The Paracellular Channel

Biology, Physiology, and Disease

The Paracellular Channel

Biology, Physiology, and Disease

Jianghui Hou
Department of Internal Medicine
Washington University St. Louis
660 South Euclid Avenue
St. Louis, Missouri, USA

Academic Press is an imprint of Elsevier
125 London Wall, London EC2Y 5AS, United Kingdom
525 B Street, Suite 1650, San Diego, CA 92101, United States
50 Hampshire Street, 5th Floor, Cambridge, MA 02139, United States
The Boulevard, Langford Lane, Kidlington, Oxford OX5 1GB, United Kingdom

Library of Congress Cataloging-in-Publication Data
A catalog record for this book is available from the Library of Congress

British Library Cataloguing-in-Publication Data
A catalogue record for this book is available from the British Library

ISBN: 978-0-12-814635-4

For information on all Academic Press publications visit our website at
https://www.elsevier.com/books-and-journals

Publisher: John Fedor
Acquisition Editor: Mica Haley
Editorial Project Manager: Megan Ashdown
Production Project Manager: Sreejith Viswanathan
Designer: Miles Hitchen

Typeset by Thomson Digital

Dedication

To my parents and my wife
who have consistently supported
my scientific inquiry.
To my son Harvey
who will explore his curiosity.

Contents

Author Biography

Dr. Jianghui Hou is a professor of Molecular Medicine and Cell Biology in Washington University in St. Louis. Dr. Hou holds a Ph.D. degree in Molecular Biology from the University of Edinburgh, United Kingdom. Dr. Hou specialized in tight junction biology during his postdoctoral training at Harvard Medical School. Dr. Hou has been studying tight junction biology for the past 15 years and published over 45 peer-reviewed articles on this topic in leading journals. Dr. Hou's major research interests include tight junction structure and function, super-resolution measurement of paracellular conductance, pathophysiology of tight junction disease, and pharmacologic development of paracellular modulator.

Preface

When Elsevier offered me an opportunity to write a book on tight junction and claudins, I started battling against myself to decide what would be the best title for the new book. My colleagues Dr. James Anderson and Dr. Alan Yu have edited two books entitled "Tight Junction" and "Claudins" respectively. Tight junctions are essential cellular barriers that separate extracellular compartments and allow passive transport of ions, solutes, and water in a regulated manner. Claudins are vital components of the regulatory machinery for paracellular transport. By comparison, tight junction behaves like an ion channel. Claudins may well form the channel pore. I reckon, the term "Paracellular Channel" will best describe what tight junction is and what claudins do.

I have maintained a historical flavor in the book and would like to emphasize that many of today's discoveries have roots in strong biochemical work of the past. Among the forerunners in the field, Dr. Shoichiro Tsukita has made the most important contribution—the discovery of claudin in 1998. We now know, claudins play pivotal roles in almost every aspect of paracellular channel's life. This book is intended to comprehensively review the current state of knowledge on paracellular channels, and inspire an influx of new minds and approaches for future exploration. This book is conceptually organized into three areas—biology, physiology, and disease of paracellular channel. The structure and function are the most important question in paracellular channel biology. Because paracellular channels reside in tight junction, how they become assembled into an ordered molecular architecture will present additional complexity to paracellular channel biology. Similar to transcellular membrane channels, paracellular channels can permeate cations, anions, and water. The charge and size selectivities define paracellular channel physiology. We have now gained a great deal of knowledge on paracellular channel's role in organ function. Such knowledge has proven vital to our understanding of how dysfunction in paracellular channels may cause diseases. Next-generation medicine will exploit genetic or pharmacologic manipulation of paracellular channel function.

Finally, I must say paracellular channel biology is a young field. Exciting discoveries are being made at a phenomenal rate. I hope this book will enthuse graduate students, research trainees, and scientists from both academia and industry to enter the field, grow with the field, and pioneer the field into a new frontier.

Acknowledgment

I am deeply grateful to the support of many brilliant mentors, colleagues, and fellows over the years, in particular Dr. Daniel Goodenough, who introduced me to the field of tight junction biology, nourished my career development, and motivated me in countless scientific setbacks. Production of this book is not possible without the help of the highly professional editorial team in Elsevier, particularly my project manager, Megan Ashdown, and my production manager, Sreejith Viswanathan. Finally, I am sincerely indebted to National Institute of Diabetes and Digestive and Kidney Diseases, National Science Foundation, Department of Defense, and American Heart Association for their continuous support of my thinking, writing and exploring for 10 years.

Chapter 1

Introduction

1.1 A NEW CLASS OF ION CHANNEL

The cell membrane is a biologic lipid bilayer that separates the interior of the cell from the exterior environment. Ion channels are pore forming membrane proteins that allow passage of ions through the cell membrane (Fig. 1.1). When life evolves from unicellular to multicellular organisms, coordination of cell growth by a process known as morphogenesis creates complex three-dimensional structures, such as the blood vessel, renal tubule, pulmonary alveolus, and so on. These tissue structures, made of continuous layers of cells, separate the exterior environment into two independent compartments. The paracellular channel is a new class of ion channel oriented perpendicular to the plasma membrane plane and serving to join the two exterior compartments (Fig. 1.2). The paracellular channel, now believed to be made of the class of *claudin* proteins, conducts ions in a similar way to the conventional membrane channel by displaying size and charge selectivity (Tang & Goodenough, 2003; Tsukita & Furuse, 2000). However, the paracellular channel is fundamentally different from the membrane channel in many aspects.

1.2 SECRET LIFE OF CELL JUNCTION

The vertebrate epithelial cell forms a tripartite junctional complex near the apical membrane, which comprises the tight junction (TJ), adherens junction (AJ), and desmosome (Farquhar & Palade, 1963). AJ and desmosome are structurally and functionally similar organelles. Both play vital roles in cell adhesion (Delva, Tucker, & Kowalczy, 2009; Meng & Takeichi, 2009). TJ, on the other hand, has been a mysterious organelle. It was initially recognized as a barrier to impede the paracellular permeation of macromolecules (Farquhar & Palade, 1963). Soon afterwards, the transport function of TJ was discovered, which indicated that TJ provided the main route of passive ion permeation between the epithelial cells (Fromter & Diamond, 1972). More importantly, the level of TJ permeability can be correlated with the degree of its structural complexity (Claude & Goodenough, 1973). It seems certain that TJ contains the long-sought paracellular channel.

The Paracellular Channel. http://dx.doi.org/10.1016/B978-0-12-814635-4.00001-2

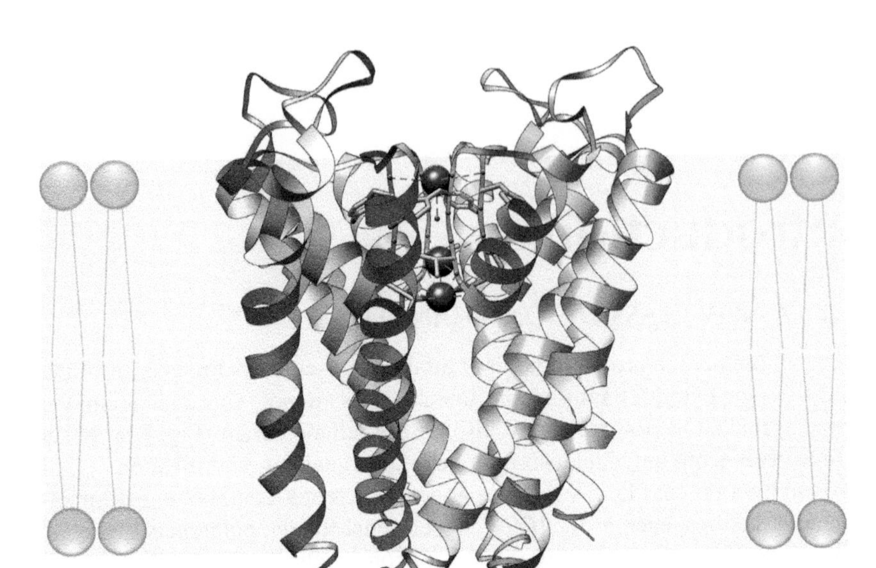

FIGURE 1.1 Membrane ion channel. The ion channel in the lipid bilayer permits passage of potassium ions (labeled in *violet*) but not sodium ions. The oxygen atoms (labeled in *red*) of the amino acid residues forming the channel pore interact with and stabilize the potassium ions by creating an environment very similar to the aqueous environment outside the lipid bilayer. Cells may open or close the channel by employing additional gating mechanisms. The depicted ion channel structure is based upon the X-ray analysis of the KcsA K^+ channel from *Streptomyces lividans* (MacKinnon, 2004).

1.3 FIRST BUT NOT LAST TIGHT JUNCTION PROTEIN

The discovery of zonula occludens-1 (ZO-1) by Goodenough and coworkers is a triumph to the field of TJ biology (Stevenson, Siliciano, Mooseker, & Goodenough, 1986). Prior to this work, there was no knowledge of the molecular makeup of TJ. Whether TJ was composed of proteins or lipids had been debated for years. Goodenough and coworkers developed a biochemical protocol to isolate TJ-enriched membrane fractions from the bile canaliculi in mouse livers (Stevenson & Goodenough, 1984). At the time when molecular cloning tools were not available, they used the crude TJ membrane fraction as primary immunogen to generate monoclonal antibodies. In the theory, each clone of antibody recognizes a unique polypeptide present in the fraction. One antibody, which specifically labeled the TJ in mouse epithelial tissues, allowed the identification of ZO-1 protein (Stevenson et al., 1986). ZO-1 is of ~225 kDa and detergent solubility assay indicates that ZO-1 is a peripheral but not integral membrane protein (Anderson, Stevenson, Jesaitis, Goodenough, & Mooseker, 1988).

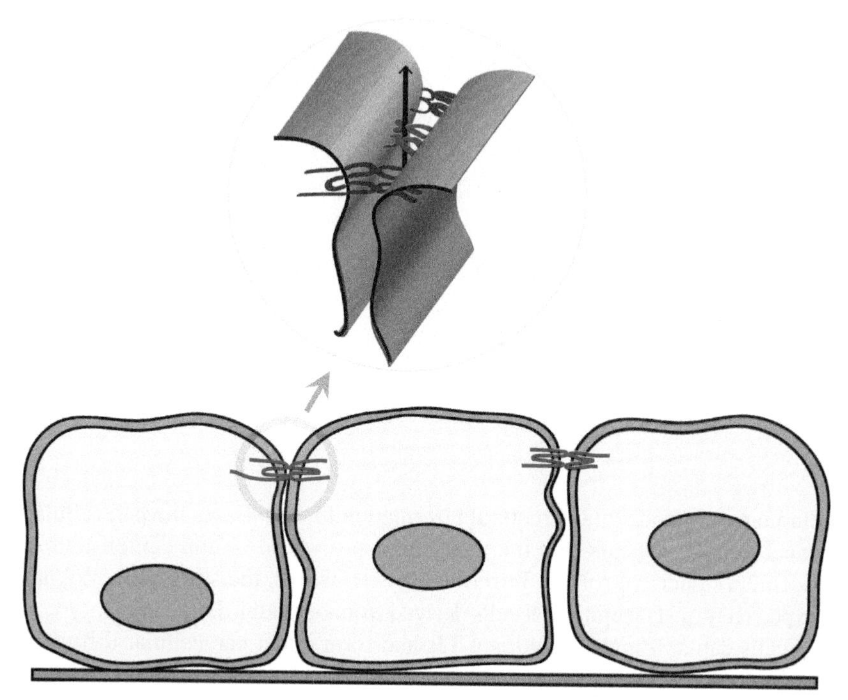

FIGURE 1.2 **Paracellular ion channel.** The paracellular channel is found within the TJ structure and is part of the paracellular diffusion barrier between the two plasma membranes. The paracellular channel conducts ions on the basis of size and charge. The paracellular channel is thought to be formed by intercellular claudin protein interactions that create pore-like structures similar to those found in membrane ion channels.

1.4 SEARCH FOR PARACELLULAR CHANNEL PROTEIN

1.4.1 First But Wrong Hit

The search for TJ integral membrane protein continued. Using a similar biochemical protocol to what Goodenough and coworkers had devised, Tsukita and coworkers prepared a TJ enriched membrane fraction from chicken livers and took a further step to remove all peripheral proteins. Then, they generated monoclonal antibodies against the proteins present in the purified membrane fraction and selected the clones of antibody based upon their ability to bind to the TJ. Three clones of antibody recognized a TJ protein of ~65 kDa, which Tsukita and coworkers named occludin (Fig. 1.3) (Furuse et al., 1993). Hydrophobicity plot reveals that occludin consists in four transmembrane domains, two extracellular loop domains and two cytoplasmic domains. The discovery of occludin immediately excited the TJ field because many would consider an integral membrane protein is essential for establishing the fundamental TJ architecture, technically known as the TJ strand. Furthermore, the extracellular domains in

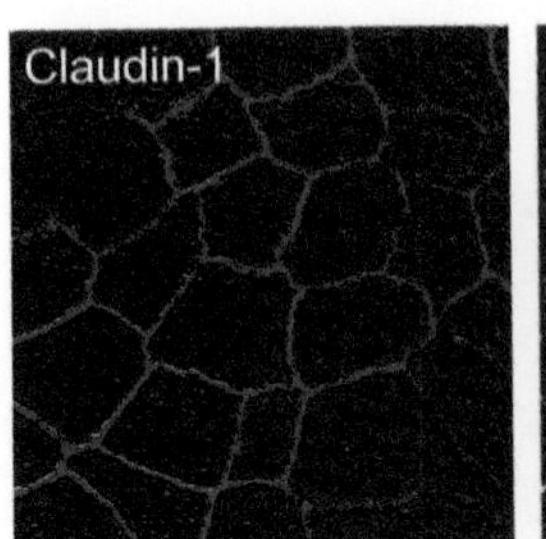

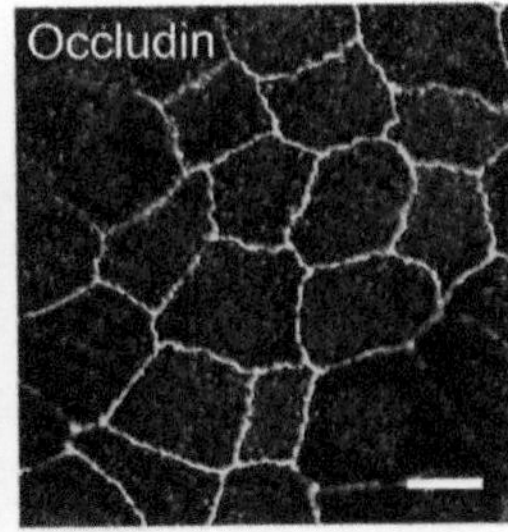

FIGURE 1.3 Localization pattern of TJ proteins. The TJ proteins: claudin-1 and occludin were immunostained in the Madin-Darby Canine Kidney (MDCK) cells to demonstrate their TJ localization. Bar: 10 μm. *(Reproduced with permission from Furuse, M., Fujita, K., Hiiragi, T., Fujimoto, K., & Tsukita, S. (1998a). Claudin-1 and -2: novel integral membrane proteins localizing at tight junctions with no sequence similarity to occludin. The Journal of Cell Biology, 141, 1539–1550.)*

occludin may encode vital structural information for stabilizing the paracellular channel pore. It appears that the next logic step would be the demonstration of occludin's decisive role in TJ physiology. However, the story soon took an unexpected turn. The epithelial cells derived from occludin-deficient embryonic stem cells can still make functional TJs and form intact paracellular diffusion barriers (Saitou et al., 1998). Occludin also failed the cell adhesion assay, which suggests that it is dispensable for the obliteration of intercellular space at the TJ (Kubota et al., 1999).

1.4.2 Never Giving Up

If occludin is not the true backbone of TJ, then there must exist another mysterious protein that somehow escaped the initial antibody-based search. Knowing that occludin is located in the TJ membrane, Tsukita and coworkers hypothesized that the new protein might interact with occludin in the TJ enriched membrane fraction. Therefore, the biochemical condition used to dissolve occludin from the membrane may allow copurification of the new protein. They succeeded in identifying two small membrane proteins of ~22 kDa from the TJ, which were termed as claudin-1 and claudin-2 (Fig. 1.3) (Furuse, Fujita, Hiiragi, Fujimoto, & Tsukita, 1998a). The discovery of claudin is a major breakthrough in TJ biology. When reconstituted into cell membrane, claudin can polymerize into TJ-like structures (Furuse, Sasaki, Fujimoto, & Tsukita, 1998b). Claudin also mediates cell adhesion in a Ca^{++}-independent manner (Kubota et al., 1999). The claudin protein family now encompasses 27 members in mammals. From the structural perspective, claudin belongs to the pfam00822 superfamily of proteins, whose members also include peripheral myelin protein 22 (PMP22), epithelial membrane protein (EMP), lens-specific membrane protein 20 (MP20), and voltage-gated calcium channel γ subunit (CACNG) (Anderson & van Itallie, 2009).

1.5 CONNECTION TO HUMAN DISEASE

Shortly after the discovery of claudin, Lifton and coworkers mapped a hereditary kidney disease, familial hypomagnesemia with hypercalciuria and nephrocalcinosis (FHHNC), to a locus on chromosome 3q, and identified the causal gene, which they named paracellin-1 (Simon et al., 1999). Sequence homology suggests that paracellin-1 is a member of the claudin family, so it has been renamed as claudin-16. Lifton's work is vital to the physiology of paracellular channel. It proves that paracellular permeability is required for normal reabsorption function of the kidney. Damages to the paracellular channel due to genetic mutations are sufficient to cause disease. No other genes are redundant or compensatory to the function of claudin-16, including the genes making transcellular channels. Diverse missense mutations, premature termination, and splice site mutations have been found in claudin-16 gene from patients with FHHNC (Simon et al., 1999; Weber et al., 2001). These mutations offer important biological insights into claudin gene transcription, protein translation, intracellular trafficking, TJ assembly, and paracellular channel permeability. Several years later, a second causal gene for FHHNC was discovered, which turned out to be another claudin gene, referred to as claudin-19 (Konrad et al., 2006). The strong connection of claudin to electrolyte imbalance disease emphasizes the rising role of paracellular channel in human disease.

1.6 PROTEIN INTERACTION

The spatial orientation of paracellular channel requires claudin to polymerize into a high-order protein complex that can seal the paracellular space. Claudin polymerization involves two types of interaction: the *cis*-interaction within cell membrane and the *trans*-interaction between cell membranes. Knowing that mutation of either claudin-16 or claudin-19 causes the same disease, Hou and coworkers probed the *cis*- and *trans*-binding affinity between these two claudins. Their work clearly demonstrates that claudin-16 interacts with claudin-19 in *cis* but not in *trans* (Hou et al., 2008). The *cis*-interaction is important for claudin assembly into the TJ. In transgenic mouse kidneys, removing either claudin protein can render the other to be dissembled from the TJ (Hou et al., 2009). The *trans*-interaction appears to play more roles than cell adhesion. Hou and coworkers revealed that the hybrid TJ made of claudin-16 and claudin-19 in cocultured cells conducted no ion (Gong et al., 2015). Their data suggest that no claudin hemichannel exists and normal paracellular channel function requires intercellular compatibility.

1.7 CRYSTAL STRUCTURE

Fujiyoshi and coworkers resolved the first crystal structure of claudin protein, the claudin-15 protein (Suzuki et al., 2014). The most intriguing feature is the β-sheet structure formed by two extracellular loop domains. The β-sheet contains charged amino acid sites that are purported to stabilize the paracellular channel pore. The β-sheet is also important for interaction with *Clostridium perfringens* enterotoxin (CPE). Claudins are the receptor proteins for CPE (Katahira et al., 1997). CPE is a powerful modulator of paracellular permeability by acting to disintegrate the TJ architecture (Sonoda et al., 1999). The binding of CPE distorts the conformation of the β-sheet according to the crystal structure of CPE bound claudin-19 (Saitoh et al., 2015). As a result, both *cis*- and *trans*-claudin interactions are disrupted. For years, whether TJ can be druggable is a tantalizing question. These structural studies lay the foundation for future pharmacological manipulation of paracellular permeability.

1.8 WATER IN TRICELLULAR JUNCTION

Cell membrane utilizes a unique class of channels to transport water, known as aquaporins. The separation of water from ion permeation across the membrane allows the cell to independently adjust salinity and osmolality. The paracellular channel, if behaving analogously to the transcellular channel, may also separate water from ion permeation. Whether claudin permeates water has been under debate since its discovery. Hou and coworkers have tendered a new hypothesis that the tricellular TJ (tTJ) permits water permeation via a different pathway from the bicellular TJ (bTJ). Hou's work reveals that manipulating a tTJ protein, known as angulin, alters the paracellular permeability to water without affecting claudin's function (Gong et al., 2017). The structure and function of tTJ significantly differ from those of bTJ. The molecular components of these two junctions appear not interchangeable. Perhaps, tTJ and bTJ evolve divergently to confer heterogeneity to the paracellular pathway. As a result, a variety of molecules with size and charge differences can be handled by the paracellular channel.

1.9 RESOLUTION RACE

The most common way to measure paracellular conductance is by applying Ohm's law to a circuit in which the epithelium can be viewed as a conductor. The ion conductive pathways in the epithelium can be described by the transepithelial conductance, that is, the inverse quantity of transepithelial resistance (TER). TER in essence is influenced by both transcellular and paracellular pathways. Therefore, how to isolate the paracellular conductance from the transepithelial conductance is a major challenge to the field. The traditional way is by blocking the transcellular channels with various pharmacological inhibitors. Frömter first introduced the principle of conductance scanning, with which the paracellular pathway can be analyzed in situ by an electrode positioned directly above the TJ (Fromter, 1972). In this type of recording, an optical microscope is needed for manual manipulation of the electrode. System errors are inevitable. Baker and coworkers significantly improved the resolution of conductance scanning by employing a new technique known as scanning ion conductance microscopy

(SICM). SICM can resolve transepithelial conductance differences in a nominal area of less than 1 μm in diameter (Chen et al., 2013). Although patch clamp is widely used in transcellular channel recording, whether paracellular channels can be patch-clamped is still highly debatable, because the membrane architecture in the TJ makes it difficult to establish giga-ohm seals for patch clamp.

REFERENCES

Anderson, J. M., Stevenson, B. R., Jesaitis, L. A., Goodenough, D. A., & Mooseker, M. S. (1988). Characterization of ZO-1, a protein component of the tight junction from mouse liver and Madin-Darby canine kidney cells. *The Journal of Cell Biology*, *106*, 1141–1149.

Anderson, J. M., & van Itallie, C. M. (2009). Physiology and function of the tight junction. *Cold Spring Harbor Perspectives in Biology*, *1*, a002584.

Chen, C. C., Zhou, Y., Morris, C. A., Hou, J., & Baker, L. A. (2013). Scanning ion conductance microscopy measurement of paracellular channel conductance in tight junctions. *Analytical Chemistry*, *85*, 3621–3628.

Claude, P., & Goodenough, D. A. (1973). Fracture faces of zonulae occludentes from "tight" and "leaky" epithelia. *The Journal of Cell Biology*, *58*, 390–400.

Delva, E., Tucker, D. K., & Kowalczyk, A. P. (2009). The desmosome. *Cold Spring Harbor Perspectives in Biology*, *1*, a002543.

Farquhar, M. G., & Palade, G. E. (1963). Junctional complexes in various epithelia. *The Journal of Cell Biology*, *17*, 375–412.

Fromter, E. (1972). The route of passive ion movement through the epithelium of Necturus gallbladder. *The Journal of Membrane Biology*, *8*, 259–301.

Fromter, E., & Diamond, J. (1972). Route of passive ion permeation in epithelia. *Nature*, *235*, 9–13.

Furuse, M., Fujita, K., Hiiragi, T., Fujimoto, K., & Tsukita, S. (1998a). Claudin-1 and -2: novel integral membrane proteins localizing at tight junctions with no sequence similarity to occludin. *The Journal of Cell Biology*, *141*, 1539–1550.

Furuse, M., Hirase, T., Itoh, M., Nagafuchi, A., Yonemura, S., Tsukita, S., & Tsukita, S. (1993). Occludin: a novel integral membrane protein localizing at tight junctions. *The Journal of Cell Biology*, *123*, 1777–1788.

Furuse, M., Sasaki, H., Fujimoto, K., & Tsukita, S. (1998b). A single gene product, claudin-1 or -2, reconstitutes tight junction strands and recruits occludin in fibroblasts. *The Journal of Cell Biology*, *143*, 391–401.

Gong, Y., Himmerkus, N., Sunq, A., Milatz, S., Merkel, C., Bleich, M., & Hou, J. (2017). ILDR1 is important for paracellular water transport and urine concentration mechanism. *Proceedings of the National Academy of Sciences of the United States of America*, *114*, 5271–5276.

Gong, Y., Renigunta, V., Zhou, Y., Sunq, A., Wang, J., Yang, J., Renigunta, A., Baker, L. A., & Hou, J. (2015). Biochemical and biophysical analyses of tight junction permeability made of claudin-16 and claudin-19 dimerization. *Molecular Biology of the Cell*, *26*, 4333–4346.

Hou, J., Renigunta, A., Gomes, A. S., Hou, M., Paul, D. L., Waldegger, S., & Goodenough, D. A. (2009). Claudin-16 and claudin-19 interaction is required for their assembly into tight junctions and for renal reabsorption of magnesium. *Proceedings of the National Academy of Sciences of the United States of America*, *106*, 15350–15355.

Hou, J., Renigunta, A., Konrad, M., Gomes, A. S., Schneeberger, E. E., Paul, D. L., Waldegger, S., & Goodenough, D. A. (2008). Claudin-16 and claudin-19 interact and form a cation-selective tight junction complex. *The Journal of Clinical Investigation*, *118*, 619–628.

Katahira, J., Inoue, N., Horiguchi, Y., Matsuda, M., & Sugimoto, N. (1997). Molecular cloning and functional characterization of the receptor for *Clostridium perfringens* enterotoxin. *The Journal of Cell Biology*, *136*, 1239–1247.

Konrad, M., Schaller, A., Seelow, D., Pandey, A. V., Waldegger, S., Lesslauer, A., Vitzthum, H., Suzuki, Y., Luk, J. M., Becker, C., et al. (2006). Mutations in the tight-junction gene claudin 19 (CLDN19) are associated with renal magnesium wasting, renal failure, and severe ocular involvement. *American Journal of Human Genetics*, *79*, 949–957.

Kubota, K., Furuse, M., Sasaki, H., Sonoda, N., Fujita, K., Nagafuchi, A., & Tsukita, S. (1999). Ca(2 +)-independent cell-adhesion activity of claudins, a family of integral membrane proteins localized at tight junctions. *Current Biology*, *9*, 1035–1038.

MacKinnon, R. (2004). Potassium channels and the atomic basis of selective ion conduction (Nobel Lecture). *Angewandte Chemie*, *43*, 4265–4277.

Meng, W., & Takeichi, M. (2009). Adherens junction: molecular architecture and regulation. *Cold Spring Harbor Perspectives in Biology*, *1*, a002899.

Saitoh, Y., Suzuki, H., Tani, K., Nishikawa, K., Irie, K., Ogura, Y., Tamura, A., Tsukita, S., & Fujiyoshi, Y. (2015). Tight junctions. Structural insight into tight junction disassembly by *Clostridium perfringens* enterotoxin. *Science*, *347*, 775–778.

Saitou, M., Fujimoto, K., Doi, Y., Itoh, M., Fujimoto, T., Furuse, M., Takano, H., Noda, T., & Tsukita, S. (1998). Occludin-deficient embryonic stem cells can differentiate into polarized epithelial cells bearing tight junctions. *The Journal of Cell Biology*, *141*, 397–408.

Simon, D. B., Lu, Y., Choate, K. A., Velazquez, H., Al-Sabban, E., Praga, M., Casari, G., Bettinelli, A., Colussi, G., Rodriguez-Soriano, J., et al. (1999). Paracellin-1, a renal tight junction protein required for paracellular Mg2+ resorption. *Science*, *285*, 103–106.

Sonoda, N., Furuse, M., Sasaki, H., Yonemura, S., Katahira, J., Horiguchi, Y., & Tsukita, S. (1999). *Clostridium perfringens* enterotoxin fragment removes specific claudins from tight junction strands: Evidence for direct involvement of claudins in tight junction barrier. *The Journal of Cell Biology*, *147*, 195–204.

Stevenson, B. R., & Goodenough, D. A. (1984). Zonulae occludentes in junctional complex-enriched fractions from mouse liver: preliminary morphological and biochemical characterization. *The Journal of Cell Biology*, *98*, 1209–1221.

Stevenson, B. R., Siliciano, J. D., Mooseker, M. S., & Goodenough, D. A. (1986). Identification of ZO-1: a high molecular weight polypeptide associated with the tight junction (zonula occludens) in a variety of epithelia. *The Journal of Cell Biology*, *103*, 755–766.

Suzuki, H., Nishizawa, T., Tani, K., Yamazaki, Y., Tamura, A., Ishitani, R., Dohmae, N., Tsukita, S., Nureki, O., & Fujiyoshi, Y. (2014). Crystal structure of a claudin provides insight into the architecture of tight junctions. *Science*, *344*, 304–307.

Tang, V. W., & Goodenough, D. A. (2003). Paracellular ion channel at the tight junction. *Biophysical Journal*, *84*, 1660–1673.

Tsukita, S., & Furuse, M. (2000). Pores in the wall: claudins constitute tight junction strands containing aqueous pores. *The Journal of Cell Biology*, *149*, 13–16.

Weber, S., Schneider, L., Peters, M., Misselwitz, J., Ronnefarth, G., Boswald, M., Bonzel, K. E., Seeman, T., Sulakova, T., Kuwertz-Broking, E., et al. (2001). Novel paracellin-1 mutations in 25 families with familial hypomagnesemia with hypercalciuria and nephrocalcinosis. *Journal of the American Society of Nephrology*, *12*, 1872–1881.

Chapter 2

Paracellular Channel Formation

2.1 TIGHT JUNCTION ULTRASTRUCTURE

2.1.1 Transmission Electron Microscopy

In 1963, Farquhar and Palade, when using transmission electron microscopy to examine ultrathin tissue sections, recognized the "cell junctional complex" near the apex of the epithelial cells where they were adjoined (Farquhar & Palade, 1963). The most apical member of this complex is the tight junction (TJ) or *zonula occludens* (ZO). At low magnification, the TJ appears between 0.1 and 0.5 μm in depth, where adjacent plasma membranes are in close apposition (Fig. 2.1). At high magnification, the TJ is seen where the exoplasmic leaflets of the adjacent membrane bilayers appear to fuse (Fig. 2.1).

2.1.2 Freeze Fracture Replica Electron Microscopy

Freeze fracture replica electron microscopy enables examining the TJ ultrastructure in its lipid environment. Most lipids do not react with conventional aldehyde fixative and thus are prone to redistribution during sample preparation. Rapid freezing at low temperatures (−115°C) physically immobilizes the membrane lipids. After fracturing the cell membrane along the interior hydrophobic plane of the lipid bilayer, the integral membrane proteins can be exposed to reveal their high-order oligomeric architecture. The fractured membrane is covered with a fine layer of platinum by a process known as shadowing to make the replica. The topographical features of the fractured membrane are then converted into variations in thickness of the deposited platinum layer of the replica (Bullivant & Ames, 1966).

Freeze fracture replica electron microscopy has revealed that the TJ is made of a continuous network of "fibrils" or "strands" on the protoplasmic (P) fracture face and complementary empty grooves on the exoplasmic (E) fracture face (Fig. 2.2). These "fibrils" or "strands" are visualized as linear parallel arrays of intramembrane particles of ~10 nm in diameter and separated by a distance of ~8 nm (Goodenough & Revel, 1970).

The Paracellular Channel. http://dx.doi.org/10.1016/B978-0-12-814635-4.00002-4

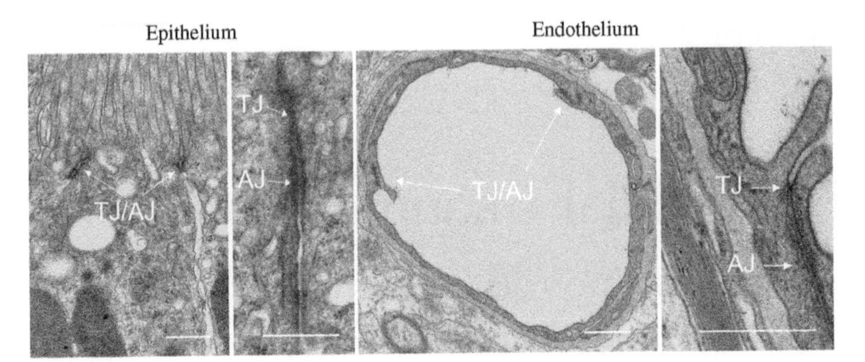

FIGURE 2.1 Epithelial and endothelial tight junctions. The TJ obliterates the paracellular space by fusing the plasma membranes of adjacent cells. The adherens junction (AJ) is located beneath the TJ and connects the adjacent plasma membranes with the electron-dense fibrillary material known as cadherin. On the cytoplasmic sides of both junctions are the electron-dense plaques that consist of the scaffold proteins such as zonula occludens-1 (ZO-1) and anchor the junctions onto the cytoskeleton. The epithelium is from the mouse kidney proximal tubule. The endothelium is from the mouse cerebral cortex. Bar: 500 nm. *(Reproduced with permission from Cain, M. D., Salimi, H., Gong, Y., Yang, L., Hamilton, S. L., Heffernan, J. R., Hou, J., Miller, M. J., Klein, R. S. (2017). Virus entry and replication in the brain precedes blood-brain barrier disruption during intranasal alphavirus infection. Journal of Neuroimmunology, 308, 118–130.)*

2.2 LIPID VERSUS PROTEIN MODELS

Because the TJ creates a unique microenvironment where the proteins and the lipids can interact, it is highly likely that both the proteins and the lipids are important for making its ultrastructural architecture. The lipid model proposes that the junctional membrane hemifusions give the characteristic appearance of the TJ under the ultrathin section transmission electron microscopy. The

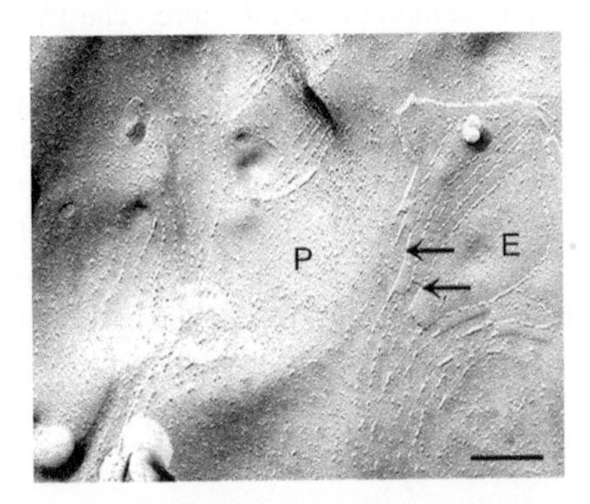

FIGURE 2.2 Tight junction ultrastructure. Freeze fracture replica electron micrograph reveals the TJ strands in the mouse L cells transfected with a claudin protein. A P-to-E fracture face transition can be seen where the TJ strands align in parallel (*arrow*). Bar: 1 μm. *(Reproduced with permission from Hou, J., Renigunta, A., Konrad, M., Gomes, A. S., Schneeberger, E. E., Paul, D. L., Waldegger, S., & Goodenough, D.A. (2008). Claudin-16 and claudin-19 interact and form a cation-selective tight junction complex. The Journal of Clinical Investigation, 118, 619–628).*

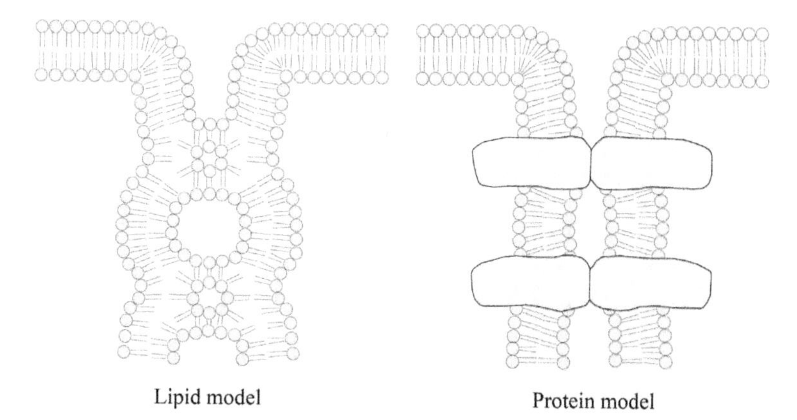

FIGURE 2.3 **Lipid versus protein models.** In the lipid model, the inverted cylindrical lipid micelles make the tight junction strands. In the protein model, the integral transmembrane proteins constitute the tight junction strands via *cis* and *trans* interactions.

TJ "fibrils" or "strands" revealed by freeze fracture replica electron microscopy are in fact made of cylinders of inverted lipid micelles, that is, the hexagonal transitory lipid arrangement where the polar head groups of the lipids are oriented inwards (Fig. 2.3). This model is based upon the observation that protein-free liposomes can form TJ-like strands (Kachar & Reese, 1982; Meyer, 1983). Cell-to-cell lipid diffusion experiments have provided some support to the lipid model. Using fluorescently labeled lipids, researchers have found that lipids with small polar head groups can diffuse freely from one cell to another even when the TJ is intact, suggesting that the exoplasmic leaflets of the neighboring cell plasma membranes are fused at the TJ, as proposed by the lipid model (Grebenkamper & Galla, 1994). However, the lipid model cannot explain why lipids with large polar head groups such as the Forssman glycolipids failed the cell-to-cell diffusion assay and were restricted to individual cells (van Meer, Gumbiner, & Simons, 1986).

The protein model proposes that the linear arrays of transmembrane proteins polymerize via intracellular interactions to form the TJ "fibrils" or "strands" and that these "fibrils" or "strands" from apposing cells pair via intercellular interactions to obliterate the paracellular space (Fig. 2.3). At least three classes of TJ proteins have been identified. Occludin and claudins are the integral transmembrane proteins that form the backbone of the TJ (Furuse, Fujita, Hiiragi, Fujimoto, & Tsukita, 1998; Furuse et al., 1993). Transfection experiments in mouse L fibroblasts that normally lack the TJ have demonstrated that ectopically expressed claudin proteins can reconstitute the TJ-like "fibrils" or "strands" (Furuse, Sasaki, Fujimoto, & Tsukita, 1998). While occludin itself cannot make such TJ-like structures, when introduced into the claudin-expressing L fibroblasts, occludin became incorporated into the claudin-based TJ strands (Furuse, Sasaki et al., 1998). Junctional adhesion molecules (JAMs) are the immunoglobulin-type adhesion proteins of the

TJ (Martin-Padura et al., 1998). JAM does not participate in the formation of TJ strands, but plays a key role in the initial phase of TJ assembly via interaction with the adherens junction. The zonula occludens (ZO) proteins are the cytosolic plaque components of the TJ (Stevenson, Siliciano, Mooseker, & Goodenough, 1986). The junctional plaque is a complex protein network that anchors the junctional membrane proteins such as occludin, claudins, and JAMs onto the cytoskeleton, that is, actin filaments and microtubules.

2.3 THE MOLECULAR STRUCTURE OF CLAUDIN

Claudins are tetraspan proteins consisting of a family with at least 27 members ranging in molecular mass from 20 to 28 kDa (Mineta et al., 2011). Claudins have four transmembrane domains, two extracellular loops (ECLs: ECL1 and ECL2), amino- and carboxyl-terminal cytoplasmic domains, and a short cytoplasmic turn (Hou, Rajagopal, & Yu, 2013). The ECL1 of claudin consists of ~50 amino acids with a common motif (-GLWCC; PROSITE ID: PS01346) (Krause et al., 2008), and intercalating negative and positive charges that contribute to paracellular ion selectivity (Hou, Paul, & Goodenough, 2005; Van Itallie, Rahner, & Anderson, 2001). The GLWCC motif functions as a receptor for Hepatitis C virus entry (Cukierman, Meertens, Bertaux, Kajumo, & Dragic, 2009). The charges in ECL1 may regulate paracellular ion selectivity by electrostatic interactions (Yu et al., 2009). The ECL2 consists of ~25 amino acids that mediate intercellular claudin interactions (Piontek et al., 2008). The carboxyl-terminal domain of claudin contains a PDZ (*p*ostsynaptic density 95/*d*iscs large/ZO-1) binding motif (YV) that is critical for interaction with the TJ plaque protein ZO-1 and correct localization in the TJ (Itoh, Furuse et al., 1999).

The first crystal structure of claudin, mouse claudin-15 protein (PDB: 4P79), has been resolved at a resolution of 2.4 Å (Suzuki et al., 2014). The transmembrane domains (TM: TM1–TM4) of claudin-15 form typical left-handed α-helices, and large portions of the two ECLs form a β-sheet structure (Fig. 2.4). The β-sheet domain extends from the membrane surface and comprises five β strands (β1–β5), four contributed by ECL1 and one by ECL2. The β-sheet structure is stabilized by a disulfide bond between the cysteine residues at β3 and β4 (Cys52 and Cys62, respectively), which are conserved among all members of the claudin family as part of the consensus motif (-GLWCC). In the structure of claudin-15, all of the charged residues in ECL1 extend away from the β-sheet surface. Among them, two negatively charged residues, Asp55 and Asp64, are located to the distal edge of the β-sheet domain (Fig. 2.4). Mutations of these two loci to positively charged residues can reverse the paracellular ion selectivity of claudin-15 (Colegio, Van Itallie, McCrea, Rahner, & Anderson, 2002). Two additional claudin structures are now available: mouse claudin-19 (PDB: 3X29) (Saitoh et al., 2015) and human claudin-4 (PDB: 5B2G) (Shinoda et al., 2016). The overall conformations of all three claudin structures are highly conserved.

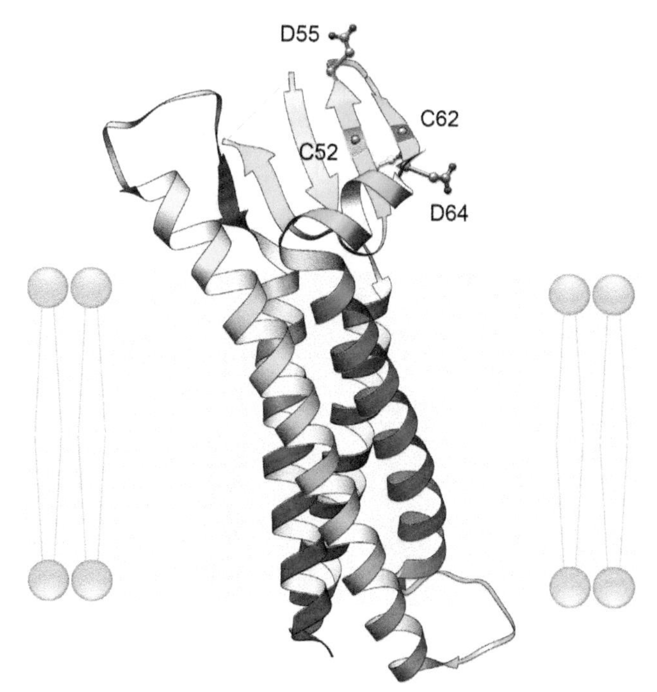

FIGURE 2.4 Molecular structure of claudin. Crystal structure of claudin-15 in ribbon representation when viewed from the side of the membrane. The color changes gradually from the N terminus (*blue*) to the C terminus (*red*). A disulfide bond (*yellow*) is made between the cysteine residues C52 and C62. Two negatively charged residues, D55 and D64, are also labeled. The depicted structure is based upon the X-ray analysis by Suzuki H., Nishizawa T., Tani K., Yamazaki Y., Tamura A., Ishitani R., et al., Crystal structure of a claudin provides insight into the architecture of tight junctions, Science (New York, NY) 344, 2014, 304–307.

2.4 INTRACELLULAR AND INTERCELLULAR INTERACTION OF CLAUDIN

2.4.1 Models of Claudin Interaction

In concept, two claudins can interact during TJ formation in a number of ways (Fig. 2.5). In Model A, interactions are only permitted with the same type of claudin, both within the plasma membrane of one cell (homomeric, *cis*) and between the plasma membranes of adjacent cells (homotypic, *trans*). In Model B, only homomeric interaction is permitted within a single plasma membrane, but mixed (heterotypic, *trans*) interactions are allowed between cells. Model C diagrams the complementary possibility, where mixed (heteromeric, *cis*) interactions are permitted within cells but only homotypic interactions between cells. Finally, Model D illustrates a situation where both heteromeric and heterotypic interactions are allowed.

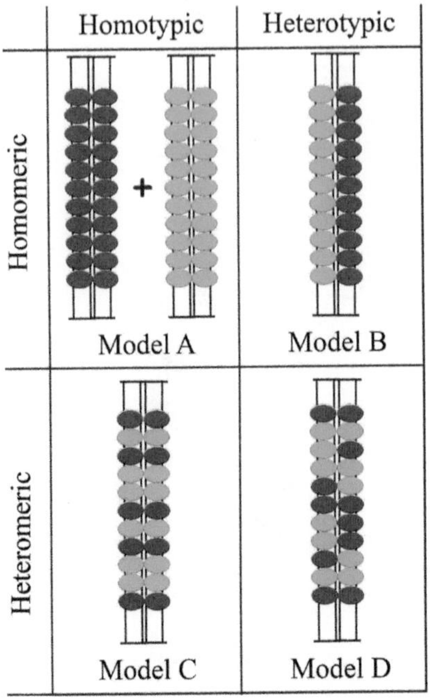

FIGURE 2.5 Models of claudin interaction. Four possible models of claudin interaction in each paired TJ strand are illustrated.

2.4.2 Intracellular Claudin Interaction

Freeze fracture replica electron microscopy can be used to examine the oligomeric structure of integral membrane proteins. The cross-sectional area of the freeze fractured protein particles varies linearly with the number of membrane spanning α-helices in the protein structure (Eskandari, Wright, Kreman, Starace, & Zampighi, 1998). On average, each helix occupies ~1.4 nm^2. This approach has accurately estimated that connexins, which make the gap junction (GJ), form a hexamer containing ~24 helices (Eskandari et al., 1998). Because claudins and connexins are tetraspan membrane proteins of similar molecular weight, and because TJ and GJ are seen as intramembrane particles of similar size (~10 nm in diameter) by freeze fracture replica electron microscopy, it has been hypothesized that claudin hexamer is the fundamental structural unit for making the TJ. This hypothesis, however, is now being challenged by several new lines of evidence. When extracted from the Madin-Darby canine kidney (MDCK) epithelial cells, claudin-2 preferentially exists as a homodimer (Van Itallie, Mitic, & Anderson, 2011). Mutational analyses followed by cross-linking experiments have further demonstrated that the second transmembrane domains in claudin-2 molecules are arranged in close proximity in the homodimer. Recombinant protein purification experiments combined with fluorescence resonance energy transfer assays have revealed that claudin-5 also adopts a dimeric structure in human embryonic kidney 293 (HEK293) cell membrane (Rossa, Ploeger et al., 2014). Amino acid loci in the third and fourth transmembrane domains of claudin-5 appear to mediate the homomeric interaction (Rossa, Protze et al., 2014). The heteromeric claudin interaction is best characterized by studies of claudin-16 and claudin-19. A yeast 2-hybrid assay first revealed the *cis* interaction between claudin-16 and claudin-19 (Hou et al., 2008). Several biochemical approaches including sucrose gradient sedimentation, cross-linking, and chromatography were used to capture the oligomeric state of claudin-16 and claudin-19 association. In fall armyworm (*Spodoptera frugiperda*) 9 (Sf9) cell membrane and HEK293 cell membrane, the only stable claudin oligomer is a heterodimer made of claudin-16 and claudin-19 (Gong et al., 2015). Alanine-insertion mutagenesis has identified amino acid loci in the third and fourth transmembrane domains of claudin-16 and claudin-19 that are important for their *cis* dimerization (Gong et al., 2015).

2.4.3 Intercellular Claudin Interaction

Trans claudin interactions depend upon both the ECL1 and the ECL2 domains. In human HeLa cells, claudin-5 can heterotypically interact with claudin-3, but not with claudin-4. Genetically engineered claudin-4 proteins expressing the ECL1 or the ECL2 domain of claudin-3 in place of the corresponding domain of claudin-4 showed binding affinity to claudin-5 (Daugherty, Ward, Smith, Ritzenthaler, & Koval, 2007). Mutagenesis has identified amino acid loci in these two ECL domains important for *trans* claudin interactions (Angelow & Yu, 2009; Daugherty et al., 2007; Piontek et al., 2008). The intercellular claudin interaction can be regulated by extracellular proteases. It has been shown that trypsin or prostasin acts upon the ECL2 domain of claudin-4 to break its homotypic interaction in the HEK293 cells (Gong et al., 2014).

2.4.4 Assembly of Claudin Molecules

The molecular packing arrangement of claudin-15 molecules in the crystal lattice offers by far the most important hint of how claudins are assembled in the TJ (Suzuki et al., 2014). The tandem intermolecular interactions align the claudin-15 protomers into a linear polymer. A conserved hydrophobic residue (Met68) from the ECL1 domain in one protomer fits into the hydrophobic pocket formed by the conserved residues (Phe146, Phe147, and Leu158) in the third transmembrane and the ECL2 domains of the adjacent protomer (Fig. 2.6). Mutations of these key residues prevented claudin-15 from reconstituting the TJ strands in the Sf9 cells (Suzuki et al., 2014). Structures of claudins bound to the *Clostridium perfringens* enterotoxin (CPE) provide additional support to the assembling model derived from claudin-15. The interaction of CPE with the ECL domains in claudin-4 or -19 causes a conformational change in the helical

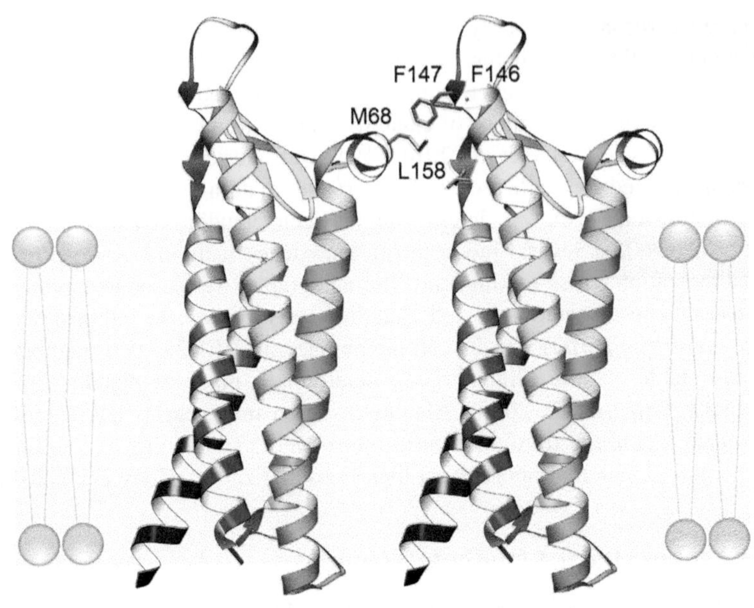

FIGURE 2.6 Linear polymerization of claudin. Ribbon representation of two linearly aligned claudin-15 protomers in the crystal when viewed in parallel to the membrane. Critical residues for the lateral interaction are labeled and shown in stick representation. The depicted structure is based upon the X-ray analysis by Suzuki H., Nishizawa T., Tani K., Yamazaki Y., Tamura A., Ishitani R., et al., Crystal structure of a claudin provides insight into the architecture of tight junctions, Science (New York, NY) 344, 2014, 304–307.

region containing Met68, therefore disrupting the putative polymerization process (Saitoh et al., 2015; Shinoda et al., 2016).

2.5 TJ PLAQUE PROTEINS

2.5.1 PDZ Domain-Containing Proteins

2.5.1.1 ZO-1, ZO-2, and ZO-3

The TJ plaque contains a large number of adaptor proteins that form a protein network via multiple protein interactions to anchor the TJ onto the cytoskeleton (Fig. 2.7). An important group of TJ plaque proteins is characterized by the presence of the so-called PDZ domain. The PDZ domain mediates protein interactions with other PDZ-containing proteins, and with membrane proteins containing specific sequence motifs, most often ending in Val at the carboxyl-terminal end (Fanning & Anderson, 1996). Intriguingly, all known claudins, with the exception of claudin-12, consist in a Val at their carboxyl termini.

The amino-terminal half of ZO-1 contains three PDZ domains, an SRC homology 3 (SH3) domain, and a yeast guanylate kinase homology (GUK) domain. The carboxyl-terminal half of ZO-1 interacts with F-actin (Itoh,

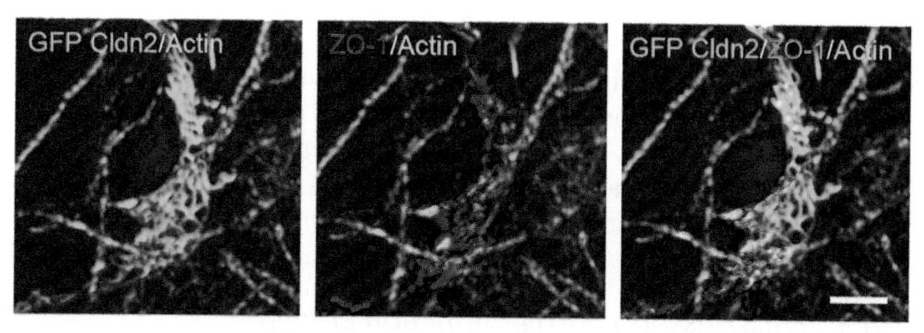

FIGURE 2.7 Localization of claudin-2, ZO-1, and actin in Rat-1 fibroblast cells. Super-resolution optical micrographs taken with structured illumination microscopy (SIM) reveal that claudin-2 forms a network of strands that are associated with ZO-1. ZO-1 is sandwiched between the claudin strands and the underlying actin filaments. Bar: 1.5 μm. *(Reproduced with permission from Van Itallie, C. M., Tietgens, A. J., & Anderson, J. M. (2017). Visualizing the dynamic coupling of claudin strands to the actin cytoskeleton through ZO-1. Molecular Biology of the Cell, 28, 524–534).*

Nagafuchi, Moroi, & Tsukita, 1997). The PDZ domains in ZO-1 interact with claudins (Itoh, Furuse et al., 1999), with ZO-2 (Itoh, Morita, & Tsukita, 1999), with ZO-3 (Haskins, Gu, Wittchen, Hibbard, & Stevenson, 1998), and with JAM (Ebnet, Schulz, Meyer Zu Brickwedde, Pendl, & Vestweber, 2000). The SH3 domain in ZO-1 binds to ZO-1-associated nucleic acid binding protein (ZONAB) (Balda & Matter, 2000). Although no GUK activity can be detected for ZO-1, the GUK domain has been found to interact with occludin in the MDCK cells (Fanning, Jameson, Jesaitis, & Anderson, 1998). ZO-2 and ZO-3 are homologous to ZO-1, and coimmunoprecipitate with ZO-1 (Gumbiner, Lowenkopf, & Apatira, 1991; Haskins et al., 1998). Like ZO-1, ZO-2 and ZO-3 interact with occludin and claudins via the amino-terminal domains, whereas they interact with F-actin via the carboxyl-terminal domains (Haskins et al., 1998; Itoh, Furuse et al., 1999; Itoh, Morita et al., 1999; Wittchen, Haskins, & Stevenson, 1999). When ZO-1 and ZO-2 were deleted from the mouse mammary gland tumor Eph4 cells, no TJ can form despite normal claudin gene expression (Umeda et al., 2006). Ectopically expressed ZO-1 or ZO-2 was able to recruit claudins to the junctional areas where they were polymerized (Umeda et al., 2006).

2.5.1.2 Membrane-Associated Guanylate Kinase Inverted Proteins

The membrane-associated guanylate kinase inverted (MAGI) proteins have the same types of domains as the ZO proteins but in an inverted arrangement (Dobrosotskaya, Guy, & James, 1997). MAGI-1 was found colocalizing with ZO-1 in the TJ plaque of the MDCK cells (Ide et al., 1999). While MAGI-1 has not been demonstrated to interact with claudins, it can interact with two important actin-binding proteins, synaptopodin and α-actinin-4, which may suggest that the MAGI proteins play an additional role in actin cytoskeleton dynamics (Patrie, Drescher, Welihinda, Mundel, & Margolis, 2002).

2.5.1.3 *Multiple PDZ Domain Protein*

Multiple PDZ domain (MPDZ) protein, also known as MUPP1, is another PDZ domain-containing protein localized to the TJ plaque (Hamazaki, Itoh, Sasaki, Furuse, & Tsukita, 2002). MPDZ can interact with claudins and JAMs (Hamazaki et al., 2002; Jeansonne, Lu, Goodenough, & Chen, 2003). Deletion of MPDZ in the mouse inner medullary collecting duct 3 (mIMCD3) cells increased the paracellular channel permeability (Lanaspa et al., 2007).

2.5.1.4 *PALS1-Associated Tight Junction Protein*

Protein associated with Lin seven 1 (PALS1, also known as Stardust in *Drosophila melanogaster*) and PALS1-associated tight junction (PATJ) protein (also known as DiscsLost in *D. melanogaster*) belong to an evolutionarily conserved protein complex important for the apicobasal polarity establishment (Hurd, Gao, Roh, Macara, & Margolis, 2003). PATJ is similar to MPDZ in protein structure, and both share several common binding partners, including JAM-A, ZO-3, and PALS1 (Adachi et al., 2009). In two epithelial cell models, MDCK and Eph4, deletion of PATJ but not MPDZ delayed the process of TJ formation (Adachi et al., 2009; Shin, Straight, & Margolis, 2005).

2.5.2 Non-PDZ Domain Proteins

2.5.2.1 *Cingulin*

Cingulin is a coiled coil domain containing protein localized to the TJ plaque (Citi, Sabanay, Jakes, Geiger, & Kendrick-Jones, 1988). Cingulin interacts with both the TJ and the cytoskeletal proteins, potentially serving as a cytoskeletal adaptor protein. The amino-terminal domain of cingulin binds to ZO-1, ZO-2, ZO-3, and JAM, whereas the carboxyl-terminal domain binds to myosin (Bazzoni et al., 2000; Cordenonsi et al., 1999). Cingulin plays a crucial role to transduce the mechanical force generated by the contraction of the actomyosin cytoskeleton into the functional regulation of the paracellular channel permeability (Turner, 2000).

2.5.2.2 *ZONAB Protein*

ZONAB is a member of the Y-box transcription factor family found in the nucleus and the TJ plaque (Balda & Matter, 2000). ZONAB interacts with the SH3 domain of ZO-1 (Balda & Matter, 2000). In the MDCK cells, ZONAB regulates the G_0/G_1 to S-phase transition in the cell cycle (Balda, Garrett, & Matter, 2003). ZONAB may convey the signals of paracellular channel permeability changes to the nuclear events such as gene replication and transcription via the transcriptional factor ErbB-2 (Balda & Matter, 2000).

2.5.2.3 *Symplekin*

Like ZONAB, symplekin is localized in two subcellular compartments, the nucleus and the TJ plaque (Keon, Schafer, Kuhn, Grund, & Franke, 1996).

It plays functional roles in nuclear pre-mRNA cleavage and polyadenylation (Barnard, Ryan, Manley, & Richter, 2004). In human colorectal cancer cells, deletion of symplekin altered the claudin-2 gene expression and the paracellular channel permeability (Buchert et al., 2010).

2.6 TIGHT JUNCTION-ASSOCIATED MARVEL DOMAIN-CONTAINING PROTEINS

The MARVEL [myelin and lymphocyte (MAL) and related proteins for vesicle trafficking and membrane link] domain is a tetraspan membrane structure that has originally been identified in proteins from the myelin and the lymphocytes (Sanchez-Pulido, Martin-Belmonte, Valencia, & Alonso, 2002). Tight junction-associated Marvel domain-containing proteins (TAMPs) include occludin (Furuse et al., 1993), tricellulin (MarvelD2) (Ikenouchi et al., 2005), and MarvelD3 (Steed, Rodrigues, Balda, & Matter, 2009). TAMPs belong to the class of TJ transmembrane proteins. Like claudins, the carboxyl-terminal domain of occludin binds to ZO-1 (Fanning et al., 1998). The high-resolution crystal structure of this domain has been determined (Li, Fanning, Anderson, & Lavie, 2005). Tricellulin is a component of the tricellular TJ and will be discussed in more detail in Chapter 6. MarvelD3 has a longer amino-terminal and shorter carboxyl-terminal domains relative to occludin and coimmunoprecipitates with occludin (Raleigh et al., 2010). When transfected into mouse L fibroblasts, none of the TAMPs can reconstitute the TJ strands on their own, but when cotransfected with claudins, they are recruited to the claudin-based TJ strands (Cording et al., 2013).

2.7 JUNCTIONAL ADHESION MOLECULE FAMILY

The JAM family proteins include three classical members—JAM-A, JAM-B, and JAM-C, and three distantly related members—JAM4, JAM-like (JAM-L) protein, and Coxsackie and adenovirus receptor (Luissint, Nusrat, & Parkos, 2014). JAMs are characterized by the molecular structure composed of two extracellular immunoglobulin domains, a single transmembrane domain, and a cytoplasmic domain with a PDZ-binding motif (Martin-Padura et al., 1998). The crystal structure of the extracellular domain in JAM-A has been resolved (Kostrewa et al., 2001). Although all members of the JAM family proteins have a PDZ-binding motif at the carboxyl terminus, they demonstrate variable affinities towards the ZO proteins. Only the classic JAMs, JAM-A, JAM-B, and JAM-C, have been confirmed to bind to the ZO proteins (Bazzoni et al., 2000; Ebnet et al., 2003). The JAM family proteins also mediate cell adhesion, either directly via the extracellular domains or indirectly via interaction with the adherens junction (Kostrewa et al., 2001). In contrast to claudins, JAMs do not reconstitute the TJ strands in transfected mouse L fibroblasts (Itoh et al., 2001).

2.8 DYNAMIC BEHAVIOR

2.8.1 Perijunctional Actomyosin Ring

Electron micrographs reveal that the apical junctional complex, including TJ and adherens junction, is associated with a highly ordered cytoskeletal structure, referred to as the perijunctional actomyosin ring, which comprises the antiparallel actin filaments and the conventional myosin, myosin II (Drenckhahn & Dermietzel, 1988). This structure has demonstrated contractile activities both in vitro and in vivo (Mooseker et al., 1982). The pharmacologic agents that disrupt the actin filaments in the perijunctional actomyosin ring, such as cytochalasins, can alter the paracellular channel assembly and permeability (Madara, Barenberg, & Carlson, 1986; Meza, Ibarra, Sabanero, Martinez-Palomo, & Cereijido, 1980). In contrast, the microtubule-disrupting agents have shown little effect on TJ structure or function (Gonzalez-Mariscal, Chavez de Ramirez, & Cereijido, 1985).

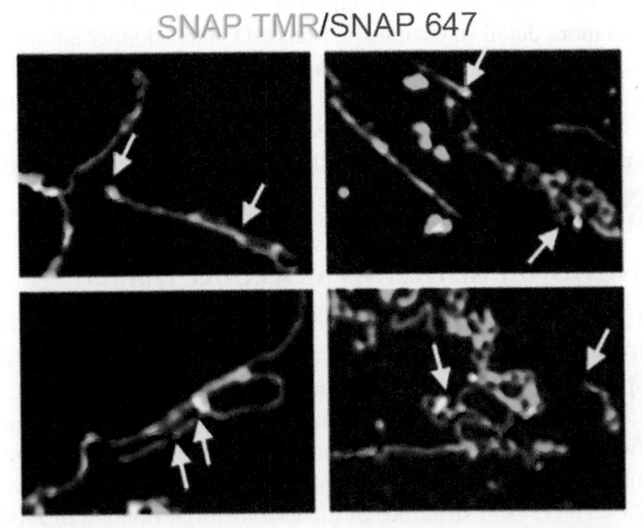

FIGURE 2.8 **Pulse-chase experiments reveal how newly synthesized claudin molecules are incorporated into the TJ strands.** The Rat-1 fibroblasts were transfected with the SNAP-tagged claudin-2 proteins. At time 0, the SNAP tags were covalently labeled with tetramethylrhodamine (TMR). At 8 h, TMR was washed and the newly synthesized SNAP-tagged claudin-2 proteins were labeled with Alexa Fluor 647. At 12 h, the images were taken and showed that the second fluorescent label was concentrated at strand ends and T-junctions (*arrows*). *(Reproduced with permission from Van Itallie, C. M., Tietgens, A. J., & Anderson, J. M. (2017). Visualizing the dynamic coupling of claudin strands to the actin cytoskeleton through ZO-1. Molecular Biology of the Cell, 28, 524–534).*

2.8.2 Molecular Mobility

Live-cell optical microscopic approaches allow analyses of fluorescently labeled TJ protein dynamics in living cells. In mouse L fibroblasts transfected with a chimeric claudin protein fused to the green fluorescent protein (GFP), the claudin-based TJ strands can be seen constantly breaking and annealing (Sasaki et al., 2003). The newly synthesized claudin molecules appear to preferentially incorporate into the break sites (Fig. 2.8) (Van Itallie, Tietgens, & Anderson, 2017). The molecular mobility of claudin, as assessed by the technique of fluorescence recovery after photobleaching (FRAP), is extremely low within the TJ strands (Sasaki et al., 2003). Occludin and ZO-1 are, in contrast, highly mobile in the TJ (Fig. 2.9) (Shen, Weber, & Turner, 2008). Unlike occludin that only diffuses into the lateral membrane, ZO-1 traffics between the TJ and the cytoplasm (Shen et al., 2008).

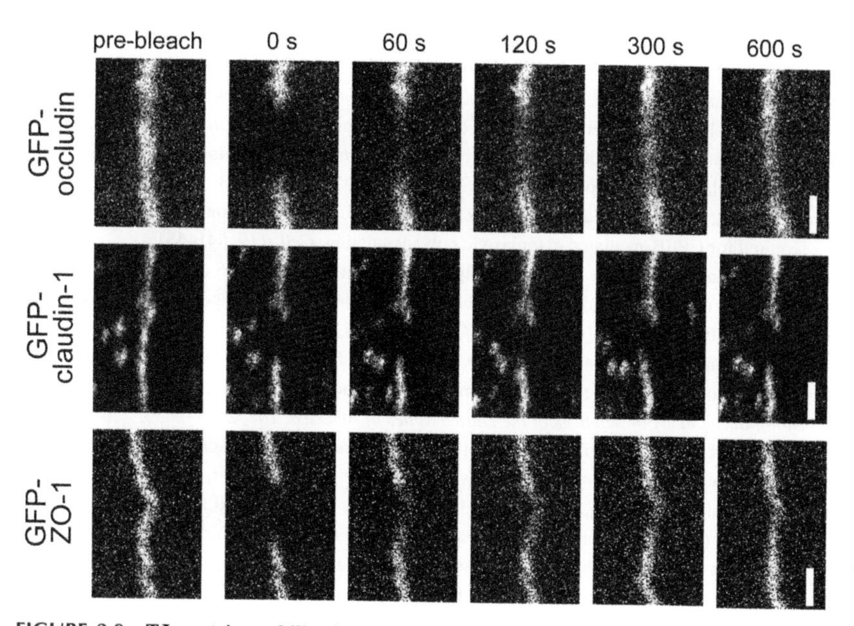

FIGURE 2.9 TJ protein mobility in polarized epithelia. GFP-occludin, GFP-claudin-1, and GFP-ZO-1 proteins were studied in MDCK cells by FRAP. High magnification images of TJ strands before and at the indicated time points after photobleaching are shown. Note that the photobleached occludin and ZO-1 proteins were rapidly replenished by neighboring unbleached proteins. However, there was little recovery of claudin-1 protein after photobleaching. Bar: 2 μm. *(Reproduced with permission from Shen L., Weber C.R. and Turner J.R., The tight junction protein complex undergoes rapid and continuous molecular remodeling at steady state, The Journal of Cell Biology 181, 2008, 683–695).*

2.8.3 Myosin Light Chain Kinase

Myosin light chain kinase (MLCK), which induces contraction of the perijunctional actomyosin ring through myosin II regulatory light chain phosphorylation, has emerged as a key regulator of the paracellular channel permeability. When the MDCK cells were transfected with an active form of MLCK, the paracellular permeability was significantly increased (Hecht et al., 1996). In the human colorectal tumor (Caco-2) cells, pharmacologic inhibition of MLCK reduced the paracellular permeability (Turner et al., 1997). MLCK is able to regulate the dynamic behavior of TJ proteins. In the Caco-2 cells, MLCK inhibition stabilized the ZO-1 protein in the TJ by reducing its molecular mobility (Yu et al., 2010).

REFERENCES

Adachi, M., Hamazaki, Y., Kobayashi, Y., Itoh, M., Tsukita, S., Furuse, M., et al. (2009). Similar and distinct properties of MUPP1 and Patj, two homologous PDZ domain-containing tight-junction proteins. *Molecular and Cellular Biology, 29*, 2372–2389.

Angelow, S., & Yu, A. S. (2009). Structure-function studies of claudin extracellular domains by cysteine-scanning mutagenesis. *The Journal of Biological Chemistry, 284*, 29205–29217.

Balda, M. S., Garrett, M. D., & Matter, K. (2003). The ZO-1-associated Y-box factor ZONAB regulates epithelial cell proliferation and cell density. *The Journal of Cell Biology, 160*, 423–432.

Balda, M. S., & Matter, K. (2000). The tight junction protein ZO-1 and an interacting transcription factor regulate ErbB-2 expression. *The EMBO Journal, 19*, 2024–2033.

Barnard, D. C., Ryan, K., Manley, J. L., & Richter, J. D. (2004). Symplekin and xGLD-2 are required for CPEB-mediated cytoplasmic polyadenylation. *Cell, 119*, 641–651.

Bazzoni, G., Martinez-Estrada, O. M., Orsenigo, F., Cordenonsi, M., Citi, S., & Dejana, E. (2000). Interaction of junctional adhesion molecule with the tight junction components ZO-1, cingulin, and occludin. *The Journal of Biological Chemistry, 275*, 20520–20526.

Buchert, M., Papin, M., Bonnans, C., Darido, C., Raye, W. S., Garambois, V., et al. (2010). Symplekin promotes tumorigenicity by up-regulating claudin-2 expression. *Proceedings of the National Academy of Sciences of the United States of America, 107*, 2628–2633.

Bullivant, S., & Ames, A., 3rd. (1966). A simple freeze-fracture replication method for electron microscopy. *The Journal of Cell Biology, 29*, 435–447.

Cain, M. D., Salimi, H., Gong, Y., Yang, L., Hamilton, S. L., Heffernan, J. R., Hou, J., Miller, M. J., Klein, R. S. (2017). Virus entry and replication in the brain precedes blood-brain barrier disruption during intranasal alphavirus infection. *Journal of Neuroimmunology,* 308, 118–130.

Citi, S., Sabanay, H., Jakes, R., Geiger, B., & Kendrick-Jones, J. (1988). Cingulin, a new peripheral component of tight junctions. *Nature, 333*, 272–276.

Colegio, O. R., Van Itallie, C. M., McCrea, H. J., Rahner, C., & Anderson, J. M. (2002). Claudins create charge-selective channels in the paracellular pathway between epithelial cells. *American Journal of Physiology Cell Physiology, 283*, C142–C147.

Cordenonsi, M., D'Atri, F., Hammar, E., Parry, D. A., Kendrick-Jones, J., Shore, D., et al. (1999). Cingulin contains globular and coiled-coil domains and interacts with ZO-1, ZO-2, ZO-3, and myosin. *The Journal of Cell Biology, 147*, 1569–1582.

Cording, J., Berg, J., Kading, N., Bellmann, C., Tscheik, C., Westphal, J. K., et al. (2013). In tight junctions, claudins regulate the interactions between occludin, tricellulin and marvelD3, which, inversely, modulate claudin oligomerization. *Journal of Cell Science, 126*, 554–564.

Cukierman, L., Meertens, L., Bertaux, C., Kajumo, F., & Dragic, T. (2009). Residues in a highly conserved claudin-1 motif are required for hepatitis C virus entry and mediate the formation of cell-cell contacts. *Journal of Virology*, *83*, 5477–5484.

Daugherty, B. L., Ward, C., Smith, T., Ritzenthaler, J. D., & Koval, M. (2007). Regulation of heterotypic claudin compatibility. *The Journal of Biological Chemistry*, *282*, 30005–30013.

Dobrosotskaya, I., Guy, R. K., & James, G. L. (1997). MAGI-1, a membrane-associated guanylate kinase with a unique arrangement of protein-protein interaction domains. *The Journal of Biological Chemistry*, *272*, 31589–31597.

Drenckhahn, D., & Dermietzel, R. (1988). Organization of the actin filament cytoskeleton in the intestinal brush border: A quantitative and qualitative immunoelectron microscope study. *The Journal of Cell Biology*, *107*, 1037–1048.

Ebnet, K., Aurrand-Lions, M., Kuhn, A., Kiefer, F., Butz, S., Zander, K., et al. (2003). The junctional adhesion molecule (JAM) family members JAM-2 and JAM-3 associate with the cell polarity protein PAR-3: A possible role for JAMs in endothelial cell polarity. *Journal of Cell Science*, *116*, 3879–3891.

Ebnet, K., Schulz, C. U., Meyer Zu Brickwedde, M. K., Pendl, G. G., & Vestweber, D. (2000). Junctional adhesion molecule interacts with the PDZ domain-containing proteins AF-6 and ZO-1. *The Journal of Biological Chemistry*, *275*, 27979–27988.

Eskandari, S., Wright, E. M., Kreman, M., Starace, D. M., & Zampighi, G. A. (1998). Structural analysis of cloned plasma membrane proteins by freeze-fracture electron microscopy. *Proceedings of the National Academy of Sciences of the United States of America*, *95*, 11235–11240.

Fanning, A. S., & Anderson, J. M. (1996). Protein-protein interactions: PDZ domain networks. *Current Biology: CB*, *6*, 1385–1388.

Fanning, A. S., Jameson, B. J., Jesaitis, L. A., & Anderson, J. M. (1998). The tight junction protein ZO-1 establishes a link between the transmembrane protein occludin and the actin cytoskeleton. *The Journal of Biological Chemistry*, *273*, 29745–29753.

Farquhar, M. G., & Palade, G. E. (1963). Junctional complexes in various epithelia. *The Journal of Cell Biology*, *17*, 375–412.

Furuse, M., Fujita, K., Hiiragi, T., Fujimoto, K., & Tsukita, S. (1998a). Claudin-1 and -2: Novel integral membrane proteins localizing at tight junctions with no sequence similarity to occludin. *The Journal of Cell Biology*, *141*, 1539–1550.

Furuse, M., Hirase, T., Itoh, M., Nagafuchi, A., Yonemura, S., Tsukita, S., et al. (1993). Occludin: A novel integral membrane protein localizing at tight junctions. *The Journal of Cell Biology*, *123*, 1777–1788.

Furuse, M., Sasaki, H., Fujimoto, K., & Tsukita, S. (1998b). A single gene product, claudin-1 or -2, reconstitutes tight junction strands and recruits occludin in fibroblasts. *The Journal of Cell Biology*, *143*, 391–401.

Gong, Y., Renigunta, V., Zhou, Y., Sunq, A., Wang, J., Yang, J., et al. (2015). Biochemical and biophysical analyses of tight junction permeability made of claudin-16 and claudin-19 dimerization. *Molecular Biology of the Cell*, *26(24)*, 4333–4346.

Gong, Y., Yu, M., Yang, J., Gonzales, E., Perez, R., Hou, M., et al. (2014). The Cap1-claudin-4 regulatory pathway is important for renal chloride reabsorption and blood pressure regulation. *Proceedings of the National Academy of Sciences of the United States of America*, *111*, E3766–E3774.

Gonzalez-Mariscal, L., Chavez de Ramirez, B., & Cereijido, M. (1985). Tight junction formation in cultured epithelial cells (MDCK). *The Journal of Membrane Biology*, *86*, 113–125.

Goodenough, D. A., & Revel, J. P. (1970). A fine structural analysis of intercellular junctions in the mouse liver. *The Journal of Cell Biology*, *45*, 272–290.

Grebenkamper, K., & Galla, H. J. (1994). Translational diffusion measurements of a fluorescent phospholipid between MDCK-I cells support the lipid model of the tight junctions. *Chemistry and Physics of Lipids*, *71*, 133–143.

Gumbiner, B., Lowenkopf, T., & Apatira, D. (1991). Identification of a 160-kDa polypeptide that binds to the tight junction protein ZO-1. *Proceedings of the National Academy of Sciences of the United States of America*, *88*, 3460–3464.

Hamazaki, Y., Itoh, M., Sasaki, H., Furuse, M., & Tsukita, S. (2002). Multi-PDZ domain protein 1 (MUPP1) is concentrated at tight junctions through its possible interaction with claudin-1 and junctional adhesion molecule. *The Journal of Biological Chemistry*, *277*, 455–461.

Haskins, J., Gu, L., Wittchen, E. S., Hibbard, J., & Stevenson, B. R. (1998). ZO-3, a novel member of the MAGUK protein family found at the tight junction, interacts with ZO-1 and occludin. *The Journal of Cell Biology*, *141*, 199–208.

Hecht, G., Pestic, L., Nikcevic, G., Koutsouris, A., Tripuraneni, J., Lorimer, D. D., et al. (1996). Expression of the catalytic domain of myosin light chain kinase increases paracellular permeability. *The American Journal of Physiology*, *271*, C1678–C1684.

Hou, J., Paul, D. L., & Goodenough, D. A. (2005). Paracellin-1 and the modulation of ion selectivity of tight junctions. *Journal of Cell Science*, *118*, 5109–5118.

Hou, J., Rajagopal, M., & Yu, A. S. (2013). Claudins and the kidney. *Annual Review of Physiology*, *75*, 479–501.

Hou, J., Renigunta, A., Konrad, M., Gomes, A. S., Schneeberger, E. E., Paul, D. L., et al. (2008). Claudin-16 and claudin-19 interact and form a cation-selective tight junction complex. *The Journal of Clinical Investigation*, *118*, 619–628.

Hurd, T. W., Gao, L., Roh, M. H., Macara, I. G., & Margolis, B. (2003). Direct interaction of two polarity complexes implicated in epithelial tight junction assembly. *Nature Cell Biology*, *5*, 137–142.

Ide, N., Hata, Y., Nishioka, H., Hirao, K., Yao, I., Deguchi, M., et al. (1999). Localization of membrane-associated guanylate kinase (MAGI)-1/BAI-associated protein (BAP) 1 at tight junctions of epithelial cells. *Oncogene*, *18*, 7810–7815.

Ikenouchi, J., Furuse, M., Furuse, K., Sasaki, H., Tsukita, S., & Tsukita, S. (2005). Tricellulin constitutes a novel barrier at tricellular contacts of epithelial cells. *The Journal of Cell Biology*, *171*, 939–945.

Itoh, M., Furuse, M., Morita, K., Kubota, K., Saitou, M., & Tsukita, S. (1999a). Direct binding of three tight junction-associated MAGUKs, ZO-1, ZO-2, and ZO-3, with the COOH termini of claudins. *The Journal of Cell Biology*, *147*, 1351–1363.

Itoh, M., Morita, K., & Tsukita, S. (1999b). Characterization of ZO-2 as a MAGUK family member associated with tight as well as adherens junctions with a binding affinity to occludin and alpha catenin. *The Journal of Biological Chemistry*, *274*, 5981–5986.

Itoh, M., Nagafuchi, A., Moroi, S., & Tsukita, S. (1997). Involvement of ZO-1 in cadherin-based cell adhesion through its direct binding to alpha catenin and actin filaments. *The Journal of Cell Biology*, *138*, 181–192.

Itoh, M., Sasaki, H., Furuse, M., Ozaki, H., Kita, T., & Tsukita, S. (2001). Junctional adhesion molecule (JAM) binds to PAR-3: A possible mechanism for the recruitment of PAR-3 to tight junctions. *The Journal of Cell Biology*, *154*, 491–497.

Jeansonne, B., Lu, Q., Goodenough, D. A., & Chen, Y. H. (2003). Claudin-8 interacts with multi-PDZ domain protein 1 (MUPP1) and reduces paracellular conductance in epithelial cells. *Cellular and Molecular Biology (Noisy-le-Grand, France)*, *49*, 13–21.

Kachar, B., & Reese, T. S. (1982). Evidence for the lipidic nature of tight junction strands. *Nature*, *296*, 464–466.

Keon, B. H., Schafer, S., Kuhn, C., Grund, C., & Franke, W. W. (1996). Symplekin, a novel type of tight junction plaque protein. *The Journal of Cell Biology, 134*, 1003–1018.

Kostrewa, D., Brockhaus, M., D'Arcy, A., Dale, G. E., Nelboeck, P., Schmid, G., et al. (2001). X-ray structure of junctional adhesion molecule: Structural basis for homophilic adhesion via a novel dimerization motif. *The EMBO Journal, 20*, 4391–4398.

Krause, G., Winkler, L., Mueller, S. L., Haseloff, R. F., Piontek, J., & Blasig, I. E. (2008). Structure and function of claudins. *Biochimica et biophysica acta, 1778*, 631–645.

Lanaspa, M. A., Almeida, N. E., Andres-Hernando, A., Rivard, C. J., Capasso, J. M., & Berl, T. (2007). The tight junction protein, MUPP1, is up-regulated by hypertonicity and is important in the osmotic stress response in kidney cells. *Proceedings of the National Academy of Sciences of the United States of America, 104*, 13672–13677.

Li, Y., Fanning, A. S., Anderson, J. M., & Lavie, A. (2005). Structure of the conserved cytoplasmic C-terminal domain of occludin: Identification of the ZO-1 binding surface. *Journal of Molecular Biology, 352*, 151–164.

Luissint, A. C., Nusrat, A., & Parkos, C. A. (2014). JAM-related proteins in mucosal homeostasis and inflammation. *Seminars in Immunopathology, 36*, 211–226.

Madara, J. L., Barenberg, D., & Carlson, S. (1986). Effects of cytochalasin D on occluding junctions of intestinal absorptive cells: Further evidence that the cytoskeleton may influence paracellular permeability and junctional charge selectivity. *The Journal of Cell Biology, 102*, 2125–2136.

Martin-Padura, I., Lostaglio, S., Schneemann, M., Williams, L., Romano, M., Fruscella, P., et al. (1998). Junctional adhesion molecule, a novel member of the immunoglobulin superfamily that distributes at intercellular junctions and modulates monocyte transmigration. *The Journal of Cell Biology, 142*, 117–127.

Meyer, H. W. (1983). Tight junction strands are lipidic cylinders. *Die Naturwissenschaften, 70*, 251–252.

Meza, I., Ibarra, G., Sabanero, M., Martinez-Palomo, A., & Cereijido, M. (1980). Occluding junctions and cytoskeletal components in a cultured transporting epithelium. *The Journal of Cell Biology, 87*, 746–754.

Mineta, K., Yamamoto, Y., Yamazaki, Y., Tanaka, H., Tada, Y., Saito, K., et al. (2011). Predicted expansion of the claudin multigene family. *FEBS Letters, 585*, 606–612.

Mooseker, M. S., Bonder, E. M., Grimwade, B. G., Howe, C. L., Keller, T. C., 3rd, Wasserman, R. H., et al. (1982). Regulation of contractility, cytoskeletal structure, and filament assembly in the brush border of intestinal epithelial cells. *Cold Spring Harbor Symposia on Quantitative Biology, 46*(Pt 2), 855–870.

Patrie, K. M., Drescher, A. J., Welihinda, A., Mundel, P., & Margolis, B. (2002). Interaction of two actin-binding proteins, synaptopodin and alpha-actinin-4, with the tight junction protein MAGI-1. *The Journal of Biological Chemistry, 277*, 30183–30190.

Piontek, J., Winkler, L., Wolburg, H., Muller, S. L., Zuleger, N., Piehl, C., et al. (2008). Formation of tight junction: Determinants of homophilic interaction between classic claudins. *FASEB Journal, 22*, 146–158.

Raleigh, D. R., Marchiando, A. M., Zhang, Y., Shen, L., Sasaki, H., Wang, Y., et al. (2010). Tight junction-associated MARVEL proteins marveld3, tricellulin, and occludin have distinct but overlapping functions. *Molecular Biology of the Cell, 21*, 1200–1213.

Rossa, J., Ploeger, C., Vorreiter, F., Saleh, T., Protze, J., Gunzel, D., et al. (2014a). Claudin-3 and claudin-5 protein folding and assembly into the tight junction are controlled by non-conserved residues in the transmembrane 3 (TM3) and extracellular loop 2 (ECL2) segments. *The Journal of Biological Chemistry, 289*, 7641–7653.

Rossa, J., Protze, J., Kern, C., Piontek, A., Gunzel, D., Krause, G., et al. (2014b). Molecular and structural transmembrane determinants critical for embedding claudin-5 into tight junctions reveal a distinct four-helix bundle arrangement. *The Biochemical Journal, 464*, 49–60.

Saitoh, Y., Suzuki, H., Tani, K., Nishikawa, K., Irie, K., Ogura, Y., et al. (2015). Tight junctions. Structural insight into tight junction disassembly by *Clostridium perfringens* enterotoxin. *Science (New York, NY), 347*, 775–778.

Sanchez-Pulido, L., Martin-Belmonte, F., Valencia, A., & Alonso, M. A. (2002). MARVEL: A conserved domain involved in membrane apposition events. *Trends in Biochemical Sciences, 27*, 599–601.

Sasaki, H., Matsui, C., Furuse, K., Mimori-Kiyosue, Y., Furuse, M., & Tsukita, S. (2003). Dynamic behavior of paired claudin strands within apposing plasma membranes. *Proceedings of the National Academy of Sciences of the United States of America, 100*, 3971–3976.

Shen, L., Weber, C. R., & Turner, J. R. (2008). The tight junction protein complex undergoes rapid and continuous molecular remodeling at steady state. *The Journal of Cell Biology, 181*, 683–695.

Shin, K., Straight, S., & Margolis, B. (2005). PATJ regulates tight junction formation and polarity in mammalian epithelial cells. *The Journal of Cell Biology, 168*, 705–711.

Shinoda, T., Shinya, N., Ito, K., Ohsawa, N., Terada, T., Hirata, K., et al. (2016). Structural basis for disruption of claudin assembly in tight junctions by an enterotoxin. *Scientific Reports, 6*, 33632.

Steed, E., Rodrigues, N. T., Balda, M. S., & Matter, K. (2009). Identification of MarvelD3 as a tight junction-associated transmembrane protein of the occludin family. *BMC Cell Biology, 10*, 95.

Stevenson, B. R., Siliciano, J. D., Mooseker, M. S., & Goodenough, D. A. (1986). Identification of ZO-1: A high molecular weight polypeptide associated with the tight junction (zonula occludens) in a variety of epithelia. *The Journal of Cell Biology, 103*, 755–766.

Suzuki, H., Nishizawa, T., Tani, K., Yamazaki, Y., Tamura, A., Ishitani, R., et al. (2014). Crystal structure of a claudin provides insight into the architecture of tight junctions. *Science (New York, NY), 344*, 304–307.

Turner, J. R. (2000). 'Putting the squeeze' on the tight junction: Understanding cytoskeletal regulation. *Seminars in Cell & Developmental Biology, 11*, 301–308.

Turner, J. R., Rill, B. K., Carlson, S. L., Carnes, D., Kerner, R., Mrsny, R. J., et al. (1997). Physiological regulation of epithelial tight junctions is associated with myosin light-chain phosphorylation. *The American Journal of Physiology, 273*, C1378–C1385.

Umeda, K., Ikenouchi, J., Katahira-Tayama, S., Furuse, K., Sasaki, H., Nakayama, M., et al. (2006). ZO-1 and ZO-2 independently determine where claudins are polymerized in tight-junction strand formation. *Cell, 126*, 741–754.

Van Itallie, C., Rahner, C., & Anderson, J. M. (2001). Regulated expression of claudin-4 decreases paracellular conductance through a selective decrease in sodium permeability. *The Journal of Clinical Investigation, 107*, 1319–1327.

Van Itallie, C. M., Mitic, L. L., & Anderson, J. M. (2011). Claudin-2 forms homodimers and is a component of a high molecular weight protein complex. *The Journal of Biological Chemistry, 286*, 3442–3450.

Van Itallie, C. M., Tietgens, A. J., & Anderson, J. M. (2017). Visualizing the dynamic coupling of claudin strands to the actin cytoskeleton through ZO-1. *Molecular Biology of the Cell, 28*, 524–534.

van Meer, G., Gumbiner, B., & Simons, K. (1986). The tight junction does not allow lipid molecules to diffuse from one epithelial cell to the next. *Nature, 322*, 639–641.

Wittchen, E. S., Haskins, J., & Stevenson, B. R. (1999). Protein interactions at the tight junction. Actin has multiple binding partners, and ZO-1 forms independent complexes with ZO-2 and ZO-3. *The Journal of Biological Chemistry, 274*, 35179–35185.

Yu, A. S., Cheng, M. H., Angelow, S., Gunzel, D., Kanzawa, S. A., Schneeberger, E. E., et al. (2009). Molecular basis for cation selectivity in claudin-2-based paracellular pores: Identification of an electrostatic interaction site. *The Journal of General Physiology, 133*, 111–127.

Yu, D., Marchiando, A. M., Weber, C. R., Raleigh, D. R., Wang, Y., Shen, L., et al. (2010). MLCK-dependent exchange and actin binding region-dependent anchoring of ZO-1 regulate tight junction barrier function. *Proceedings of the National Academy of Sciences of the United States of America, 107*, 8237–8241.

Chapter 3

Paracellular Channel Recording

3.1 THEORETIC CONSIDERATIONS

3.1.1 Equivalent Electric Circuit of an Epithelium

Electric current across the biologic membrane is caused by ion flow. Ion conductance and ion permeability are related, but not equivalent. Permeability is an intrinsic property of the membrane, independent of ion concentration; whereas ion conductance is concentration-dependent. Transepithelial electric current includes contributions from two pathways: the transcellular pathway and the paracellular pathway (Fig. 3.1A). In the equivalent circuit of the epithelium (Fig. 3.1B), R denotes the electric resistance and E denotes the equivalent electromotive force (EMC, also known as zero-current voltage). EMC is caused by the electrogenic diffusion of permeant ions. The subscripts denote: a, apical membrane; b, basolateral membrane; and s, shunt or paracellular.

3.1.2 Transepithelial Resistance

The transepithelial resistance (TER or R_{te}) depends upon the transcellular ($R_a + R_b$) and the paracellular (R_s) resistance according to the following equation.

$$\frac{1}{R_{te}} = \frac{1}{R_a + R_b} + \frac{1}{R_s} \tag{3.1}$$

The degree of epithelial leakiness is best assessed by the ratio of transcellular resistance versus paracellular resistance, that is, $(R_a + R_b)/R_s$. The larger this ratio is, the leakier the epithelium becomes. It is clear that the conductance (G) is proportional to area, and consequently the resistance ($R = 1/G$) decreases with increasing area. This is the reason why R_{te} is expressed as $\Omega\cdot cm^2$; whereas G_{te} as S/cm^2. In general, R_{te} is measured by applying a current pulse, ΔI, across the epithelium and recording the resultant voltage change ΔV. According to Ohm's law, the resistance is calculated as the voltage difference across the epithelium divided by the amplitude of the current pulse through the epithelium ($\Delta V/\Delta I$).

$$R_{te} = \frac{\Delta V}{\Delta I} \tag{3.2}$$

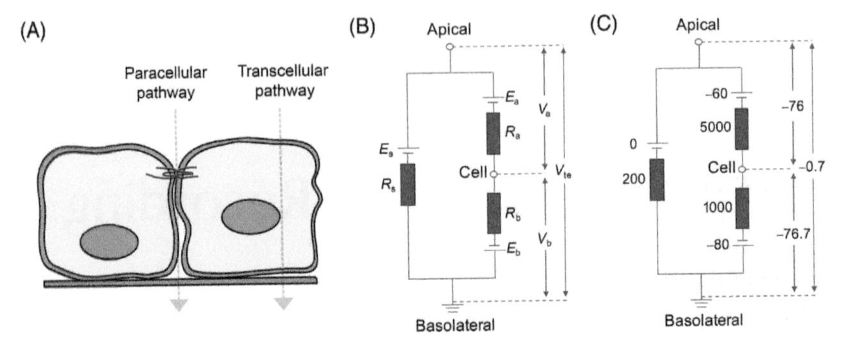

FIGURE 3.1 Equivalent electric circuit of an epithelium. (A) Ions can permeate through the tight junction (the paracellular pathway) or through the cell (the transcellular pathway) by crossing the apical and the basolateral membrane. (B) Each element in the circuit (*a*, apical membrane; *b*, basolateral membrane; *s*, paracellular shunt pathway) is represented by an electric equivalent, that is, an electromotive force, *E*, in series with a resistor, *R*. (C) Membrane voltages (in mV) for a realistic epithelium with the assumed values of *E* (in mV) and *R* (in $\Omega\cdot\text{cm}^2$). Note that the membrane voltages differ from the respective *E* values. (Reproduced *with permission from Reuss, L., Bello-Reuss, E., & Grady, T. P. (1979). Effects of ouabain on fluid transport and electrical properties of* Necturus *gallbladder. Evidence in favor of a neutral basolateral sodium transport mechanism. The Journal of General Physiology, 73, 385–402.)*

3.1.3 Transepithelial Voltage

The transepithelial voltage depends upon *E* and *R* of all circuit elements. The reason is that there is an intraepithelial loop current due to the presence of the paracellular pathway. The loop current (I_e) is given by:

$$I_e = \frac{E_b + E_s - E_a}{R_a + R_b + R_s} \quad (3.3)$$

and the membrane voltage is given by:

$$V = E - I_e R \quad (3.4)$$

Inserting the expression for I_e into the membrane voltage equation yields the following three equations for the apical membrane voltage (V_a), the basolateral membrane voltage (V_b), and the transepithelial voltage (V_{te}) or the shunt voltage (V_s).

$$V_a = \frac{E_a\left(R_a + R_s\right) + \left(E_b - E_s\right)R_a}{R_a + R_b + R_s} \quad (3.5)$$

$$V_b = \frac{E_b\left(R_a + R_s\right) + \left(E_a + E_s\right)R_b}{R_a + R_b + R_s} \quad (3.6)$$

$$V_{te}, V_s = \frac{(E_b - E_a)R_s + E_s(R_a + R_b)}{R_a + R_b + R_s} \tag{3.7}$$

Fig. 3.1C shows the solution of Eqs. (3.5)–(3.7) for assumed values of E and R in realistic cell membranes. The voltages of the two membrane domains depend upon each other because they are connected by the paracellular pathway, which acts as an electric shunt. The changes in the properties of one of the membranes alter not only its own voltage, but also the voltage across the opposite membrane and the voltage across the entire epithelium. The electric shunting process requires a low junctional resistance (R_S) due to the presence of the paracellular channel. If the value of R_S approaches infinity, then the cell membrane voltage approaches the value of the respective E, and I_e approaches zero.

3.1.4 Transepithelial Flux Assay

Fluxes of various hydrophilic tracers can be measured to determine the permeability properties across the epithelium. Fick's law relates the diffusive flux to the concentration of the tracer under the assumption that the concentration profile of the tracer is linear in the epithelium, and that the electric potential is zero across the epithelium if the tracer carries charge. Considering at a given point χ, the concentration of the tracer j is a function of $C(\chi)$.

$$J(x) = -D\frac{dC(x)}{dx} \qquad \text{Fick's law}$$

where J is the *flux* or the number of tracer molecules that pass a unit area per unit time; D is called the *diffusion coefficient* of the tracer.

By definition,

$$P = \frac{D}{h}$$

where P is the permeability coefficient; h is the thickness of the epithelium.

Because the concentration profile of the tracer is linear in the epithelium,

$$\frac{dC}{dx} = \frac{C^{(a)} - C^{(b)}}{h}$$

where $C^{(b)}$ and $C^{(a)}$ are the concentration of the tracer j in the bathing buffer of basal and apical sides; assuming $C^{(b)} > C^{(a)}$, which, when inserted in Fick's law, gives

$$J = -D\frac{C^{(a)} - C^{(b)}}{h}$$

When written in P, the equation aforementioned gives

$$J = P[C^{(b)} - C^{(a)}]$$

and relates the permeability of the tracer j with its flux rate and concentration at basal and apical sides.

Let $V^{(b)}$ and $V^{(a)}$ represent the volume of basal and apical compartments. $C^{(b)}(t)$ and $C^{(a)}(t)$ are the concentration of the tracer molecule in the bathing buffer of basal and apical compartments at time t; $m^{(b)}(t)$ and $m^{(a)}(t)$ are the number of the tracer molecule in the bathing buffer of basal and apical compartments at time t.

$$J(t) = P(C^{(b)}(t) - C^{(a)}(t))$$

After time dt, the apical compartment receives an amount of tracer molecule $dm^{(a)}$, that is,

$$dm^{(a)} = AJ(t)dt = AP(C^{(b)}(t) - C^{(a)}(t))dt \tag{3.8}$$

where A is the area of the epithelium. The law of mass conservation requires,

$$m_t = m^{(b)}(0) + m^{(a)}(0) = m^{(b)}(t) + m^{(a)}(t)$$

where m_t is the total amount of tracer molecule.

The concentration $C(t)$ are then written as

$$C^{(b)}(t) = \frac{m^{(b)}(t)}{V^{(b)}} = \frac{m_t - m^{(a)}(t)}{V^{(b)}} \qquad C^{(a)}(t) = \frac{m^{(a)}(t)}{V^{(a)}}$$

Thus,

$$\frac{dm^{(a)}}{dt} = AP\{\frac{m_t - m^{(a)}(t)}{V^{(b)}} - \frac{m^{(a)}(t)}{V^{(a)}}\} = AP\{\frac{m_t}{V^{(b)}} - (\frac{1}{V^{(b)}} + \frac{1}{V^{(a)}})m^{(a)}(t)\}$$

$$= AP\frac{m_t}{V^{(b)}} - AP\frac{V^{(b)} + V^{(a)}}{V^{(b)}V^{(a)}}m^{(a)}(t)$$

Dividing by $V^{(a)}$ gives the concentration equation,

$$\frac{dC^{(a)}}{dt} + kC^{(a)} = b$$

where

$$k = AP\frac{V^{(b)} + V^{(a)}}{V^{(b)}V^{(a)}} \qquad b = AP\frac{m_t}{V^{(b)}V^{(a)}}$$

Note that

$$\frac{b}{k} = \frac{m_t}{V^{(b)} + V^{(a)}} = C(\infty)$$

Rearranging the concentration equation gives,

$$\frac{d(b - kC^{(a)})}{b - kC^{(a)}} = -kdt$$

Integration from $t = 0$ gives,

$$\int_0^t \ln(b - kC^{(a)}) = -k\int_0^t t$$

$$\ln\frac{b - kC^{(a)}(t)}{b - kC^{(a)}(0)} = -kt$$

Replacing b gives,

$$\ln\frac{C(\infty) - C^{(a)}(t)}{C(\infty) - C^{(a)}(0)} = -kt = -APt\frac{V^{(b)} + V^{(a)}}{V^{(b)}V^{(a)}}$$

Thus,

$$P = -\frac{V^{(b)}V^{(a)}}{tA(V^{(b)} + V^{(a)})}\ln\frac{m_t - C^{(a)}(t)(V^{(b)} + V^{(a)})}{m_t - C^{(a)}(0)(V^{(b)} + V^{(a)})} \tag{3.9}$$

for calculation of the permeability coefficient, provided that the total amount of tracer molecule m_t and the volume of basal and apical compartments are known.

If there is no tracer in the apical compartment at $t = 0$ or $C^{(a)}(0) = 0$, the equation aforementioned is simplified as,

$$P = -\frac{V^{(b)}V^{(a)}}{tA(V^{(b)} + V^{(a)})} \ln \frac{m_t - C^{(a)}(t)(V^{(b)} + V^{(a)})}{m_t}$$

$$P = -\frac{V^{(b)}V^{(a)}}{tA(V^{(b)} + V^{(a)})} \ln \frac{C^{(b)}(0)V^{(b)} - C^{(a)}(t)(V^{(b)} + V^{(a)})}{C^{(b)}(0)V^{(b)}}$$

where $m_t = C^{(b)}(0)V^{(b)}$

When experiments are conducted in the *Ussing* chamber with $V^{(b)} = V^{(a)} = V$,

$$P = -\frac{V}{2tA} \ln \frac{C^{(b)}(0) - 2C^{(a)}(t)}{C^{(b)}(0)}$$

If $C^{(b)}(0) > > C^{(a)}(t)$ or diffusion is a slow process, Taylor approximation applies with

$$\ln(1 + x) = x - \frac{1}{2}x^2 + \frac{1}{3}x^3 - \cdots \qquad -1 < \chi < 1$$

Thus,

$$P = -\frac{V}{2tA} \frac{-2C^{(a)}(t)}{C^{(b)}(0)} = \frac{VC^{(a)}(t)}{tAC^{(b)}(0)} \tag{3.10}$$

3.1.5 Diffusion Potential and Ion Selectivity

The ion selectivity of an epithelium can be estimated from the diffusion potential that builds up across the epithelium if an ion gradient is applied. The *Goldman–Hodgkin–Katz* equation relates the ion selectivity to the diffusion potential. Its derivation requires the following assumptions:

1. The membrane has the thickness of h.
2. All the ions are monovalent.

3. The concentrations of an ion of type j inside the epithelium are at the boundaries at $\chi = 0$ and $\chi = h$ equal to $C_j(0)$ and $C_j(h)$. Their connections to the outside concentrations $C^{(b)}$ (basal side) and $C^{(a)}$ (apical side) are given by $C_j(0) = \alpha_j C^{(b)}$, $C_j(h) = \alpha_j C^{(a)}$, where α_j is the partition coefficient between the epithelium and the aqueous solution for the ion j.
4. The potentials at $\chi = 0$ and $\chi = h$ equal to $\phi(0)$ and $\phi(h)$. For reasons of convenience, $\phi(h)$ is regarded as zero (apical). The electric field in the epithelium is assumed to be constant. Hence, $\frac{d\phi}{dx} = \frac{\phi(h) - \phi(0)}{h} = -\frac{\phi(0)}{h} = -\frac{V}{h}$, where V is the transepithelial potential.
5. The total current I across the epithelium is zero at equilibrium.

   According to the *Nernst–Planck* equation, $J_j = -D_j \frac{dC_j}{dx} - z_j e B_j C_j \frac{d\phi}{dx}$

where J_j is the flux of the ion j at the position χ with the diffusion coefficient D_j and mobility B_j and concentration $C_j(\chi)$ under the influence of the concentration gradient $dC_j(\chi)/d\chi$ and the potential gradient $d\phi/d\chi$, and e is the elementary charge.

The Einstein relation gives, $D_j = kTB_j$, in which k is the *Boltzmann* constant and T is the *Kelvin* temperature.

Eliminating B_j gives,

$$-\frac{J_j}{D_j} = \frac{dC_j}{dx} + \frac{z_j e C_j}{kT}\frac{d\phi}{dx}$$

A new variable ψ is defined as

$$\psi = \frac{\phi}{kT/e}$$

Thus,

$$-\frac{J_j}{D_j} = \frac{dC_j}{dx} + z_j C_j \frac{d\psi}{dx}$$

Applying *Kramers*' transformation gives,

$$J_j e^{z_j\psi} = -D_j \frac{d}{dx}\left(C_j e^{z_j\psi}\right)$$

Integration from $\chi = 0$ to $\chi = h$ gives,

$$\int_0^h J_j e^{z_j\psi(x)}dx = -\int_{C(0),\psi(0)}^{C(h),\psi(h)} D_j d\left(C_j(x)e^{z_j\psi(x)}\right)$$

At the steady state, $dJ_j/d\chi = 0$. Thus,

$$J_j = -D_j \frac{C_j(h)e^{z_j\psi(h)} - C_j(0)e^{z_j\psi(0)}}{\int_0^h e^{z_j\psi(x)dx}}$$

Making use of

$$d\psi = \frac{d\phi}{kT/e}$$

$$\frac{d\psi}{dx} = \frac{\frac{d\phi}{dx}}{kT/e} = \frac{-\frac{V}{h}}{kT/e} = -\frac{v}{h}$$

where $v = \frac{V}{kT/e}$, and $\psi(h) = 0$.

$$\int_0^h e^{z_j\psi(x)}dx = \int_0^h e^{z_j\psi(x)}(\frac{dx}{d\psi})d\psi = -\frac{h}{v}\int_v^0 e^{z_j\psi(x)}d\psi = -\frac{h}{z_jv}[1-e^{z_jv}]$$

Hence,

$$J_j = -D_j \frac{C_j(h)e^{z_j\psi(h)} - C_j(0)e^{z_j\psi(0)}}{\int_0^h e^{z_j\psi(x)dx}} = z_jv\frac{D_j}{h}\frac{C_j(h) - C_j(0)e^{z_jv}}{1-e^{z_jv}}$$

As $C_j(0) = \alpha_jC^{(b)}$ and $C_j(h) = \alpha_jC^{(a)}$,

$$J_j = z_jv\frac{ajD_j}{h}\frac{C^{(a)} - C^{(b)}e^{z_jv}}{1-e^{z_jv}} = z_jP_jv\frac{C^{(a)} - C^{(b)}e^{z_jv}}{1-e^{z_jv}}$$

where, by definition, $P_j = \frac{a_jD_j}{h}$. If the apical and the basal solutions contain only NaCl, and by definition, $C^{(a)}$ and $C^{(b)}$ are the apical and the basal NaCl concentrations, $C = C^{(a)}$, $\varepsilon = C^{(b)}/C^{(a)}$ (dilution factor), then

$$J_{Na} = P_{Na}vC\frac{1-\varepsilon e^{v}}{1-e^{v}}$$

$$J_{Cl} = -P_{Cl}\nu C\frac{1-\varepsilon e^{-\nu}}{1-e^{-\nu}}$$

The currents are,

$$I_{Na} = z_j F J_{Na} = FP_{Na}\nu C\frac{1-\varepsilon e^{\nu}}{1-e^{\nu}}$$

$$I_{Cl} = z_j F J_{Cl} = FP_{Cl}\nu C\frac{1-\varepsilon e^{-\nu}}{1-e^{-\nu}}$$

$$= FP_{Cl}\nu C\frac{\varepsilon - e^{\nu}}{1-e^{\nu}}$$

where F is the *Faraday* constant.

As $I = I_{Na} + I_{Cl} = 0$,

$$FP_{Na}\nu C\frac{1-\varepsilon e^{\nu}}{1-e^{\nu}} + FP_{Cl}\nu C\frac{\varepsilon - e^{\nu}}{1-e^{\nu}} = 0$$

$$P_{Na}(1-\varepsilon e^{\nu}) + P_{Cl}(\varepsilon - e^{\nu}) = 0$$

By definition, $\eta = P_{Na}/P_{Cl}$,

$$\eta = \frac{-(\varepsilon - e^{\nu})}{(1-\varepsilon e^{\nu})} \tag{3.11}$$

When $\varepsilon = 1$, the dilution potential $\nu = 0$.

3.1.6 A Simplified Regimen to Calculate Na^+ and Cl^- Permeability

By *Ohm's* law, the total ion conductance G (1/TER or $1/R_{te}$) through the epithelium can be deduced from Eq. (3.2).

By definition, the partial conductance of ion j is,

$$\frac{1}{G_j} = \int_0^h \frac{1}{z_j^2 e B_j F C_j(x)} dx$$

When there is no concentration gradient, $C_j(x) = a_j C^{(b)} = a_j C^{(a)} = a_j C$.

$$G_j = \frac{z_j^2 e B_j F a_j C}{h} = \frac{z_j^2 e D_j F a_j C}{kTh} = \frac{(z_j F)^2 C D_j a_j}{RTh} = \frac{(z_j F)^2 C}{RT} P_j$$

where $D_j = kTB_j$ (*Einstein* relation), $k/e = R/F$ (R is the gas constant), and $P_j = \frac{a_j D_j}{h}$

$$G = G_{Na} + G_{Cl} = \frac{F^2 C}{RT} P_{Na} + \frac{F^2 C}{RT} P_{Cl} = \frac{F^2 C}{RT}(P_{Na} + P_{Cl})$$

Making use of $\eta = P_{Na}/P_{Cl}$

$$P_{Na} = \frac{RT}{F^2}\frac{G}{C}\frac{\eta}{(\eta+1)} \tag{3.12}$$

$$P_{Cl} = \frac{RT}{F^2}\frac{G}{C}\frac{1}{(\eta+1)} \tag{3.13}$$

3.1.7 Transepithelial Water Permeability

According to Fick's law and Eq. (3.8), the equation for solute flow can be written as,

$$dm^{(a)} = AJ(t)dt = AP(C^{(b)}(t) - C^{(a)}(t))dt$$

Analogously, the equation for volume flow when no hydrostatic pressure difference exists between the apical and basal compartments is based upon the proportionality of the flow of volume $dV^{(a)}/dt$ (due to the flow of water), to the difference of osmotic pressure ($\pi^{(a)} - \pi^{(b)}$) between the apical and basal solutions (Kedem & Katchalsky, 1958).

$$\frac{dV^{(a)}}{dt} = AP_{os}(\pi^{(a)} - \pi^{(b)}) \tag{3.14}$$

where P_{os} denotes the permeability coefficient of water. Inserting $\pi = RTC$ to Eq. (3.14) gives,

$$\frac{dV^{(a)}}{dt} = ARTP_{os}(C^{(a)} - C^{(b)})$$

where C is the osmotic concentration and denotes the sum of the concentrations of all the solutes no matter whether the epithelium is permeable to them or not; R is the gas constant; and T is the *Kelvin* temperature.

If the system only contains nonpermeable solutes, N_m, and the basal solution is continuously perfused to maintain its osmotic concentration, then $C^{(b)}$ is a constant, and $C^{(a)} = N_m^{(a)} / V^{(a)}$, where $N_m^{(a)}$ is also a constant. The aforementioned equation can be written as,

$$\frac{dV^{(a)}}{dt} = ARTP_{os}(\frac{N_m^{(a)}}{V^{(a)}} - C^{(b)}) \tag{3.15}$$

Solving this differential equation, however, has exceeded the scope of this book.

3.2 PRACTICAL APPLICATIONS

3.2.1 The Ussing Chamber

3.2.1.1 Basic Configuration

In the 1950s, Dr. Hans Ussing made the first breakthrough in the measurement of transepithelial ion transport. Dr. Ussing developed an apparatus, later known as the Ussing chamber, with paired compartments between which an epithelial tissue can be mounted and analyzed (Fig. 3.2) (Ussing, 1953). By putting the same solution on both sides and short-circuiting the transepithelial voltage, Dr. Ussing was able to simultaneously measure electric current and Na^+ flux across the frog skin (Ussing & Zerahn, 1951). The fluids on both sides of the tissue are stirred by a circulating bubble lift, which, at the same time, provides equilibration of these fluids with O_2 and CO_2. Thermostated water can be infused into the water jackets of the chamber to maintain a desired temperature. Salt bridges are used both to eliminate the liquid junction potential and to prevent exposure of the epithelium to the toxic effects of Ag from the Ag-AgCl electrodes. The salt bridges typically consist of 3% agar melted in 3 M KCl solution that is congealed in the tubing.

3.2.1.2 Transepithelial Resistance Measurement

During transepithelial resistance measurement, Ringer's solution is filled into both compartments of the Ussing chamber. Two pairs of electrodes are placed into the compartments and separated by the epithelial tissue (Fig. 3.2). When an electric current step (ΔI) generated by the current source passes through the

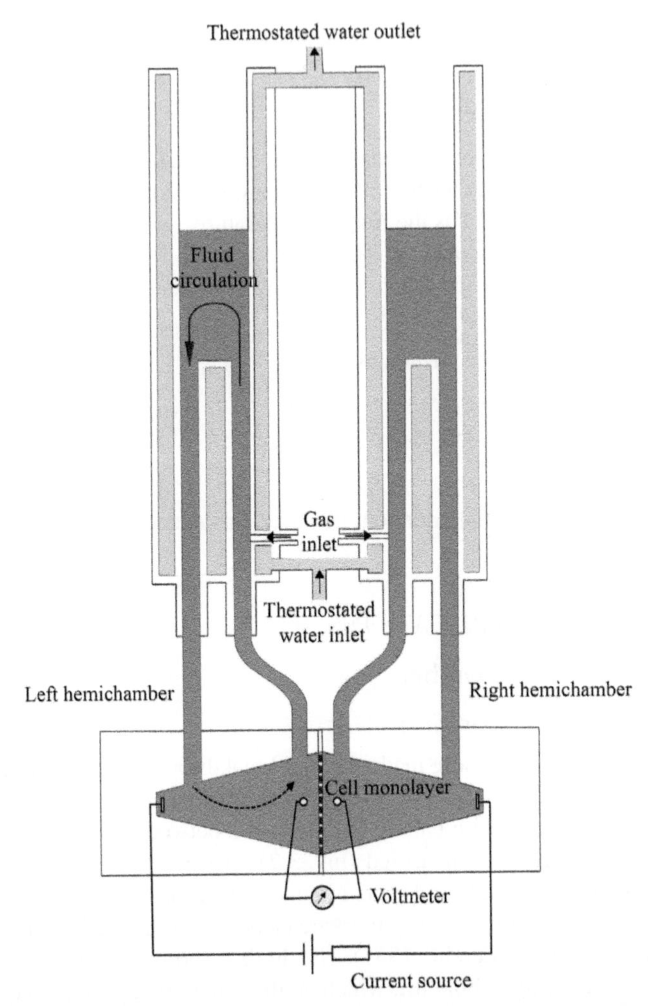

FIGURE 3.2 Schematic diagram of the Ussing chamber. The Ussing chamber is named after the Danish physiologist Dr. Hans Ussing and consists of two fluid-filled hemichambers separated by an epithelial tissue in the center. Two fluid reservoirs with gas lifts provide circulation and equilibration with O_2 and CO_2. Electric current is applied via two current electrodes connected to a current generator. Transepithelial voltage is recorded by two voltage electrodes connected to a voltmeter.

tissue from the pair of current electrodes, a voltage change (ΔV) across the tissue due to the transepithelial resistance, TER or R_{te}, can be recorded by the pair of voltage electrodes connected to a voltmeter. TER or R_{te} can then be calculated according to Eq. (3.2).

In a simplified form of the Ussing chamber, an epithelial tissue can be prepared on a permeable filter support and analyzed in a culture dish known as the Transwell. TER or R_{te} is conveniently and repetitively measured by the commercially available "chopstick electrode" system. While the principle for determining TER or R_{te} holds the same for the Ussing chamber and the chopstick system, the measurements made by the chopstick system often yield very different results from those by the Ussing chamber. First, due to a nonuniform current field produced by the chopstick electrodes across the epithelial tissue, the TER or R_{te} value may vary depending on the position of the electrodes relative to the tissue (Jovov, Wills, & Lewis, 1991). Second, because the fluids are not stirred in the Transwell, an unstirred layer may develop in the apical or basal side of the epithelium to cause local changes in the concentration of ions and solutes, which will in turn affect TER or R_{te}. Finally, the chopstick system is usually driven by alternating electric currents, which may short-circuit the resistor elements of the cell membrane to cause an underestimation of TER or R_{te}.

3.2.1.3 Diffusion Potential Measurement

The ion selectivity of transepithelial transport can be deduced from the diffusion potential that builds up across the epithelium according to Eq. (3.11). To determine, for example, whether the Na^+ and Cl^- permeabilities (P_{Na}, P_{Cl}) across the epithelium differ, part of the NaCl in the bathing solution on one side of the epithelium is replaced iso-osmotically by an uncharged substance, for example, mannitol. Thus, NaCl is "diluted" on one side of the epithelium. If $P_{Na} \neq P_{Cl}$, a transepithelial potential difference (diffusion or dilution potential, ΔV) will develop, which, for an ion permeability ratio (P_{Na}/P_{Cl}) > 1, is positive on the diluted side, and vice versa. The absolute values of P_{Na} and P_{Cl} are calculated from the ion selectivity (P_{Na}/P_{Cl}) and the transepithelial resistance (TER or R_{te}) using Eqs. (3.12) and (3.13).

3.2.1.4 Impedance Measurement

Simple TER or R_{te} measurements with a direct current (DC) circuit do not take cell membrane capacitance into account. Part of the electric current passing through the circuit is in fact diverted to charge the cell membrane capacitor (Fig. 3.3A). Impedance spectroscopy makes a more accurate measurement of TER or R_{te} by using an alternating current (AC) circuit. An alternating current (I) with an angular frequency (ω) generates an oscillating potential (V) across the epithelium with the same frequency but different phase. The impedance (Z_{te}), deriving from V/I, reflects the transepithelial resistance when ω approaches zero (Fig. 3.3B).

3.2.1.5 Differentiation of Paracellular Pathway From Transcellular Pathway

It is important to note that both the paracellular pathway and the transcellular pathway can influence the value of TER or R_{te}, no matter whether it is measured with the DC circuit (Fig. 3.1B) or the AC circuit (Fig. 3.3A). In a leaky epithelium, that is, $R_s << R_a + R_b$ (Fig. 3.1C), TER or R_{te} approaches R_s according to Eq. (3.1). Blocking the transcellular pathway with various ion channel inhibitors further increases $R_a + R_b$ to ensure R_s dominates R_{te} even in a tight epithelium.

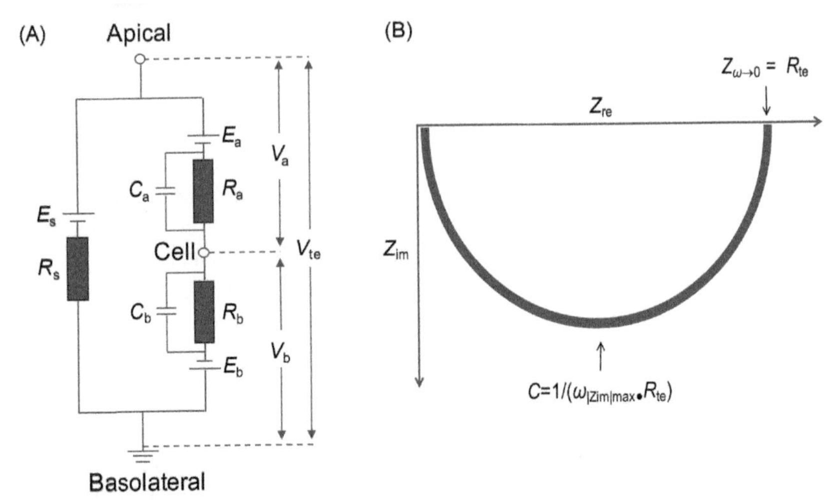

FIGURE 3.3 **Equivalent electric circuit for impedance measurement.** (A) An AC circuit takes cell membrane capacitance into account. Each element in the circuit (*a*, apical membrane; *b*, basolateral membrane; *s*, paracellular shunt pathway) is represented by an electric equivalent, that is, an electromotive force, *E*, in series with a resistor, *R*, and with a capacitor, *C*. (B) Nyquist diagram (plot of the real and the imaginary axis of the impedance, Z_{re}, Z_{im}). At low frequencies ($\omega \to 0$), Z_{re} approaches R_{te}. The capacitance *C* can be calculated from the frequency at which $|Z_{im}|$ reaches a maximum value: $C = 1/(\omega|Z_{im}|_{max} \cdot R_{te})$.

3.2.2 Conductance Scanning

3.2.2.1 Basic Concepts and Initial Attempts

Conductance scanning employs a scanning electrode to measure local variations in current density close to a cell surface and has been applied to characterize the underlying conductive pathways. In 1972, Frömter first introduced the principle for conductance scanning, with which heterogeneous electric fields above different conductive pathways of an epithelium were observed (Frömter & Diamond, 1972). Continued efforts by Fromm and coworkers have improved the localized measurement of transcellular and paracellular conductance over a variety of epithelia (Gitter, Bendfeldt, Schulzke, & Fromm, 2000; Gitter, Bertog, Schulzke, & Fromm, 1997). In these recordings, one pair of electrodes applied a transepithelial potential across an epithelial tissue, whereas a second pair of electrodes measured the resultant changes in electric field at discrete locations above the sample. The electric field was measured for each location at two depths (close to and far away from the surface) (Cereijido, Meza, & Martinez-Palomo, 1981; Gitter, Bertog, Schulzke, & Fromm, 1997; Kockerling, Sorgenfrei, & Fromm, 1993). Local conductance can then be calculated from the following equation (Cereijido et al., 1981; Gitter et al., 1997).

$$G = \frac{E}{\rho V_{te}} = \frac{(\Delta V_{close\ point} - \Delta V_{distant\ point}) / \Delta z}{\rho V_{te}} \tag{3.16}$$

where E is the electrical field, and equals the potential difference, ΔV measured at two depths relative to the vertical displacement of the pipet, Δz, ρ is the bathing solution resistivity and V_{te} is the applied transepithelial potential. In these recordings, an optical microscope was utilized to position the pipet over cell junction or cell body with the help of a micromanipulator to record the paracellular conductance or the transcellular conductance respectively. Vertical position was determined when the pipet first touched the cell surface as a virtual zero and then was manually withdrawn from the surface.

3.2.2.2 Scanning Ion Conductance Microscopy

3.2.2.2.1 Instrumentation

Scanning ion conductance microscopy (SICM) is a technique based on scanning probe microscopy (SPM), which raster-scans nonconductive samples to record ion currents and generate topographic images. First introduced by Hansma and coworkers in 1989, SICM utilizes a hollow pipet with nanoscale tip dimensions as a probe to image samples (Hansma, Drake, Marti, Gould, & Prater, 1989). Pipets are pulled from glass capillaries with methods commonly used in patch-clamp studies. Through the manipulation of pulling parameters, such as temperature, pulling velocity, and delay, pipets with a variety of tip diameters and geometries can be fabricated. After fabrication, the pipet is filled with electrolyte solution and a Ag/AgCl electrode is inserted into the pipet. A potential difference is applied between the pipet electrode and a reference electrode placed in the bathing solution to generate an ion current, which flows from the reference electrode to the pipet electrode through the pipet tip opening. When the pipet is positioned close to a surface, the amplitude of ion current correlates to the pipet-surface distance. Therefore, the pipet position relative to a surface can be controlled precisely with a piezo-electric positioner. This phenomenon forms the basis of current feedback regulation in SICM and allows SICM to dynamically scan surfaces in a noninvasive fashion.

Baker and coworkers have modified SICM specifically for measurements of transepithelial conductance (Chen, Zhou, Morris, Hou, & Baker, 2013; Zhou, Chen, & Baker, 2012). To incorporate conductance scanning into SICM, a dual-barrel pipet has been designed, in which the first barrel is for topographic imaging and positioning of the pipet; the second barrel is for recording the electric field over the epithelium (Fig. 3.4A). In a perfusion cell culture system (e.g., Transwell) where epithelial cells grow to form a monolayer, the dual-barrel pipet is placed into the upper chamber. The current electrode (PE) in one barrel detects current passing from the PE to a reference electrode (RE), serving as the conventional SICM probe to record the pipet-surface distance and to obtain the topographic information of the cell membrane. The potential electrode (UE) in the second barrel serves to measure localized potential differences at cell surfaces relative to the RE. A transepithelial potential (V_{te}) between the RE and a working electrode (WE, bottom chamber) is applied over heterogeneous conductive pathways in the epithelium (Fig. 3.4A). The equivalent electric circuit of the dual-barrel SICM configuration is shown in Fig. 3.4B.

To evaluate local conductance changes with SICM, potential deflections at the pipet tip are measured at two fixed pipet-surface distances (D_{ps}), which is precisely controlled by a piezo-electric positioner. One position is close to the surface (0.2 μm), which is controlled through the robust feedback signal [detailed SICM feedback mechanism can be found in reviews: (Chen, Zhou, & Baker, 2012; Happel, Thatenhorst, & Dietzel, 2012)]. Here, D_{ps} is kept constant during imaging and conductance measurements, and can be determined experimentally through approach curves. A second position is 12.5 μm above the sample, which in this case is half of the range of vertical piezo movement. A longer vertical movement of the pipet can be achieved with the combination of piezo and Z-stepper motor, but local potential variations are limited within approximately 10 μm above the cell surface. Therefore, these two positions are ideal for accurate evaluation of cell surface conductance. The transepithelial conductance can be described by the following equation:

$$G = \frac{E}{\rho V^e} = \frac{(\Delta V_{0.2\mu m} - \Delta V_{12.5\mu m}) / \Delta z}{\rho V^e} \tag{3.17}$$

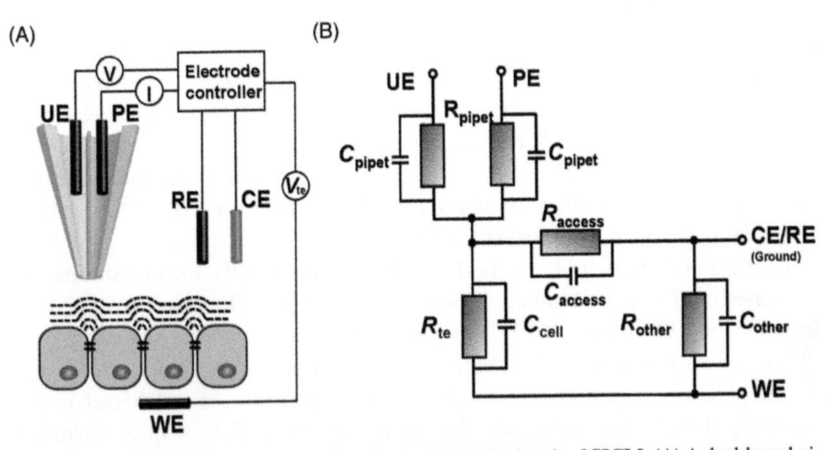

FIGURE 3.4 Schematic diagram and equivalent electric circuit of SICM. (A) A dual-barrel pipette is designed to obtain topographic information and measure local changes in transepithelial conductance. *CE*, Counter electrode; D_{ps}, pipet-surface distance; *PE*, current electrode; *RE*, reference electrode; *UE*, potential electrode; *WE*, working electrode. (B) The electric circuit from PE to RE monitors the access resistance (R_{access}); the electric circuit from WE to RE applies the transepithelial potential; the electric circuit from UE to WE records the transepithelial conductance. *(Reproduced with permission from Gong, Y., Renigunta, V., Zhou, Y., Sunq, A., Wang, J., Yang, J., Renigunta, A., Baker, L. A., & Hou, J. (2015). Biochemical and biophysical analyses of tight junction permeability made of claudin-16 and claudin-19 dimerization. Molecular biology of the cell, 26(24), 4333–4346.)*

in which, E, the electrical field is obtained by measuring potential deflection (ΔV) induced by transepithelial potential (V_{te}) at two pipet-surface distances (D_{ps}); V^e is the range of transepithelial potential sweeping (V^e = 120 mV, −60 mV to +60 mV); ρ is the cell medium resistivity; and Δz is the vertical displacement of the pipet. To avoid depolarization of cell plasma membrane, an alternating V_{te} is applied, the frequency of which is determined from impedance measurement to minimize the capacitive contribution. The magnitude of the transepithelial potential is less than 60 mV to avoid damage to the cell membrane.

3.2.2.2.2 Modeling With Nanopore

The polyimide membranes with nanopores of ~500 nm were used to model the conductive pathways in the epithelium. These membranes displayed an electrical resistance (TER or R_{te}) of ~ 94 $\Omega \cdot cm^2$ (in 0.1 M KCl), similar to that of typical epithelial cells. The potential responses over two closely spaced nanopores (<5 μm) are shown in Fig. 3.5A. The signal-to-noise ratio from potential measurement is significantly higher (S/N: 17.7) than that from current measurement (S/N: 4.8) (Fig. 3.5B), because of the large access resistance (on the order of MΩ) (Fig. 3.4B). Fig. 3.5C shows the relationship between the applied transepithelial potential (V_{te}) and the resultant potential deflection (ΔV) when the pipet is positioned over the center of a nanopore (pore 1 in Fig. 3.5A).

3.2.2.2.3 Point Conductance Measurement of Biologic Sample

Baker and coworkers have applied the SICM technique to record the paracellular conductance in the Madin-Darby Canine Kidney (MDCK) cells (Chen

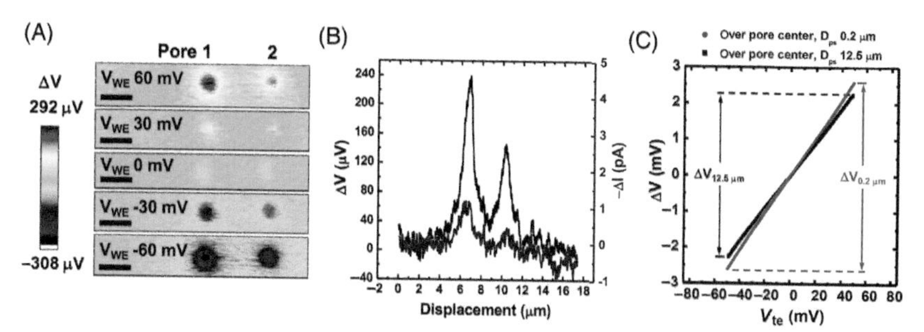

FIGURE 3.5 Characterization of porous membrane with SICM. (A) Images of potential variations for two closely spaced nanopores induced by a series of transmembrane potentials applied to the working electrode, WE. Scale bar: 1 μm. (B) The line scan across the two pores exemplifies the improvement in signal-to-noise ratio gained by potential measurement relative to current measurement. (C) The potential deflections at two pipet-surface distances (D_{ps}), far (12.5 μm) and close (0.2 μm), are recorded over the center of pore 1. *(Reproduced with permission from Chen, C. C., Zhou, Y., Morris, C. A., Hou, J., & Baker, L. A. (2013). Scanning ion conductance microscopy measurement of paracellular channel conductance in tight junctions. Analytical Chemistry, 85, 3621–3628.)*

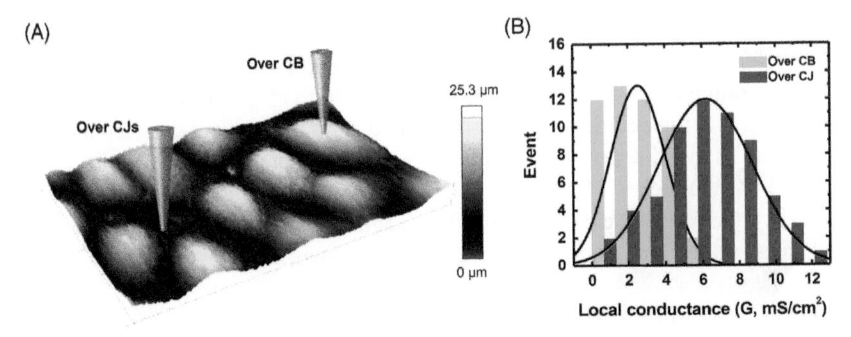

FIGURE 3.6 Characterization of epithelial cell monolayer with SICM. (A) A topographic image of the apical surface of the MDCK cell monolayer shows the location of cell body (CB) and cell junction (CJ). (B) Histograms of conductance measurements obtained over cell bodies ($N = 49$) and cell junctions ($N = 62$). *(Reproduced with permission from Chen, C. C., Zhou, Y., Morris, C. A., Hou, J., & Baker, L. A. (2013). Scanning ion conductance microscopy measurement of paracellular channel conductance in tight junctions. Analytical Chemistry, 85, 3621–3628.)*

et al., 2013). First, a topographic image is obtained with SICM imaging mode in absence of transepithelial potential (Fig. 3.6A). The locations of cell body (CB, transcellular pathway) and cell junctions (CJ, paracellular pathway) can then be pinpointed from the image to extract their spatial coordinates. The recording pipet is positioned over CB or CJ according to these coordinates to record the potential deflection (ΔV) induced by the applied transepithelial potential (V_{te}) at two discrete pipet-surface distances (D_{ps}, 0.2 μm and 12.5 μm), respectively. The recorded conductance over CBs (transcellular) or CJs (paracellular) displays Gaussian distributions with means of 2.53 ± 1.49 mS/cm^2 and 6.20 ± 2.54 mS/cm^2, respectively (Fig. 3.6B). The paracellular conductance significantly differs from the transcellular conductance. The spatial resolution for topographic imaging is under 100 nm. The lateral distribution of ΔV displays submicron spatial resolution, which is limited largely by the scanning parameters employed, especially the geometry of the pipet tip, D_{ps}, and V_{te}. While the location of CJ or CB can be pinpointed by SICM, the continuous distribution of the multitude of competing conductive pathways surrounding CJ or CB may prevent the isolation of paracellular- or transcellular-specific conductance.

3.2.2.2.4 Two-Dimensional Conductance Scan of Biological Sample

Compared to fixed-position point measurement, two-dimensional conductance scan by SICM can provide real-time visualization of the conductance distribution over heterogeneous pathways in an epithelium (Zhou et al., 2017). This automated SICM approach combines the hopping function of SICM topographic scanning configuration with the stationary conductance measurement described in the previous section (Fig. 3.7A). A conductance heat map can now be generated for an epithelial cell monolayer, for example, the MDCK cell monolayer (Fig. 3.7B). It is clear from the conductance heat map that the

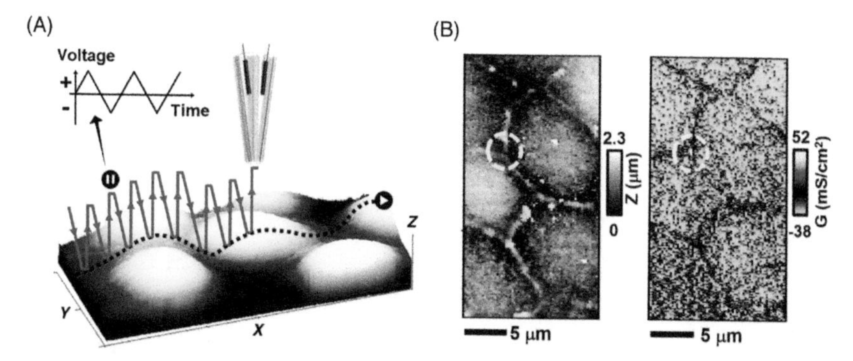

FIGURE 3.7 Two-dimensional conductance mapping. (A) Schematic diagram of the hopping SICM. Nanopipet first approaches the surface and then withdraws by a pre-set vertical displacement at each coordinate to map the surface topography with "pause" periods added to both the beginning and the end of each hopping movement to allow application of transepithelial potential and conductance recording. (B) Topographic image (*left*) and conductance heat map (*right*) of an MDCK cell monolayer recorded under physiologic condition. Bar: 5 μm. *(Reproduced with permission from Zhou. L., Gong, Y., Hou, J., Baker, L.A., (2017). Quantitative Visualization of Nanoscale Ion Transport. Anal Chem, 89(24),13603–13609.)*

paracellular pathway is more conductive than the transcellular pathway. There are conductance heterogeneities along individual cell junctions. The conductance hot spots are often found in the tricellular junctions (*yellow circle* in Fig. 3.7B).

3.2.3 Patch Clamp

Until recently, the patch clamp technique has been considered unsuitable to paracellular conductance recording because of the difficulty in achieving a gigaohm seal over the tight junction. Turner and coworkers have reported for the first time a successful recording of the paracellular conductance in the MDCK cells with patch clamp (Weber et al., 2015). There appears to be a channel of 90 pS in the tight junction when claudin-2 is present. The gating behavior of this putative paracellular channel is very similar to that of the conventional membrane channel (Weber et al., 2015).

3.2.4 Optical Microscopic Measurement of Transjunctional Water Permeability

Spring and coworkers developed an optical microscopic approach to measure transjunctional water permeability (Kovbasnjuk, Leader, Weinstein, & Spring, 1998). It is based upon the assumption that water flow into the lateral intercellular space (LIS) can be visualized by trapping a high-molecular-weight fluorescent dye in the LIS and observing its concentration profile along the LIS from the tight junction to the basement membrane (Fig. 3.8). If

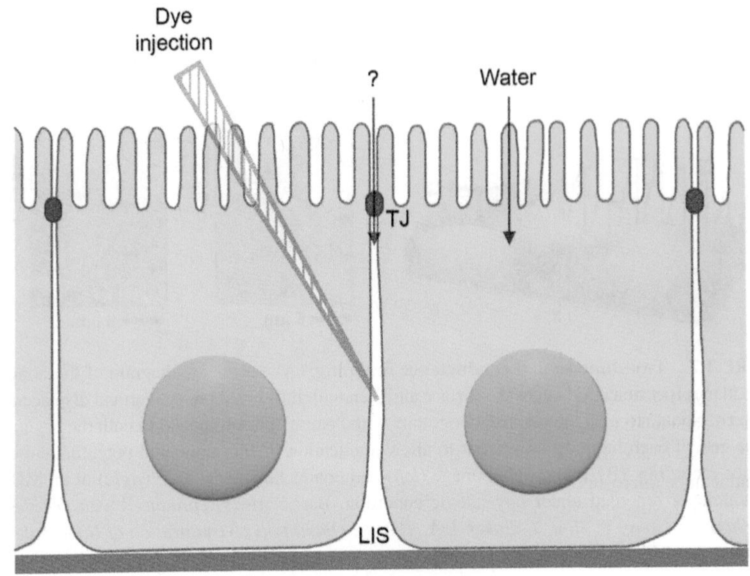

FIGURE 3.8 Illustration of the principle of water flux measurement across the tight junction. The epithelium is grown on a plastic permeable support (e.g., Transwell). A high-molecular-weight fluorescein dextran dye is injected into the lateral intercellular space (LIS) and trapped by the plastic support. An inverted confocal microscope is used to measure the fluorescent intensity at focal planes along the LIS.

significant water flow occurs across the tight junction, then the dye concentration will fall precipitously in the region of the LIS immediately adjacent to the tight junction. The water flow rate, F_v, is calculated from the following equation:

$$F_v = DA\frac{dln(C)}{dx} \tag{3.18}$$

where C is the dye concentration, D is the dye diffusion coefficient, A is the LIS cross-sectional area, and x is the displacement from tight junction. The dye concentration can be determined from the intensity of fluorescence at focal planes along the LIS, which is normalized relative to the measured value closest to the tight junction.

REFERENCES

Cereijido, M., Meza, I., & Martinez-Palomo, A. (1981). Occluding junctions in cultured epithelial monolayers. *American Journal of Physiology Cell Physiology*, *240*, C96–C102.

Chen, C. -C., Zhou, Y., & Baker, L. A. (2012). Scanning ion conductance microscopy. *Annual Review of Analytical Chemistry*, *5*, 207–228.

Chen, C. C., Zhou, Y., Morris, C. A., Hou, J., & Baker, L. A. (2013). Scanning ion conductance microscopy measurement of paracellular channel conductance in tight junctions. *Analytical Chemistry*, *85*, 3621–3628.

Frömter, E. (1972). The route of passive ion movement through the epithelium of *Necturus* gallbladder. *Journal of Membrane Biology*, *8*, 259–301.

Frömter, E., & Diamond, J. (1972). Route of passive ion permeation in epithelia. *Nature New Biology*, *235*, 9–13.

Gitter, A. H., Bendfeldt, K., Schulzke, J. D., & Fromm, M. (2000). Trans/paracellular, surface/crypt, and epithelial/subepithelial resistances of mammalian colonic epithelia. *Pflugers Archiv*, *439*, 477–482.

Gitter, A. H., Bertog, M. -D, Schulzke, J., & Fromm, M. (1997). Measurement of paracellular epithelial conductivity by conductance scanning. *Pflugers Archiv*, *434*, 830–840.

Gong, Y., Renigunta, V., Zhou, Y., Sunq, A., Wang, J., Yang, J., Renigunta, A., Baker, L. A., & Hou, J. (2015). Biochemical and biophysical analyses of tight junction permeability made of claudin-16 and claudin-19 dimerization. *Molecular biology of the cell*, *26*(24), 4333–4346.

Hansma, P. K., Drake, B., Marti, O., Gould, S. A. C., & Prater, C. B. (1989). The scanning ion-conductance microsope. *Science*, *243*, 641–643.

Happel, P., Thatenhorst, D., & Dietzel, I. (2012). Scanning ion conductance microscopy for studying biological samples. *Sensors*, *12*, 14983–15008.

Jovov, B., Wills, N. K., & Lewis, S. A. (1991). A spectroscopic method for assessing confluence of epithelial cell cultures. *The American Journal of Physiology*, *261*, C1196–C1203.

Kedem, O., & Katchalsky, A. (1958). Thermodynamic analysis of the permeability of biological membranes to non-electrolytes. *Biochimica et Biophysica Acta*, *27*, 229–246.

Kockerling, A., Sorgenfrei, D., & Fromm, M. (1993). Electrogenic Na^+ absorption of rat distal colon is confined to surface epithelium: a voltage-scanning study. *American Journal of Physiology Cell Physiology*, *264*, C1285–C1293.

Kovbasnjuk, O., Leader, J. P., Weinstein, A. M., & Spring, K. R. (1998). Water does not flow across the tight junctions of MDCK cell epithelium. *Proceedings of the National Academy of Sciences of the United States of America*, *95*, 6526–6530.

Ussing, H. H. (1953). Transport through biological membranes. *Annual Review of Physiology*, *15*, 1–20.

Ussing, H. H., & Zerahn, K. (1951). Active transport of sodium as the source of electric current in the short-circuited isolated frog skin. *Acta Physiologica Scandinavica*, *23*, 110–127.

Weber, C. R., Liang, G. H., Wang, Y., Das, S., Shen, L., Yu, A. S., Nelson, D. J., & Turner, J. R. (2015). Claudin-2-dependent paracellular channels are dynamically gated. *eLife*, *4*, e09906.

Zhou, Y., Chen, C. C., & Baker, L. A. (2012). Heterogeneity of multiple-pore membranes investigated with ion conductance microscopy. *Analytical Chemistry*, *84*, 3003–3009.

Zhou, L., Gong, Y., Hou, J., & Baker, L. A. (2017). Quantitative Visualization of Nanoscale Ion Transport. *Anal Chem*, 89(24), 13603–13609.

Chapter 4

Paracellular Cation Channel

4.1 CHANNEL-LIKE PROPERTIES OF TIGHT JUNCTION

Paracellular ion transport through the tight junction (TJ) is passive, driven by electrochemical gradients, and demonstrates channel-like properties such as electric conductance, charge, and size selectivity similar to those found in membrane ion channels. The paracellular conductance, often expressed by its inverse value, the paracellular resistance (R_p or R_s, the shunt resistance), is the most important parameter for describing the paracellular permeability through the TJ. Epithelia can be classified as "leaky" or "tight" based upon the R_p or R_s. The leaky epithelia are those with the R_p or R_s smaller than the cell membrane resistance (R_c), and the total transepithelial resistance (TER or R_{te}) less than 1000 $\Omega \cdot cm^2$, and vice versa. The leaky epithelia include the proximal tubule of the kidney, the small intestine, and the colon, whereas the tight epithelia include the distal tubule of the kidney, the gastric mucosa, and the bladder (Powell, 1981). The TJ can also been seen as a barrier perforated by aqueous pores capable of discriminating charge and size in the paracellular pathway. The electrochemical potential created by placing unequal ion concentrations on either side of the epithelium reflects the charge selectivity of the paracellular pathway. The charge selectivity is often expressed as the permeability ratio of Na^+ to Cl^- (P_{Na}/P_{Cl}). The paracellular pathways in most epithelia are cation selective (Powell, 1981). The anion selective paracellular pathways have been found in the skin and the colon (Frizzell, Koch, & Schultz, 1976; Mandel & Curran, 1972). The paracellular permeability of hydrophilic nonelectrolyte is inversely related to its size. The size selectivity among epithelia is relatively narrow (7–15 Å in diameter) (Tang & Goodenough, 2003), whereas the paracellular pathway in the endothelial cell appears able to permit passage of much larger solutes of 40–60 Å in diameter owing to the atypical TJs that incompletely seal the paracellular space (Firth, 2002).

4.2 THE FUNCTIONAL DIVERSITY OF CLAUDIN

Among the proteins making the TJ, the claudin family can confer charge and size selectivity to the paracellular pathway, resulting in differences in TER and paracellular permeabilities. It appears that each claudin protein has its own unique property of ion permeability (Table 4.1). The claudins making paracellular cation channels include claudin-2, -10b, -12, -15, -16, and -21. The

The Paracellular Channel. http://dx.doi.org/10.1016/B978-0-12-814635-4.00004-8

TABLE 4.1 Claudin Permeabilities Evaluated in Different Epithelial Cells

Species	Permeability	Cell background	References
Claudin-1	Cation and anion barrier	MDCK-II	Inai, Kobayashi, and Shibata (1999)
Claudin-2	Cation channel	MDCK-I, -II, -C7, LLC-PK1, Caco2	Amasheh et al. (2002), Fujita et al. (2008), Hou, Gomes, Paul, and Goodenough (2006), Van Itallie et al. (2003), Yu et al. (2009)
	Divalent cation channel	MDCK-I, Caco2	Fujita et al. (2008), Yu et al. (2009)
Claudin-3	Cation and anion barrier	MDCK-II	Milatz et al. (2010)
Claudin-4	Cation barrier	MDCK-II	Van Itallie, Rahner, and Anderson (2001)
	Anion channel	LLC-PK1, M-1, mIMCD3	Hou et al. (2006), Hou, Renigunta, Yang, and Waldegger (2010)
Claudin-5	Cation barrier	MDCK-II	Wen, Watry, Marcondes, and Fox (2004)
Claudin-6	Anion barrier	MDCK-II	Sas, Hu, Moe, and Baum (2008)
Claudin-7	Cation channel and anion barrier	LLC-PK1	Alexandre, Lu, and Chen (2005)
	Anion channel	LLC-PK1	Hou et al. (2006)
Claudin-8	Cation barrier	MDCK-II	Yu, Enck, Lencer, and Schneeberger (2003)
	Anion channel	M-1, mIMCD3	Hou et al. (2010)
Claudin-9	Anion barrier	MDCK-II	Sas et al. (2008)
Claudin-10a	Anion channel	MDCK-II	Van Itallie et al. (2006)
Claudin-10b	Cation channel	LLC-PK1	Van Itallie et al. (2006)
Claudin-11	Cation barrier	MDCK-II	Van Itallie et al. (2003)
Claudin-12	Divalent cation channel	Caco2	Fujita et al. (2008)
Claudin-14	Cation barrier	MDCK-II	Ben-Yosef et al. (2003)
Claudin-15	Cation channel	LLC-PK1	Van Itallie et al. (2003)
Claudin-16	Cation channel	LLC-PK1	Hou et al. (2005)
	Divalent cation channel	MDCK-C7	Kausalya et al. (2006)

TABLE 4.1 Claudin Permeabilities Evaluated in Different Epithelial Cells *(cont.)*

Species	Permeability	Cell background	References
Claudin-17	Anion channel	MDCK-C7, LLC-PK1	Krug et al. (2012)
Claudin-18	Cation barrier	MDCK-II	Jovov et al. (2007)
Claudin-19	Cation barrier	MDCK-II	Angelow, El-Husseini, Kanzawa, and Yu (2007)
	Anion barrier	LLC-PK1	Hou et al. (2008)
Claudin-21	Cation channel	MDCK-I	Tanaka et al. (2016)

Caco2, Human Colon Carcinoma Cell Line 2; LLC-PK1, Lilly Laboratories Cell - Porcine Kidney 1; MDCK-I, -II or -C7, Madin-Darby Canine Kidney type I, type II or type C7 cell; M-1, Mouse cortical collecting duct type 1 cell; mIMCD3, mouse Inner Medullary Collecting Duct type 3 cell.

claudins making paracellular anion channels include claudin-4, -7, -8, -10a, and -17. Notably, the measurement of claudin permeability depends upon the background of the endogenous claudins expressed in the epithelium. Because the transport property of the TJ is determined by the combination of all the claudin proteins present in the cells, studying any individual claudin will need to take into account the permeabilities of the remaining claudins.

4.3 THE STRUCTURAL BASIS OF CATION SELECTIVITY

4.3.1 The First Extracellular Loop of Claudin

Anderson and coworkers performed a key experiment to demonstrate the pivotal role of the first extracellular loop (ECL1) domain of claudin in dictating the charge selectivity of the paracellular pathway. Domain swapping of the first, but not the second ECL between claudin-2, a cation selective claudin, and claudin-4, an anion selective claudin, rendered the paracellular pathway made of claudin-2 to become more anion selective, and the paracellular pathway made of claudin-4 more cation selective (Fig. 4.1) (Colegio, Van Itallie, Rahner, & Anderson, 2003).

4.3.2 The Selectivity Filter in the First Extracellular Loop of Claudin

The charged amino acid residues in the ECL1 of claudin may create a favorable electrostatic environment to facilitate the permeation of ions. According to the crystal structure of claudin-15 (Fig. 2.4), the 4th β-strand in ECL1 faces the extracellular space and may contain the amino acid loci that act as the selectivity filter to dictate the paracellular charge selectivity (Suzuki et al., 2014). Experimental data are compatible

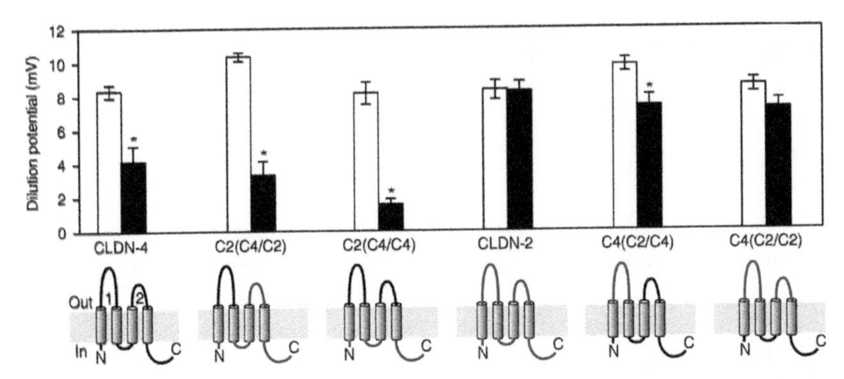

FIGURE 4.1 The first extracellular loop domain in claudin determines the paracellular charge selectivity. *Lower panel*: domain swapping of claudin extracellular loops (*black segments*, claudin-4; *red segments*, claudin-2). *Upper panel*: dilution potentials compared between the MDCK-II cells that expressed (*filled bars*) and unexpressed (*open bars*) the chimeric claudin protein. Positive dilution potential indicates cation selectivity, that is, $P_{Na}/P_{Cl} > 1$. *$p < 0.05$ compared to unexpressed cells. *(Reproduced with permission from Colegio, O. R., Van Itallie, C., Rahner, C., & Anderson, J. M. (2003). Claudin extracellular domains determine paracellular charge selectivity and resistance but not tight junction fibril architecture. American Journal of Physiology Cell Physiology, 284. C1346–C1354.)*

with the structural insights. There are two negatively charged residues, Asp55 and Asp64, located in the 4th β-strand of the ECL1 of claudin-15 (-D_{55}SLGVSNCWD_{64}-, highlighted in Fig. 4.2). Mutation at either locus to a positively charged amino acid, Lys or Arg, abolished the cation selectivity of claudin-15 (Fig. 4.3) (Colegio, Van Itallie, McCrea, Rahner, & Anderson, 2002). The charged residues outside the selectivity filter may also play important roles. In the ECL1 of claudin-16, there are ten negatively charged residues scattered in all four β-strand domains (Fig. 4.2). Mutagenesis analyses have identified five out of these ten negatively charged sites (D34, D35, E49, D56, and E70) that affect the cation selectivity of claudin-16 (Hou, Paul, & Goodenough, 2005). Among them, only D56 is within the putative selectivity filter.

4.3.3 The Electrostatic Field Strength Model

The net charge in the 4th β-strand of ECL1 may dictate the charge selectivity by exerting an electrostatic field. Alignment of the ECL1 domains in claudins allows deducing the amino acid sequences of 4th β-strand domains according to the crystal structure of claudin-15 (Fig. 4.2). Among all known claudins, the net charge in the 4th β-strand from claudin-2, -10b, -12, -15, -16, or -21 is negative. Functionally, all these claudins form cation channels (Table 4.1). The increase in charge density in the 4th β-strand appears to apply synergistic effects on paracellular charge selectivity due to higher electrostatic field strength. For example,

```
                                                            4th β-strand
Claudin-1     27 LPQWKIYSYAG--DNIVTAQAIYEGLWMSCVSQS-TGQIQCKVFDSLLN-LNS 75
Claudin-2     27 LPNWRTSSYVG--ASIVTAVGFSKGLWMECATHS-TGITQCDIYSTLLG-LPA 75
Claudin-3     26 LPMWRVSAFIG--SSIITAQITWEGLWMNCVVQS-TGQMQCKMYDSLLA-LPQ 74
Claudin-4     27 LPMWRVTAFIG--SNIVTAQTSWEGLWMNCVVQS-TGQMQCKMYDSMLA-LPQ 75
Claudin-5     27 LPMWQVTAFLD--HNIVTAQTTWKGLWMSCVVQS-TGHMQCKVYESVLA-LSA 75
Claudin-6     27 LPMWKVTAFIG--NSIVVAQMVWEGLWMSCVVQS-TGQMQCKVYDSLLA-LPQ 75
Claudin-7     27 IPQWQMSSYAG--DNIITAQAMYKGLWMECVTQS-TGMMSCKMYDSVLA-LPG 75
Claudin-8     27 MPQWRVSAFIE--SNIVVFENRWEGLWMNCMRHA-NIRMQCKVYDSLLA-LSP 75
Claudin-9     27 LPLWKVTAFIG--NSIVVAQVVWEGLWMSCVVQS-TGQMQCKVYDSLLA-LPQ 75
Claudin-10a   25 SNEWKVTTR-A--SSVITATWVYQGLWMNCAGNA-LGSFHCRPHFTIFK-VEG 72
Claudin-10b   27 TDYWKVSTI-D--GTVITTATYFANLWKICVTDS-TGVANCKEFPSMLA-LDG 74
Claudin-11    27 TNDWVVTCSYTIPTCRKMDELGSKGLWADCVM-A-TGLYHCKPLVDILI-LPG 76
Claudin-12    30 LPNWRKLRLITF-NRNEKNLTIYTGLWVKCARY--DGSSDCLMYDRTWY-LSV 78
Claudin-13    27 LPVWRVT-FPD--DETDPDATIWEGLWHICQVRE-NRWIQCTLYDTRIL-VAQ 74
Claudin-14    27 LPHWRRTAHVG--TNILTAVSYLKGLWMECVWHS-TGIYQCQIYRSLLA-LPR 75
Claudin-15    26 NSYWRVSTV-H--GNVITTNTIFENLWYSCATDS-LGVSNCWDFPSMLA-LSG 73
Claudin-16    26 TDCWMVNAD-----DSLEVSTKCRGLWWECVTNAFDGIRTCDEYDSIYAEHPL 73
Claudin-17    27 LPQWRVSAFIG--SNIIIFERIWEGLWMNCIQQA-MVTLQCKFYNSILA-LPP 75
Claudin-18    27 MDMWSTQDL-Y--DNPVTAVFQYEGLWRSCVQQS-SGFTECRPYFTILG-LPA 74
Claudin-19    27 LPQWKQSSYAG--DAIITAVGLYEGLWMSCASQS-TGQVQCKLYDSLLA-LDG 75
Claudin-20    27 LPNWKVNAYAG--PNIVTAVVQVQGLWVDCTWYS-TGMFSCTLKYSILS-LPV 75
Claudin-21    30 LPQWKTLTLDL--N---EMETWVSGLWEACVNQE-EAGTVCKAFESFLS-LPQ 75
                     *                           .**  *          *
```

FIGURE 4.2 Alignment of the first extracellular loop domain in claudin. The mouse claudin sequences are shown here. Negatively charged acidic residues are labeled in *red*; positively charged basic residues in *blue*. The amino acid loci making the 4th β-strand in claudin-15 structure are highlighted in *yellow*. The claudin-21 gene is also known as the claudin-25 pseudogene in the mouse genome.

when Asp55 and Asp64 in claudin-15 were both mutated to positively charged amino acids, the magnitude of change in paracellular charge selectivity was higher than the combined effect of two single mutations (Fig. 4.3) (Colegio et al., 2002).

4.3.4 The Electrostatic Interaction Site Model

According to Eisenman's theory, the charge selectivity of an ion channel is determined by the difference between the energy cost of dehydration of the ion and the energy gain from binding to the charged sites within the channel pore (Eisenman & Horn, 1983). The higher the electrostatic interaction between ion and channel pore, the more selective the channel will be for smaller ions (e.g., Li^+) and the lower the interaction, the more selective for larger ions (e.g., Cs^+). The free water mobility sequence of alkali metal monovalent cations is described by the Eisenman selectivity sequence I: $Cs^+ > Rb^+ > K^+ > Na^+ > Li^+$. The permeability sequence of alkali metal monovalent cations through the claudin-2 paracellular channel resembles the Eisenman selectivity sequence VI–VIII: $K^+ > Rb^+ > Na^+ > Li^+ > Cs^+$ (Fig. 4.4A) (Yu et al., 2009), which suggests that the electrostatic interaction in the channel pore is of moderate strength and can only allow partial dehydration of the permeant cations (Eisenman & Horn, 1983). Asp65 is the only charged residue in the 4th β-strand in the ECL1

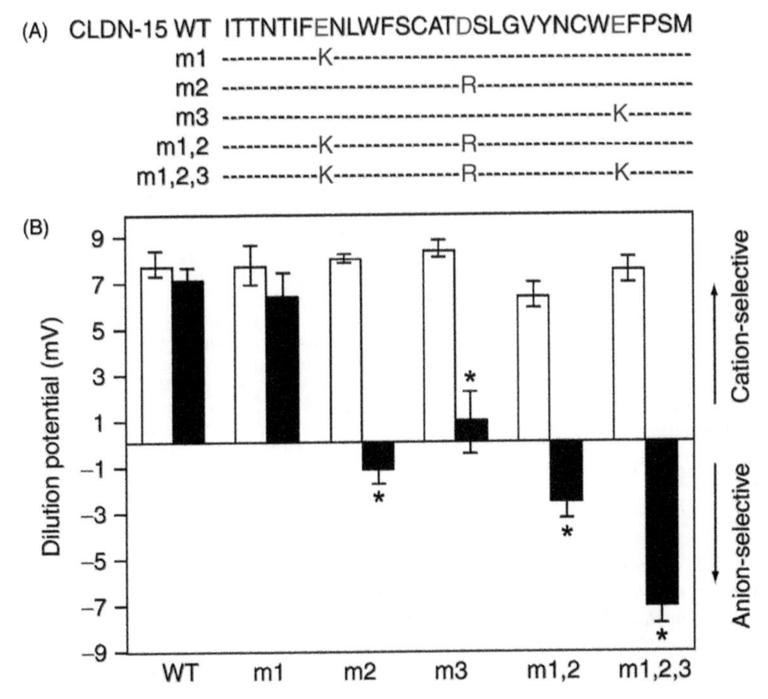

FIGURE 4.3 The charge sites in the first extracellular loop domain of claudin determines the paracellular charge selectivity. (A) Amino acid sequence alignment of the first extracellular domain of human claudin-15 and charge-reversal mutants: m1 (E46K); m2 (D55R); m3 (E64K); m1,2 (E46K, D55R); and m1,2,3 (E46K, D55R, E64K) and (B) dilution potential measurements in MDCK-II cells expressing wild type or mutant claudin-15 (*solid bars*) compared with unexpressed cells (*open bars*). Note that the increase in amino acid charge density correlates with the increase in paracellular ion selectivity. $*p < 0.05$ compared to unexpressed cells. *(Reproduced with permission from Colegio, O. R., Van Itallie, C. M., McCrea, H. J., Rahner, C., & Anderson, J. M. (2002). Claudins create charge-selective channels in the paracellular pathway between epithelial cells. American journal of physiology Cell physiology,* 283. *C142–C147.)*

of claudin-2, which makes the putative selectivity filter in the channel pore (Fig. 4.2). Mutation of Asp65 to a neutral amino acid, Asn65, not only reduced the permeabilities of claudin-2 to the alkali metal monovalent cations, but also shifted the permeability sequence of claudin-2 to a lower order matching the Eisenman selectivity sequence V: $K^+ > Rb^+ > Na^+ > Cs^+ > Li^+$ (Fig. 4.4A) (Yu et al., 2009). Moreover, low extracellular pH can inhibit the cation permeability and reduce the charge selectivity of wild type claudin-2 but not the D65N mutant isoform (Fig. 4.4B) (Yu et al., 2009). These findings suggest that the carboxylate group of Asp65 stabilizes the permeant cation in claudin-2 channel pore by electrostatic interaction and that removing or neutralizing this charged group diminishes the electrostatic interaction strength required for establishing the paracellular charge selectivity.

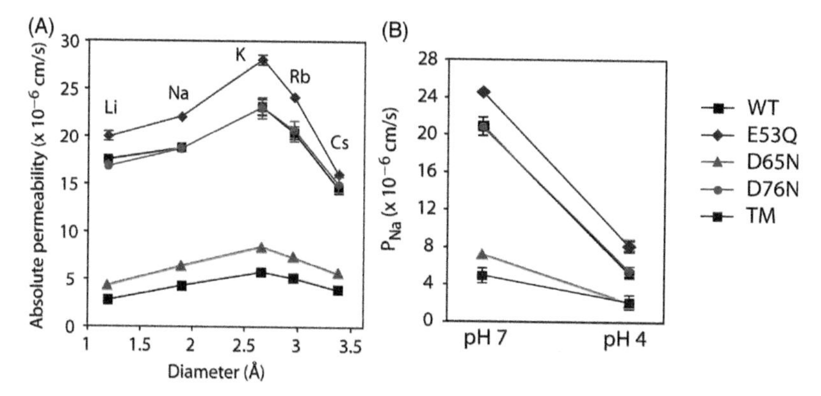

FIGURE 4.4 The electrostatic interaction in claudin-2 channel pore. (A) Absolute permeability to alkali metal cations of wild type and mutant claudin-2 proteins in MDCK-I cells and (B) effect of acidification to pH 4 on paracellular Na^+ permeability (P_{Na}) of wild type and mutant claudin-2 proteins in MDCK-I cells. *(Reproduced with permission from Yu, A. S., Cheng, M. H., Angelow, S., Gunzel, D., Kanzawa, S. A., Schneeberger, et al.. (2009). Molecular basis for cation selectivity in claudin-2-based paracellular pores: Identification of an electrostatic interaction site. The Journal of General Physiology,* 133. *111–127.)*

4.3.5 The Site Number Model

The paracellular channel is not selective among alkali metal monovalent cations (Hou et al., 2005; Tanaka et al., 2016; Yu et al., 2009). The ratio between the permeabilities of the most permeable cation (K^+) and the least permeable cation (Cs^+) is below 1.6 for the claudin-2 channel (Yu et al., 2009). A type of transmembrane channel, known as the nonselective NaK channel, shares similar permeability profiles (Shi, Ye, Alam, Chen, & Jiang, 2006). Paradoxically, the NaK channel is highly homologous to the KcsA channel, a K^+ selective transmembrane channel (Fig. 1.1). Structural comparison of these two channels reveals that the number of cation binding sites may underlie their selectivity difference. The selectivity filter of the NaK channel preserves two cation-binding sites equivalent to sites 3 and 4 in the selectivity filter of the KcsA channel, whereas the region corresponding to sites 1 and 2 in the KcsA channel becomes a vestibule, in which ions can diffuse but not bind (Fig. 4.5) (Shi et al., 2006). It appears that the fewer the cation binding sites are, the less selective the transmembrane channel becomes (Derebe et al., 2011). The paracellular channel may obey the same principle to establish the charge selectivity that allows differentiating cations from anions but not among various cations.

4.4 THE CONDUCTANCE OF PARACELLULAR CHANNEL

The conductance of transmembrane channel depends upon many factors, such as the length of the pore (Latorre & Miller, 1983), the diameter of the pore entrance (Geng, Niu, & Magleby, 2011), the electrostatic field strength in the

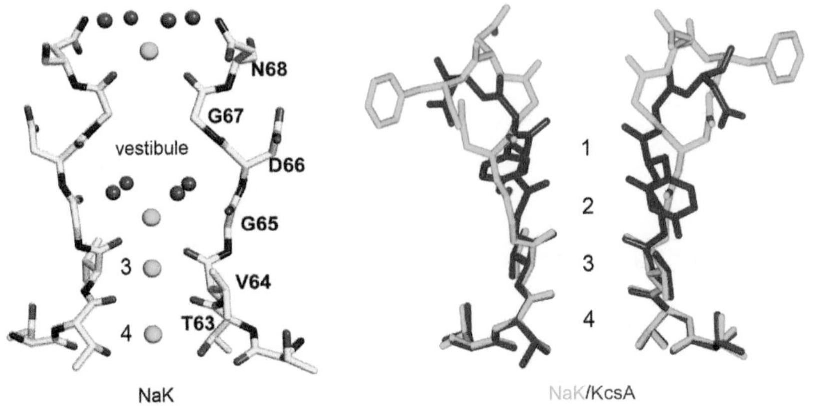

FIGURE 4.5 Principle of cation selectivity in NaK channel. The number of contiguous ion binding sites demonstrates the principle of the site number model of ion selectivity. K^+ ions are shown as *green spheres* in single file in the selectivity filter. Water molecules are shown as *red spheres* bound to K^+ in the vestibule. The nonselective NaK channel lacks #1 and #2 cation-binding sites when compared to the K^+ selective KcsA channel. The depicted ion channel structures are based upon the X-ray analyses of the NaK channel and the KcsA channel (MacKinnon, 2004; Shi et al., 2006).

pore (Nimigean, Chappie, & Miller, 2003), the ion binding sites in the selectivity filter (Zhou & MacKinnon, 2003), the rate of ion dehydration (Mullins, 1959), and the concentration of ion (Latorre & Miller, 1983). Therefore, the conductance levels may vary dramatically among the transmembrane channels owing to their structural diversity. The paracellular channel conductance, in contrast, is maintained within a narrow range by claudin-2, -10b, -15, -16, and -21 (Hou et al., 2005; Tanaka et al., 2016; Van Itallie, Fanning, & Anderson, 2003; Van Itallie et al., 2006; Yu et al., 2009). This would suggest that the paracellular channels made of these claudins are structurally alike. Patch clamp experiments have revealed single claudin-2 channel conductance of 90 pS in the MDCK cells (Fig. 4.6) (Weber et al., 2015). The MDCK cells are approximately hexagonal and have an average radius of 7 μm, and hence an area of 1.5×10^{-6} cm^2 and a cell perimeter of 42 μm (Cereijido, Gonzalez-Mariscal, & Borboa, 1983). From the macroscopic claudin-2 conductance of 10 mS/cm^2, it can be estimated that there is one claudin-2 channel every 25 nm along the cell perimeter after averaging the TJ strand number (Yu et al., 2009). The claudin channel density in the TJ strand is most likely determined by the TJ architecture. The TJ strands appear as parallel arrays of intramembrane protein particles of ~10 nm in diameter and separated by a distance of ~8 nm on freeze fracture replica electron micrograph (Fig. 2.2) (Goodenough & Revel, 1970). If each protein particle constitutes a putative paracellular channel, then the number of paracellular channels per unit length of a TJ strand will be constant. The number of TJ strands is, however, inversely correlated to the paracellular conductance (Claude

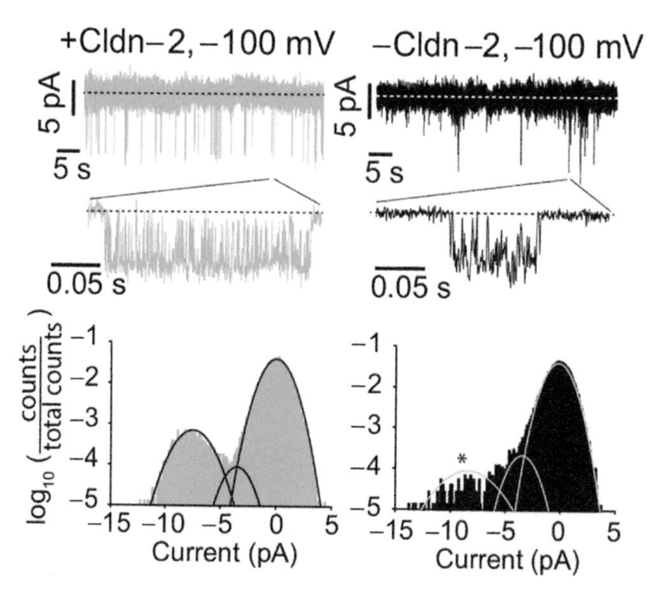

FIGURE 4.6 Putative single claudin channel conductance. Trans-TJ patch clamp revealed conductance events detected at −100 mV when claudin-2 was expressed (+Cldn-2) or not expressed (−Cldn-2) by the MDCK-I cells. These events, when fitted to Gaussian distribution, indicated the claudin-2 dependent conductance centered at ~9 pA, which might be due to a paracellular channel of ~90 pS. *(Reproduced with permission from Weber, C. R., Liang, G. H., Wang, Y., Das, S., Shen, L., Yu, A.S., et al. (2015). Claudin-2-dependent paracellular channels are dynamically gated. eLife, 4. e09906.)*

& Goodenough, 1973). It has been postulated that the paracellular channel fluctuates between open and closed states (Claude, 1978). The overall paracellular conductance (G_p) is described by the following equation:

$$G_p = G_i \times P_0^n$$

in which, G_i denotes the conductance level of individual TJ strand, P_0 is the probability of a paracellular channel staying in its open state, and n is the number of strands in the TJ (Claude, 1978).

4.5 THE SIZE SELECTIVITY OF PARACELLULAR CHANNEL

To characterize the size selectivity of paracellular channel independent of influences from the charge selectivity, Watson and coworkers first measured the "apparent permeability" (P_{app}) of a graded series of uncharged polyethylene glycols (PEGs) of increasing molecular diameter (7–15 Å) across intestinal epithelial monolayers by the transepithelial flux assay (Chapter 3, Section 3.1.4). Their results have clearly demonstrated that the size selectivity is biphasic, consisting of a high-capacity, size-restrictive pathway and a low-capacity, size-independent pathway (Fig. 4.7) (Watson, Rowland, & Warhurst, 2001). The high-capacity,

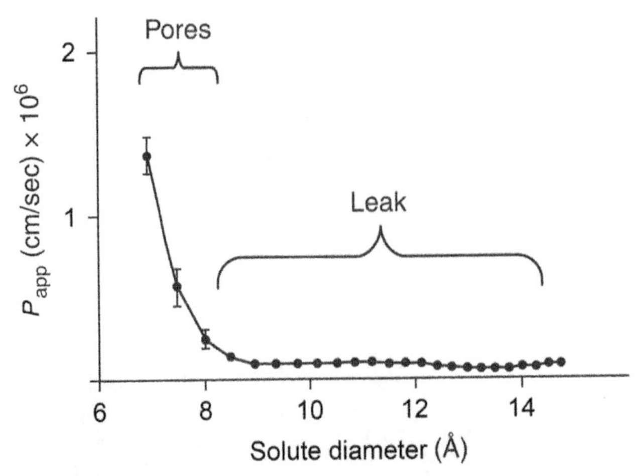

FIGURE 4.7 The two-pathway model of paracellular size selectivity. Idealized data reveal the permeability of uncharged polyethylene glycol molecules of increasing sizes across an epithelium. The pathway for molecules of <8 Å in diameter is formed by the claudin channels that are size selective. The pathway for molecules >8 Å in diameter shows no size selectivity but is dependent upon TJ dynamics or through tricellular junction. *(Reproduced with permission from Anderson, J. M., & Van Itallie, C. M. (2009). Physiology and function of the tight junction. Cold Spring Harbor Perspectives in Biology,* 1. *a002584.)*

size-restrictive pathway behaves as a system of small pores with diameter of <8 Å. Expression of claudin-2 selectively increased the small pore density in the TJ but had no effect on the permeability of PEGs that are larger than 8 Å in diameter (Van Itallie et al., 2008). The P_{app} for the small pore appears dissociated from the TER or R_{te} in several epithelial cell models (Van Itallie et al., 2008). These findings suggest that the permeability of small-uncharged solutes depends upon the small pore density, whereas the permeability of electrolytes, especially small cations, is subject to further selection by the cation-selective claudin molecules present in the TJ. The low-capacity, size-independent pathway allows permeation of molecules with diameter of >8 Å. It could represent a novel pathway through the tricellular TJ (Chapter 6) or transient breaks in the bicellular TJ as part of the TJ dynamic behaviors (Chapter 2, Section 2.8).

4.6 THE DIVALENT CATION PERMEABILITY OF PARACELLULAR CHANNEL

The paracellular channels made of claudin-2, -12, and -16 can permeate the divalent cations such as Ca^{++} and Mg^{++} (Fujita et al., 2008; Hou et al., 2005; Yu et al., 2009). The divalent cation permeability is approximately four- to five-fold lower than the monovalent cation permeability. In the claudin-2 channel, the divalent cation permeability is highly dependent upon the negative charge in the side chain of Asp65 in the ECL1 domain (Fig. 4.2). Neutralizing this charge by genetic mutation to Asn65 exerted a disproportionately larger effect on the

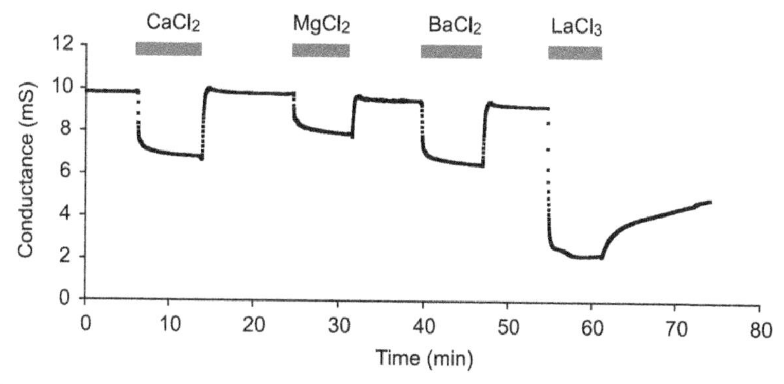

FIGURE 4.8 Inhibition of claudin-2 conductance by divalent and polyvalent cations. The MDCK-I cells were overexpressed with claudin-2 and bathed in Ringer's solution containing Ca^{++} and Mg^{++} each at 0.25 mM. At the time points indicated by the *gray bars*, the bathing solution was changed to solution containing the indicated divalent or polyvalent inorganic cation (at 5 mM). *(Reproduced with permission from Yu, A. S., Cheng, M. H., & Coalson, R. D. (2010). Calcium inhibits paracellular sodium conductance through claudin-2 by competitive binding. The Journal of Biological Chemistry,* 285. *37060–37069.)*

permeability of Ca^{++} (P_{Ca}) than Na^{+} (P_{Na}): P_{Ca} (WT)/P_{Ca} (D65N) = 22.1 ± 0.8 versus P_{Na} (WT)/P_{Na} (D65N) = 2.9 ± 0.1 (Yu et al., 2009). Because of its high valence, the divalent cation interacts more strongly than the monovalent cation with the negatively charged site of Asp65. Stronger intrapore binding would also mean slower dissociation from the charged site, which explains the lower permeability of the divalent cation. On the other hand, the divalent cation may compete with the monovalent cation for binding to the charged site in the channel pore. Yu and coworkers have found that the divalent cations including Ca^{++}, Mg^{++}, and Ba^{++} act as a reversible inhibitor of the paracellular conductance carried by Na^{+} through the claudin-2 channel, without causing changes in the TJ architecture (Fig. 4.8) (Yu, Cheng, & Coalson, 2010). Mutation of the intrapore binding site Asp65 in claudin-2 partially abrogated the inhibitory effect of divalent cations. The polyvalent cation La^{+++} caused nearly complete and irreversible inhibition of claudin-2 channel conductance (Fig. 4.8), which further strengthened the electrostatic interaction model of paracellular cation permeability (Yu et al., 2010).

4.7 THE REGULATION OF PARACELLULAR CATION CHANNEL

4.7.1 Claudin-2

4.7.1.1 Growth Factors and Hormones

4.7.1.1.1 Epidermal Growth Factor

In the MDCK-II cells, the epidermal growth factor (EGF)-induced EGF receptor (EGFR) activation significantly reduces the claudin-2 gene expression and the paracellular Na^{+} permeability (Singh & Harris, 2004). A membrane bound EGFR ligand can elicit a similar effect on claudin-2 gene expression and paracellular conductance, suggesting a juxtacrine activation mechanism in cell-autonomous manners (Singh, Sugimoto, Dhawan, & Harris, 2007). EGFR regulates the mRNA and protein levels of claudin-2 gene expression. The transcriptional factor STAT3 is rapidly phosphorylated by the kinase Src in response to EGFR activation, translocates to the nucleus, and blocks the claudin-2 mRNA transcription (Garcia-Hernandez et al., 2015). EGFR also accelerates clathrin-dependent endocytosis and lysosomal degradation of the claudin-2 protein, leading to decreased protein stability and abundance in the TJ (Ikari, Takiguchi, Atomi, & Sugatani, 2011).

4.7.1.1.2 Vitamin D

Vitamin D, via its metabolized active form, 1α,25-dihydroxyvitamin D_3, is known to increase the paracellular Ca^{++} permeability by upregulation of the claudin-2 gene expression in the intestinal epithelial cells (Fujita et al., 2008). The claudin-2 gene is a direct target of the transcriptional factor, vitamin D receptor (VDR). Genetic deletion of VDR in transgenic mouse models caused significant decreases in the claudin-2 gene expression in the intestines (Zhang et al., 2015).

4.7.1.2 Kinases

Transfection of an active form of the mitogen-activated protein kinase or the extracellular signal-regulated kinase into the MDCK-II cells inhibited the claudin-2 gene expression and increased the TER (Lipschutz, Li, Arisco, & Balkovetz, 2005). Atypical protein kinase C (aPKC), including two isoforms—ι and ζ, is the enzymatic component of the Par polarity complex localized to the TJ. In the MDCK-II cells, knockdown of PKCι resulted in increased TER and reduced claudin-2 protein abundance (Lu et al., 2015). The claudin-2 protein itself can be phosphorylated. Anderson and coworkers have shown that the Ser208 site in the cytoplasmic tail domain of claudin-2 is heavily phosphorylated when purified from the MDCK-II cells or from the mouse kidney (Fig. 4.9A) (Van Itallie et al., 2012). The phosphorylation of claudin-2 protein appears to regulate its TJ localization. The dephosphorylated form of claudin-2 is intrinsically unstable and prone to lysosomal degradation (Fig. 4.9B) (Van Itallie et al., 2012).

4.7.1.3 Osmolality

Apical hyposmolality or basolateral hyperosmolality can decrease the paracellular cation permeability in the MDCK-II cells via regulation of claudin-2 gene expression and protein localization (Ikari et al., 2015; Tokuda, Hirai, & Furuse, 2016). The promoter in the claudin-2 gene contains a GATA-2 binding site. Hyperosmolality decreases the nuclear level of GATA-2 via changes in intracellular Ca^{++} and PKCβ phosphorylation (Ikari et al., 2015). The claudin-2 protein localization is also regulated by extracellular osmolality. A basolateral to apical osmolality gradient delocalized the claudin-2 protein

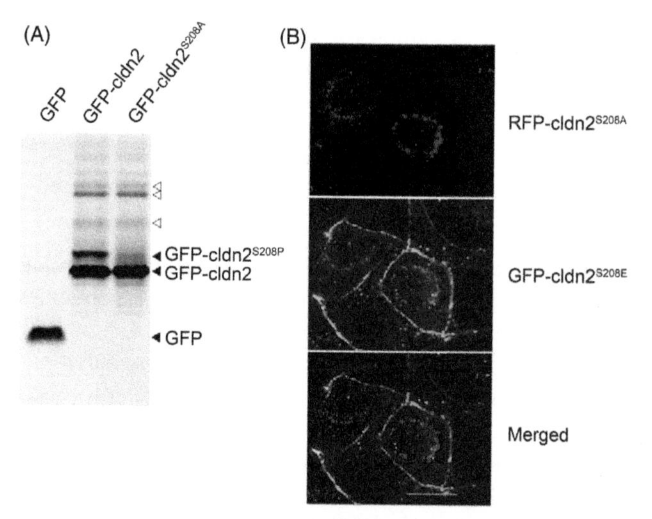

FIGURE 4.9 Claudin-2 protein phosphorylation and localization. (A) Lysate from MDCK-I cells expressing GFP alone, GFP-tagged wild type claudin-2, and GFP-tagged phosphorylation deficient mutant (S208A) claudin-2 proteins were subjected to Phos-tag electrophoresis. Unphosphorylated claudin-2 proteins were found in cells expressing either wild type or mutant claudin-2 (*lower filled arrowhead*), whereas a strongly phosphorylated band in wild type but not mutant claudin-2 (*upper filled arrowhead*) was consistent with S208 being the major site of phosphorylation. The protein bands indicated by *upper open arrowheads* were likely minor phosphorylated forms of claudin-2 and (B) when N-terminally red fluorescent protein (RFP) labelled claudin-2 S208A and GFP-labelled S208E proteins (mimicking phosphorylated state) were cotransfected into MDCK-I cells, the claudin-2 S208A proteins were found in intracellular vesicles, whereas the claudin-2 S208E proteins were found at the cell junction. Bar: 5 μm. *(Reproduced with permission from Van Itallie, C. M., Tietgens, A. J., LoGrande, K., Aponte, A., Gucek, M., & Anderson, J. M. (2012). Phosphorylation of claudin-2 on serine 208 promotes membrane retention and reduces trafficking to lysosomes. Journal of Cell Science, 125. 4902–4912.)*

from the TJ by creating membrane blebs to disorganize the TJ architecture (Tokuda et al., 2016).

4.7.1.4 Cingulin

Cingulin is a TJ plaque protein involved in the TJ formation (Chapter 2, Section 2.5.2.1). Knockdown of cingulin in the MDCK-II cells upregulated the mRNA and the protein levels of claudin-2 by activating the small GTPase RhoA (Guillemot & Citi, 2006). Transgenic knockout of cingulin in mice increased the claudin-2 protein levels in the duodenum and the kidney (Guillemot et al., 2012). No paracellular permeability difference has been reported by these two cingulin studies.

4.7.1.5 Symplekin

Symplekin is a TJ plaque protein involved in cytoplasmic RNA polyadenylation and transcriptional regulation (Chapter 2, Section 2.5.2.3). In human colorectal cancer cells, knockdown of symplekin suppressed claudin-2 gene expression on the levels of mRNA and protein, which led to a decrease in the paracellular cation permeability (Buchert et al., 2010).

4.7.2 Claudin-16

4.7.2.1 Human Genetic Mutation

Genetic mutations in the claudin-16 gene, formerly known as paracellin-1, cause a human autosomal recessive disorder—familial hypomagnesemia and hypercalciuria with nephrocalcinosis (FHHNC, OMIM #248250) (Simon et al., 1999). While most human mutations are rare missense mutations affecting the transmembrane domains in claudin-16, there are several missense mutations clustering in the ECL1 domain of claudin-16 and causing cation permeability defects, e.g. S110R (equivalent to S40R in mouse claudin-16), R114Q (R44Q in mouse), C131R (C61R in mouse), and H141D (H71D in mouse) (Fig. 4.2) (Konrad et al., 2008; Simon et al., 1999; Weber, Schneider et al., 2001). Among them, the human C131 site, equivalent to C61 in mouse claudin-16, is homologous to C62 in mouse claudin-15 and may play a critical role to stabilize the β4 strand in ECL1 that makes the selectivity filter according to the crystal structure of claudin-15 (Fig. 2.4).

4.7.2.2 Alternative Translation Initiation

The human claudin-16 gene encodes a 305-amino acid protein that possesses two in-frame start codons (ATG: encoding methionine M1 and M71, respectively). The mouse and rat claudin-16 genes lack the first start codon in the human gene and translate proteins from the ATG corresponding to the second start codon in the human gene (Weber, Schlingmann et al., 2001). Transfection of the full-length human claudin-16 gene into the MDCK cells resulted in the production of two proteins that migrated with different electrophoretic mobility (33 and 27 kDa) (Fig. 4.10A) (Hou et al., 2005). The molecular weight of 33 kDa matches the predicted molecular weight of the full-length claudin-16 protein (amino acids 1–305) and 27 kDa matches the alternatively translated claudin-16 protein (Δ70: amino acids 71–305). Hou and coworkers have revealed that only the shorter Δ70-claudin-16 protein is correctly localized at the TJ, whereas the full-length claudin-16 protein is mistargeted to the endosome or the lysosome (Fig. 4.10B) (Hou et al., 2005). Interestingly, there is a missense mutation (M71T) found in a cohort of FHHNC patients (Simon et al., 1999), which destines the mutant claudin-16 protein for lysosomal degradation due to the missing of the second start codon (Konrad et al., 2008).

4.7.2.3 Interaction With ZO-1

The carboxyl-terminal domain of the claudin-16 protein contains a PDZ domain-binding motif (TRV) that interacts with the PDZ domain of the TJ

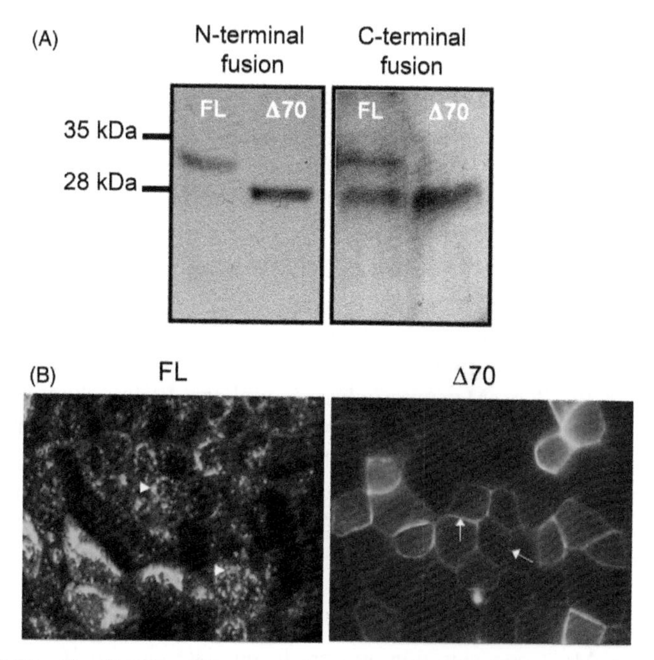

FIGURE 4.10 Claudin-16 protein translation and localization. (A) Western immunoblots of MDCK-II cells transfected with N-terminally or C-terminally HA fused full-length (FL) or truncated (Δ70) human claudin-16 proteins and labeled with anti-HA antibody. Note that the C-terminal fusion allows detection of two bands in full-length claudin-16 suggesting alternative translation initiation and (B) Immunofluorescent microscopy reveals that while the full-length (FL) claudin-16 protein is mistargeted to the lysosome (*arrowhead*), the truncated (Δ70) claudin-16 protein is localized at the cell junction (*arrow*). Bar: 10 μm. *(Reproduced with permission from Hou, J., Paul, D. L., & Goodenough, D. A. (2005). Paracellin-1 and the modulation of ion selectivity of tight junctions. Journal of Cell Science,* 118*. 5109–5118.)*

plaque protein ZO-1 (Ikari et al., 2004). Deletion of the PDZ domain-binding motif (ΔTRV) from the claudin-16 protein caused its delocalization from the TJ (Ikari et al., 2004). Muller and coworkers have identified a missense mutation (T303R, equivalent to mouse T233R) in the claudin-16 gene affecting the PDZ domain-binding motif (TRV mutated to RRV) from two cohorts of FHHNC patients (Muller et al., 2003). This mutation disrupts claudin-16 binding to ZO-1 and renders claudin-16 mistargeted to the lysosome (Muller et al., 2003).

4.7.2.4 Endocytosis and exocytosis

When transfected into the MDCK cells, the claudin-16 protein is constantly endocytosed into the clathrin-coated pits (Kausalya et al., 2006). Blocking the clathrin-mediated endocytosis by cytosolic acidification increased the cell surface abundance level of a mutated form (L203X) of mouse claudin-16 protein that lacks the entire carboxyl-terminal domain of the protein important for interaction with ZO-1 and anchoring onto TJ plaque (Muller, Kausalya, Meij, &

Hunziker, 2006). Syntaxins play important roles in the fusion of vesicles with target membranes. Syntaxin-8 has been shown to interact with claudin-16 and regulate its TJ localization (Ikari et al., 2014). Syntaxin-8 knockdown decreased the TJ abundance level of claudin-16 and the paracellular permeability to Mg^{++} in the MDCK cells. Recycling assays indicated that syntaxin-8 facilitated the trafficking of claudin-16 to the plasma membrane without affecting its endocytosis (Ikari et al., 2014).

4.7.2.5 Phosphorylation

The claudin-16 protein can be phosphorylated at its carboxyl-terminal domain (Ser217) by the protein kinase A under physiological conditions in the MDCK cells (Ikari, Matsumoto, Harada, Takagi, & Hayashi et al., 2006). Dephosphorylation of claudin-16 by PKA inhibitors disrupted the interaction of claudin-16 with ZO-1, caused claudin-16 to accumulate in the lysosome, and decreased the paracellular Mg^{++} permeability (Ikari, Matsumoto, Harada, Takagi, & Degawa et al., 2006; Ikari, Matsumoto, Harada, Takagi, & Hayashi et al., 2006).

4.7.2.6 Extracellular Mg^{++} Concentration

Extracellular Mg^{++} concentration appears to regulate claudin-16 on both the mRNA and the protein levels. Mice fed with a low Mg^{++} diet developed hypomagnesemia but expressed more claudin-16 mRNA in the kidney, whereas mice fed with a high Mg^{++} diet developed hypermagnesemia but expressed less claudin-16 mRNA in the kidney when compared with mice under normal dietary condition (Efrati et al., 2010). Mg^{++} depletion in the culture medium for the MDCK cells reduced the claudin-16 phosphorylation levels at two loci—T295 and T303 located in the carboxyl-terminal domain of the human protein, which delocalized claudin-16 from the TJ and inhibited the paracellular Mg^{++} permeability (Ikari et al., 2010).

REFERENCES

Alexandre, M. D., Lu, Q., & Chen, Y. H. (2005). Overexpression of claudin-7 decreases the paracellular Cl- conductance and increases the paracellular Na+ conductance in LLC-PK1 cells. *Journal of Cell Science*, *118*, 2683–2693.

Amasheh, S., Meiri, N., Gitter, A. H., Schoneberg, T., Mankertz, J., Schulzke, J. D., et al. (2002). Claudin-2 expression induces cation-selective channels in tight junctions of epithelial cells. *Journal of Cell Science*, *115*, 4969–4976.

Anderson, J. M., & Van Itallie, C. M. (2009). Physiology and function of the tight junction. *Cold Spring Harbor Perspectives in Biology*, *1*, a002584.

Angelow, S., El-Husseini, R., Kanzawa, S. A., & Yu, A. S. (2007). Renal localization and function of the tight junction protein, claudin-19. *American Journal of Physiology Renal Physiology*, *293*, F166–F177.

Ben-Yosef, T., Belyantseva, I. A., Saunders, T. L., Hughes, E. D., Kawamoto, K., Van Itallie, C. M., et al. (2003). Claudin 14 knockout mice, a model for autosomal recessive deafness

DFNB29, are deaf due to cochlear hair cell degeneration. *Human Molecular Genetics, 12*, 2049–2061.

Buchert, M., Papin, M., Bonnans, C., Darido, C., Raye, W. S., Garambois, V., et al. (2010). Symplekin promotes tumorigenicity by up-regulating claudin-2 expression. *Proceedings of the National Academy of Sciences of the United States of America, 107*, 2628–2633.

Cereijido, M., Gonzalez-Mariscal, L., & Borboa, L. (1983). Occluding junctions and paracellular pathways studied in monolayers of MDCK cells. *The Journal of Experimental Biology, 106*, 205–215.

Claude, P. (1978). Morphological factors influencing transepithelial permeability: A model for the resistance of the zonula occludens. *The Journal of Membrane Biology, 39*, 219–232.

Claude, P., & Goodenough, D. A. (1973). Fracture faces of zonulae occludentes from "tight" and "leaky" epithelia. *The Journal of Cell Biology, 58*, 390–400.

Colegio, O. R., Van Itallie, C. M., McCrea, H. J., Rahner, C., & Anderson, J. M. (2002). Claudins create charge-selective channels in the paracellular pathway between epithelial cells. *American Journal of Physiology Cell Physiology, 283*, C142–C147.

Colegio, O. R., Van Itallie, C., Rahner, C., & Anderson, J. M. (2003). Claudin extracellular domains determine paracellular charge selectivity and resistance but not tight junction fibril architecture. *American Journal of Physiology Cell Physiology, 284*, C1346–C1354.

Derebe, M. G., Sauer, D. B., Zeng, W., Alam, A., Shi, N., & Jiang, Y. (2011). Tuning the ion selectivity of tetrameric cation channels by changing the number of ion binding sites. *Proceedings of the National Academy of Sciences of the United States of America, 108*, 598–602.

Efrati, E., Hirsch, A., Kladnitsky, O., Rozenfeld, J., Kaplan, M., Zinder, O., et al. (2010). Transcriptional regulation of the claudin-16 gene by Mg2+ availability. *Cellular Physiology and Biochemistry: International Journal of Experimental Cellular Physiology, Biochemistry, and Pharmacology, 25*, 705–714.

Eisenman, G., & Horn, R. (1983). Ionic selectivity revisited: The role of kinetic and equilibrium processes in ion permeation through channels. *The Journal of Membrane Biology, 76*, 197–225.

Firth, J. A. (2002). Endothelial barriers: From hypothetical pores to membrane proteins. *Journal of Anatomy, 200*, 541–548.

Frizzell, R. A., Koch, M. J., & Schultz, S. G. (1976). Ion transport by rabbit colon. I. Active and passive components. *The Journal of Membrane Biology, 27*, 297–316.

Fujita, H., Sugimoto, K., Inatomi, S., Maeda, T., Osanai, M., Uchiyama, Y., et al. (2008). Tight junction proteins claudin-2 and -12 are critical for vitamin D-dependent Ca2+ absorption between enterocytes. *Molecular Biology of the Cell, 19*, 1912–1921.

Garcia-Hernandez, V., Flores-Maldonado, C., Rincon-Heredia, R., Verdejo-Torres, O., Bonilla-Delgado, J., Meneses-Morales, I., et al. (2015). EGF regulates claudin-2 and -4 expression through Src and STAT3 in MDCK cells. *Journal of Cellular Physiology, 230*, 105–115.

Geng, Y., Niu, X., & Magleby, K. L. (2011). Low resistance, large dimension entrance to the inner cavity of BK channels determined by changing side-chain volume. *The Journal of General Physiology, 137*, 533–548.

Goodenough, D. A., & Revel, J. P. (1970). A fine structural analysis of intercellular junctions in the mouse liver. *The Journal of Cell Biology, 45*, 272–290.

Guillemot, L., & Citi, S. (2006). Cingulin regulates claudin-2 expression and cell proliferation through the small GTPase RhoA. *Molecular Biology of the Cell, 17*, 3569–3577.

Guillemot, L., Schneider, Y., Brun, P., Castagliuolo, I., Pizzuti, D., Martines, D., et al. (2012). Cingulin is dispensable for epithelial barrier function and tight junction structure, and plays a role in the control of claudin-2 expression and response to duodenal mucosa injury. *Journal of Cell Science, 125*, 5005–5014.

Hou, J., Gomes, A. S., Paul, D. L., & Goodenough, D. A. (2006). Study of claudin function by RNA interference. *The Journal of Biological Chemistry, 281*, 36117–36123.

Hou, J., Paul, D. L., & Goodenough, D. A. (2005). Paracellin-1 and the modulation of ion selectivity of tight junctions. *Journal of Cell Science, 118*, 5109–5118.

Hou, J., Renigunta, A., Konrad, M., Gomes, A. S., Schneeberger, E. E., Paul, D. L., et al. (2008). Claudin-16 and claudin-19 interact and form a cation-selective tight junction complex. *The Journal of Clinical Investigation, 118*, 619–628.

Hou, J., Renigunta, A., Yang, J., & Waldegger, S. (2010). Claudin-4 forms paracellular chloride channel in the kidney and requires claudin-8 for tight junction localization. *Proceedings of the National Academy of Sciences of the United States of America, 107*, 18010–18015.

Ikari, A., Hirai, N., Shiroma, M., Harada, H., Sakai, H., Hayashi, H., et al. (2004). Association of paracellin-1 with ZO-1 augments the reabsorption of divalent cations in renal epithelial cells. *The Journal of Biological Chemistry, 279*, 54826–54832.

Ikari, A., Matsumoto, S., Harada, H., Takagi, K., Degawa, M., Takahashi, T., et al. (2006a). Dysfunction of paracellin-1 by dephosphorylation in Dahl salt-sensitive hypertensive rats. *The Journal of Physiological Sciences: JPS, 56*, 379–383.

Ikari, A., Matsumoto, S., Harada, H., Takagi, K., Hayashi, H., Suzuki, Y., et al. (2006b). Phosphorylation of paracellin-1 at Ser217 by protein kinase A is essential for localization in tight junctions. *Journal of Cell Science, 119*, 1781–1789.

Ikari, A., Kinjo, K., Atomi, K., Sasaki, Y., Yamazaki, Y., & Sugatani, J. (2010). Extracellular Mg(2+) regulates the tight junctional localization of claudin-16 mediated by ERK-dependent phosphorylation. *Biochimica et biophysica acta, 1798*, 415–421.

Ikari, A., Takiguchi, A., Atomi, K., & Sugatani, J. (2011). Epidermal growth factor increases clathrin-dependent endocytosis and degradation of claudin-2 protein in MDCK II cells. *Journal of Cellular Physiology, 226*, 2448–2456.

Ikari, A., Tonegawa, C., Sanada, A., Kimura, T., Sakai, H., Hayashi, H., et al. (2014). Tight junctional localization of claudin-16 is regulated by syntaxin 8 in renal tubular epithelial cells. *The Journal of Biological Chemistry, 289*, 13112–13123.

Ikari, A., Fujii, N., Hahakabe, S., Hayashi, H., Yamaguchi, M., Yamazaki, Y., et al. (2015). Hyperosmolarity-induced down-regulation of claudin-2 mediated by decrease in PKCbeta-dependent GATA-2 in MDCK Cells. *Journal of Cellular Physiology, 230*, 2776–2787.

Inai, T., Kobayashi, J., & Shibata, Y. (1999). Claudin-1 contributes to the epithelial barrier function in MDCK cells. *European Journal of Cell Biology, 78*, 849–855.

Jovov, B., Van Itallie, C. M., Shaheen, N. J., Carson, J. L., Gambling, T. M., Anderson, J. M., et al. (2007). Claudin-18: A dominant tight junction protein in Barrett's esophagus and likely contributor to its acid resistance. *American Journal of Physiology Gastrointestinal and Liver Physiology, 293*, G1106–G1113.

Kausalya, P. J., Amasheh, S., Gunzel, D., Wurps, H., Muller, D., Fromm, M., et al. (2006). Disease-associated mutations affect intracellular traffic and paracellular Mg2+ transport function of claudin-16. *The Journal of Clinical Investigation, 116*, 878–891.

Konrad, M., Hou, J., Weber, S., Dotsch, J., Kari, J. A., Seeman, T., et al. (2008). CLDN16 genotype predicts renal decline in familial hypomagnesemia with hypercalciuria and nephrocalcinosis. *Journal of the American Society of Nephrology: JASN, 19*, 171–181.

Krug, S. M., Gunzel, D., Conrad, M. P., Rosenthal, R., Fromm, A., Amasheh, S., et al. (2012). Claudin-17 forms tight junction channels with distinct anion selectivity. *Cellular and Molecular Life Sciences: CMLS, 69*, 2765–2778.

Latorre, R., & Miller, C. (1983). Conduction and selectivity in potassium channels. *The Journal of Membrane Biology, 71*, 11–30.

Lipschutz, J. H., Li, S., Arisco, A., & Balkovetz, D. F. (2005). Extracellular signal-regulated kinases 1/2 control claudin-2 expression in Madin-Darby canine kidney strain I and II cells. *The Journal of Biological Chemistry*, *280*, 3780–3788.

Lu, R., Dalgalan, D., Mandell, E. K., Parker, S. S., Ghosh, S., & Wilson, J. M. (2015). PKCiota interacts with Rab14 and modulates epithelial barrier function through regulation of claudin-2 levels. *Molecular Biology of the Cell*, *26*, 1523–1531.

MacKinnon, R. (2004). Potassium channels and the atomic basis of selective ion conduction (Nobel Lecture). *Angewandte Chemie (International ed in English)*, *43*, 4265–4277.

Mandel, L. J., & Curran, P. F. (1972). Response of the frog skin to steady-state voltage clamping. I. The shunt pathway. *The Journal of General Physiology*, *59*, 503–518.

Milatz, S., Krug, S. M., Rosenthal, R., Gunzel, D., Muller, D., Schulzke, J. D., et al. (2010). Claudin-3 acts as a sealing component of the tight junction for ions of either charge and uncharged solutes. *Biochimica et biophysica acta*, *1798*, 2048–2057.

Muller, D., Kausalya, P. J., Claverie-Martin, F., Meij, I. C., Eggert, P., Garcia-Nieto, V., et al. (2003). A novel claudin 16 mutation associated with childhood hypercalciuria abolishes binding to ZO-1 and results in lysosomal mistargeting. *American Journal of Human Genetics*, *73*, 1293–1301.

Muller, D., Kausalya, P. J., Meij, I. C., & Hunziker, W. (2006). Familial hypomagnesemia with hypercalciuria and nephrocalcinosis: Blocking endocytosis restores surface expression of a novel claudin-16 mutant that lacks the entire C-terminal cytosolic tail. *Human Molecular Genetics*, *15*, 1049–1058.

Mullins, L. J. (1959). An analysis of conductance changes in squid axon. *The Journal of General Physiology*, *42*, 1013–1035.

Nimigean, C. M., Chappie, J. S., & Miller, C. (2003). Electrostatic tuning of ion conductance in potassium channels. *Biochemistry*, *42*, 9263–9268.

Powell, D. W. (1981). Barrier function of epithelia. *The American Journal of Physiology*, *241*, G275–G288.

Sas, D., Hu, M., Moe, O. W., & Baum, M. (2008). Effect of claudins 6 and 9 on paracellular permeability in MDCK II cells. *American Journal of Physiology Regulatory, Integrative and Comparative Physiology*, *295*, R1713–R1719.

Shi, N., Ye, S., Alam, A., Chen, L., & Jiang, Y. (2006). Atomic structure of a Na+- and K+-conducting channel. *Nature*, *440*, 570–574.

Simon, D. B., Lu, Y., Choate, K. A., Velazquez, H., Al-Sabban, E., Praga, M., et al. (1999). Paracellin-1, a renal tight junction protein required for paracellular Mg2+ resorption. *Science (New York, NY)*, *285*, 103–106.

Singh, A. B., & Harris, R. C. (2004). Epidermal growth factor receptor activation differentially regulates claudin expression and enhances transepithelial resistance in Madin-Darby canine kidney cells. *The Journal of Biological Chemistry*, *279*, 3543–3552.

Singh, A. B., Sugimoto, K., Dhawan, P., & Harris, R. C. (2007). Juxtacrine activation of EGFR regulates claudin expression and increases transepithelial resistance. *American Journal of Physiology Cell Physiology*, *293*, C1660–C1668.

Suzuki, H., Nishizawa, T., Tani, K., Yamazaki, Y., Tamura, A., Ishitani, R., et al. (2014). Crystal structure of a claudin provides insight into the architecture of tight junctions. *Science (New York, NY)*, *344*, 304–307.

Tanaka, H., Yamamoto, Y., Kashihara, H., Yamazaki, Y., Tani, K., Fujiyoshi, Y., et al. (2016). Claudin-21 has a paracellular channel role at tight junctions. *Molecular and Cellular Biology*, *36*, 954–964.

Tang, V. W., & Goodenough, D. A. (2003). Paracellular ion channel at the tight junction. *Biophysical Journal*, *84*, 1660–1673.

Tokuda, S., Hirai, T., & Furuse, M. (2016). Effects of osmolality on paracellular transport in MDCK II cells. *PLoS One*, *11*, e0166904.

Van Itallie, C., Rahner, C., & Anderson, J. M. (2001). Regulated expression of claudin-4 decreases paracellular conductance through a selective decrease in sodium permeability. *The Journal of Clinical Investigation*, *107*, 1319–1327.

Van Itallie, C. M., Fanning, A. S., & Anderson, J. M. (2003). Reversal of charge selectivity in cation or anion-selective epithelial lines by expression of different claudins. *American Journal of Physiology Renal Physiology*, *285*, F1078–F1084.

Van Itallie, C. M., Rogan, S., Yu, A., Vidal, L. S., Holmes, J., & Anderson, J. M. (2006). Two splice variants of claudin-10 in the kidney create paracellular pores with different ion selectivities. *American Journal of Physiology Renal Physiology*, *291*, F1288–F1299.

Van Itallie, C. M., Holmes, J., Bridges, A., Gookin, J. L., Coccaro, M. R., Proctor, W., et al. (2008). The density of small tight junction pores varies among cell types and is increased by expression of claudin-2. *Journal of Cell Science*, *121*, 298–305.

Van Itallie, C. M., Tietgens, A. J., LoGrande, K., Aponte, A., Gucek, M., & Anderson, J. M. (2012). Phosphorylation of claudin-2 on serine 208 promotes membrane retention and reduces trafficking to lysosomes. *Journal of Cell Science*, *125*, 4902–4912.

Watson, C. J., Rowland, M., & Warhurst, G. (2001). Functional modeling of tight junctions in intestinal cell monolayers using polyethylene glycol oligomers. *American Journal of Physiology Cell Physiology*, *281*, C388–C397.

Weber, S., Schlingmann, K. P., Peters, M., Nejsum, L. N., Nielsen, S., Engel, H., et al. (2001a). Primary gene structure and expression studies of rodent paracellin-1. *Journal of the American Society of Nephrology: JASN*, *12*, 2664–2672.

Weber, S., Schneider, L., Peters, M., Misselwitz, J., Ronnefarth, G., Boswald, M., et al. (2001b). Novel paracellin-1 mutations in 25 families with familial hypomagnesemia with hypercalciuria and nephrocalcinosis. *Journal of the American Society of Nephrology: JASN*, *12*, 1872–1881.

Weber, C. R., Liang, G. H., Wang, Y., Das, S., Shen, L., Yu, A. S., et al. (2015). Claudin-2-dependent paracellular channels are dynamically gated. *eLife*, *4*, e09906.

Wen, H., Watry, D. D., Marcondes, M. C., & Fox, H. S. (2004). Selective decrease in paracellular conductance of tight junctions: Role of the first extracellular domain of claudin-5. *Molecular and Cellular Biology*, *24*, 8408–8417.

Yu, A. S., Enck, A. H., Lencer, W. I., & Schneeberger, E. E. (2003). Claudin-8 expression in Madin-Darby canine kidney cells augments the paracellular barrier to cation permeation. *The Journal of Biological Chemistry*, *278*, 17350–17359.

Yu, A. S., Cheng, M. H., Angelow, S., Gunzel, D., Kanzawa, S. A., Schneeberger, E. E., et al. (2009). Molecular basis for cation selectivity in claudin-2-based paracellular pores: Identification of an electrostatic interaction site. *The Journal of General Physiology*, *133*, 111–127.

Yu, A. S., Cheng, M. H., & Coalson, R. D. (2010). Calcium inhibits paracellular sodium conductance through claudin-2 by competitive binding. *The Journal of Biological Chemistry*, *285*, 37060–37069.

Zhang, Y. G., Wu, S., Lu, R., Zhou, D., Zhou, J., Carmeliet, G., et al. (2015). Tight junction CLDN2 gene is a direct target of the vitamin D receptor. *Scientific Reports*, *5*, 10642.

Zhou, Y., & MacKinnon, R. (2003). The occupancy of ions in the K+ selectivity filter: Charge balance and coupling of ion binding to a protein conformational change underlie high conduction rates. *Journal of Molecular Biology*, *333*, 965–975.

Chapter 5

Paracellular Anion Channel

5.1 TWO FACES OF ANION SELECTIVITY

The claudins making paracellular anion channels include claudin-4, -7, -8, -10a, and -17 (Table 4.1). These paracellular anion channels often function either as paracellular barriers to cations or as paracellular channels to anions, depending upon the cellular background, to modulate the ion selectivity of the tight junction (TJ). For example, gene knockdown experiments have revealed that claudin-4 and claudin-7 function as paracellular barriers to Na^+ in the MDCK-II cells but as paracellular channels to Cl^- in the LLC-PK1 cells (Hou, Gomes, Paul, & Goodenough, 2006). The claudin protein composition of the TJ may differ between these two cell types, therefore providing alternative partners with which claudin-4 and claudin-7 may interact. It is the partner claudin that will determine the absolute permeability of claudin-4 and claudin-7. More intriguingly, transfection of claudin-10a into the MDCK-II cells simultaneously reduced the paracellular permeability to Na^+ while increased the paracellular permeability to Cl^- (Van Itallie et al., 2006). Hou and coworkers have proposed a model to explain the combinatorial effect of claudin interaction on paracellular permeability (Gong et al., 2015a). In a hypothetic cell model with cation-selective background of paracellular permeability, for example, a model resembling the MDCK-II cells because of the high expression level of the cation channel—claudin-2 (Fig. 5.1A), adding claudin-10a, an anion-selective channel, will not only create anion permeation pores but also reduce the number of cation permeation pores in the TJ strand (Fig. 5.1B), assuming that the overall paracellular channel density is governed by the TJ architectural requirement to establish the strand-like pattern and likely remains constant (Fig. 2.2). This scenario also assumes that claudin-10a can interact and co-polymerize with claudin-2 into the same TJ strand. If a claudin molecule, for example, claudin-4 or claudin-7, cannot interact with claudin-2, then it may form an independent TJ strand that will impede the cation permeation (Fig. 5.1C).

5.2 THE STRUCTURAL BASIS OF ANION SELECTIVITY

5.2.1 The Net Charge in the Selectivity Filter of Claudin

According to the structural knowledge of claudin, the net charge in the selectivity filter consisting of the 4th β-strand in ECL1 may dictate the ion selectivity of the paracellular channel (Fig. 2.4). The net charge in the selectivity filter from

The Paracellular Channel. http://dx.doi.org/10.1016/B978-0-12-814635-4.00005-X

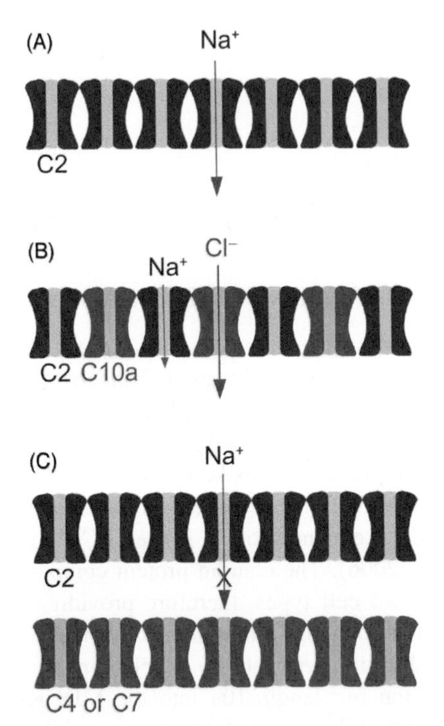

FIGURE 5.1 Hypothetic models of paracellular channel arrangement. (A) Background of cation-selective permeability due to high expression levels of claudin-2 (C2). (B) Adding claudin-10a (C10a) to the claudin-2 background will reduce the claudin-2 protein density in the TJ strand assuming that claudin-10a can interact with claudin-2. (C) Adding claudin-4 (C4) or claudin-7 (C7) to the claudin-2 background will lead to the formation of a new TJ strand made of claudin-4 or claudin-7 assuming that claudin-4 or claudin-7 cannot interact with claudin-2. The newly formed claudin-4 or claudin-7 strand is arranged in series with the claudin-2 strand and will impede cation permeation due to the anion selectivity of claudin-4 or claudin-7.

claudin-4, -7, -8, -10a, or -17 is positive, compatible with their function as anion channels (Fig. 4.2). The charged residues outside the selectivity filter may also play important roles. For example, mutation of a positively charged residue, Arg33 in the 1st β-strand of ECL1 in claudin-10a abolished its anion selectivity (Van Itallie et al., 2006).

5.2.2 The Electrostatic Interaction Site in Claudin

According to Eisenman's theory, the charge selectivity of an ion channel is determined by the difference between the energy cost of dehydration of the ion and the energy gain from binding to the charged sites within the channel pore (Eisenman & Horn, 1983). The permeability sequence to halide anions of the claudin-4 channel resembles the Eisenman selectivity sequence III: $Br^- > Cl^- \sim I^- > F^-$, which is different from the sequence of their free-water

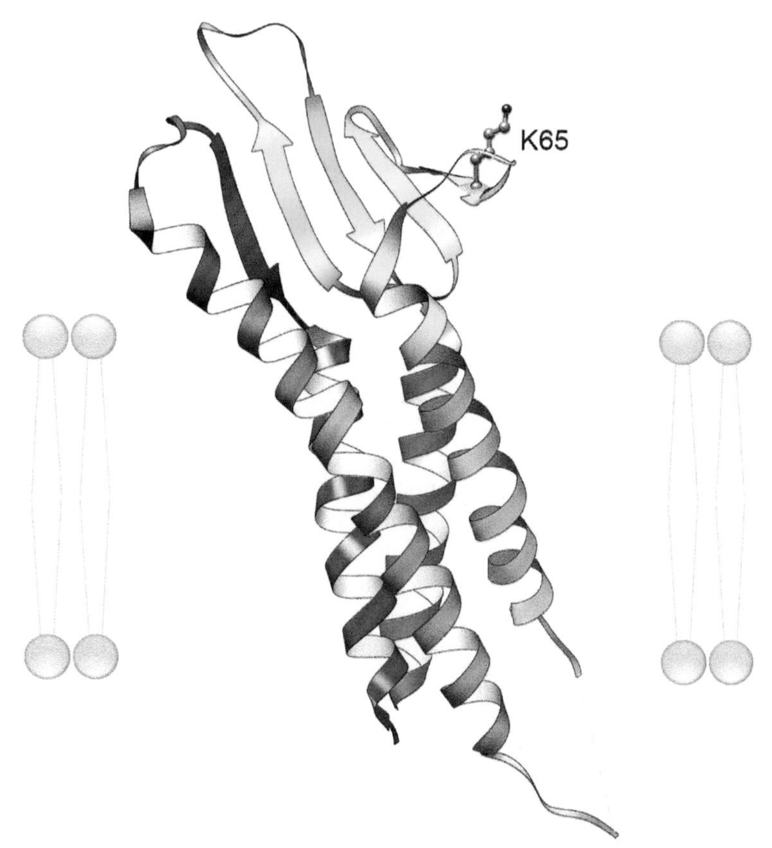

FIGURE 5.2 Molecular structure of claudin-4. Crystal structure of claudin-4 in ribbon representation when viewed from the side of the membrane. The color changes gradually from the N terminus (*green*) to the C terminus (*red*). A positively charged residue, K65, is labeled. The depicted claudin structure is based upon the X-ray analysis of human claudin-4 protein (Shinoda et al., 2016).

mobilities, that is, the Eisenman sequence I: $I^- > Br^- > Cl^- > F^-$ (Hou, Renigunta, Yang, & Waldegger, 2010). The shift of halide anion permeability sequence to a higher order indicates the presence of electrostatic interaction sites in the claudin-4 channel pore that allow partial dehydration of the permeant anions (Eisenman & Horn, 1983). The crystal structure of claudin-4 reveals that the positively charged residue, Lys65 in the 4th β-strand of ECL1 faces the extracellular space and appears to interact with the permeant anions (Fig. 5.2) (Shinoda et al., 2016). Neutralizing the positive charge on Lys65 by mutating to Thr65 in claudin-4 not only abolished the anion permeability of claudin-4 but also reversed the ion selectivity of claudin-4 to become slightly permeable to cations (Hou et al., 2010). Brownian dynamics simulations suggest that the paracellular channel conductance varies with the effective charge valence on the side chain of the charged residue in the selectivity filter of the channel pore (Yu et al., 2009). Decreases in the effective charge in claudin-4 channel pore ($+1e \rightarrow 0$) not only reduce its conductance, carried by anions, but also destabilize the electrostatic field that has now lowered the energy barrier for cation permeation. Because the K65T mutation in claudin-4 only mildly increases the cation permeability, the cation binding affinity of the mutant claudin-4 channel pore must be weak. The Arg62 site in claudin-10a and the Lys65 site in claudin-17, homologous to the Lys65 site in claudin-4 (Fig. 4.2), are also critical for the anion selectivity of the paracellular channel that claudin-10a or claudin-17 respectively makes (Krug et al., 2012; Van Itallie et al., 2006).

5.3 THE CONDUCTANCE OF PARACELLULAR ANION CHANNEL

The conductance of paracellular cation channels, for example, claudin-2, -10b, -15, -16, and -21 is at ~10 mS/cm^2 (Hou, Paul, & Goodenough, 2005; Tanaka et al., 2016; Van Itallie, Fanning, & Anderson, 2003; Van Itallie et al., 2006; Yu et al., 2009). The conductance of paracellular anion channels, for example, claudin-4 and -17 is by one magnitude lower and at ~1 mS/cm^2 (Hou et al., 2010; Krug et al., 2012). Brownian dynamics simulations predict the maximal single-channel conductance for claudin to be around 100 pS (Yu et al., 2009). The overall paracellular conductance is regulated by the single claudin channel conductance, the open probability of claudin channel, the claudin molecular density along individual TJ strand, and the number of TJ strands (Chapter 4, Section 4.4) (Claude, 1978). Assuming that cation-selective and anion-selective claudins share similar single channel conductance, open probability, and on-strand molecular density levels, then the TJ strand number will be the single most likely factor that differs between cation-selective and anion-selective paracellular pathways. The number of TJ strands is inversely correlated to the paracellular conductance (Claude & Goodenough, 1973). Interestingly, the proximal tubular epithelium in the mouse kidney expresses high levels of claudin-2, presents high levels of paracellular cation conductance, and consists in only 1–2 parallel TJ strands, whereas the distal tubular epithelium in the mouse kidney expresses high levels of claudin-4, presents low levels of paracellular anion conductance, and consists in more than 4 parallel TJ strands (Claude & Goodenough, 1973).

5.4 THE REGULATION OF PARACELLULAR ANION CHANNEL

5.4.1 Claudin-4

5.4.1.1 Phosphorylation

A rare form of human hereditary renal disease known as familial hyperkalemic hypertension (FHHt), pseudohypoaldosteronism II (PHA-II), or Gordon's syndrome, is caused by mutations in two genes encoding *w*ith *no* lysine (*K*) kinase 1 (WNK1) and WNK4 respectively (Wilson et al., 2001). Transfection of the WNK4 protein harboring the FHHt-causing mutation (D561A) into the MDCK cells increased the paracellular Cl^- permeability by phosphorylating the claudin-4 protein (Kahle et al., 2004; Yamauchi et al., 2004). Aldosterone, a type of mineralocorticoids, also upregulates the paracellular Cl^- permeability and phosphorylates the claudin-4 protein in a cortical collecting duct cell line—RCCD2 derived from the rat kidney (Le Moellic et al., 2005). In the rat salivary epithelial SMG-C6 cells, activation of the muscarinic acetylcholine receptor (mAChR) reduces the transepithelial resistance (TER or R_{te}), an inverse indicator of the paracellular conductance, by phosphorylating and endocytosing the claudin-4 protein (Cong et al., 2015). Mutagenesis studies have revealed that the S195 site in claudin-4 is phosphorylated by the ERK1/2 kinases in response to mAChR activation. The phosphorylated claudin-4 protein interacts with β-arrestin-2, which triggers claudin-4 endocytosis via the clathrin-dependent pathway (Cong et al., 2015).

5.4.1.2 *Osmolality*

Hyperosmolality can upregulate the mRNA and the protein expression levels of claudin-4 in the mouse collecting duct mIMCD3 cells or in the papilla of the mouse or the rat kidney (Lanaspa, Andres-Hernando, Rivard, Dai, & Berl, 2008). The promoter in the claudin-4 gene contains the Sp1 binding site. Hyperosmolality induces the binding of the transcriptional factors Sp1/c-Jun to the Sp1 site in the claudin-4 gene promotor, which results in the increase in the claudin-4 gene transcription (Ikari et al., 2013). The interaction between claudin-4 and the multiple PDZ domain (MPDZ) protein, also known as MUPP1, appears to be regulated by hyperosmolality too. Gene knockdown experiments have shown that MUPP1 is required to recruit claudin-4 to the TJ in response to hyperosmolality (Lanaspa et al., 2008).

5.4.1.3 *Proteases*

Trypsin, when added to the apical side of the mouse collecting duct M-1 and mIMCD3 cells, transiently increased the TER by reducing the paracellular Cl^- permeability (Liu, Hering-Smith, Schiro, & Hamm, 2002). A trypsin-like protease known as channel-activating protease 1 (CAP1) or prostasin exerts a similar effect to trypsin on paracellular Cl^- permeation in both M-1 and mIMCD3 cells (Gong et al., 2014). Mechanistically, CAP1 directly acts upon the 2nd extracellular loop (ECL2) domain of claudin-4 to disrupt its *trans*-interaction, presumably by proteolytic cleavage at the R158 site of the claudin-4 protein (Fig. 5.3). Loss of claudin-4 interaction reduces its membrane stability, causing increased endocytotic rates (Gong et al., 2014).

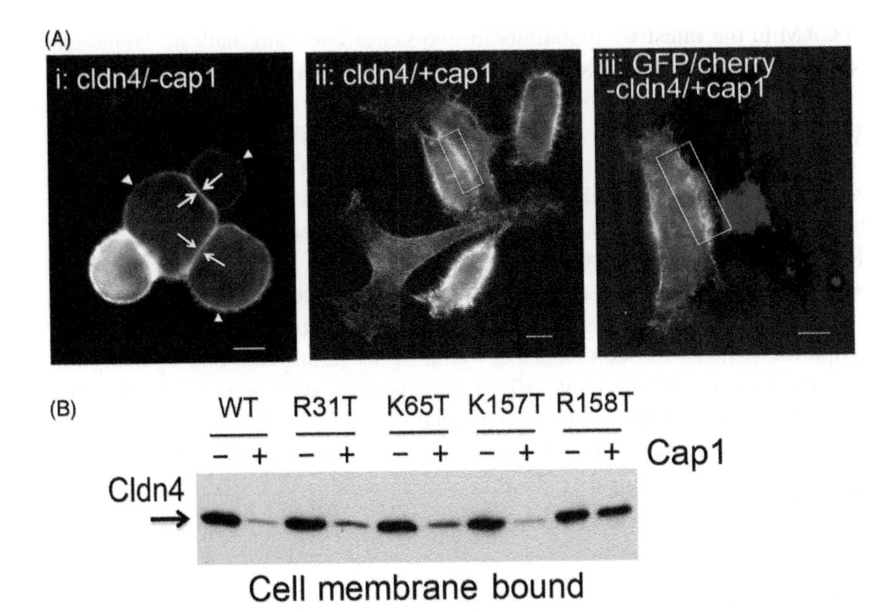

FIGURE 5.3 **Protease disrupts claudin-4 interaction.** (A) In unpolarized HEK293 cells, claudin-4 protein is predominantly localized at the cell junction (i, *arrow*) and variably at the nonjunctional plasma membrane (i, *arrowhead*). In the presence of CAP1, claudin-4 relocates to nonjunctional plasma membrane (*rectangle*) viewed by single-color (ii) and dual-color (iii) fluorescence imaging. Bar: 10 μm. (B) Cell surface abundance levels of claudin-4 protein in HEK293 cells transfected with wildtype or mutant claudin-4 proteins. Note that mutation at the R158 but not the R31, K65, or K157 site protects claudin-4 from CAP1-induced membrane destabilization. *(Reproduced with permission from Gong, Y., Yu, M., Yang, J., Gonzales, E., Perez, R., Hou, M., et al. (2014). The Cap1-claudin-4 regulatory pathway is important for renal chloride reabsorption and blood pressure regulation. Proceedings of the National Academy of Sciences of the United States of America,* 111, *E3766–3774).*

5.4.2 Claudin-7

5.4.2.1 Phosphorylation

The WNK4 kinase has been shown to phosphorylate claudin-7 at the S206 site in the carboxyl-terminus of the protein (Tatum et al., 2007). Both the wildtype WNK4 and the FHHt-causing mutant form of WNK4 increased the paracellular Cl^- permeability when transfected into the LLC–PK1 cells. More importantly, the FHHt-causing mutation potentiates the kinase activity of WNK4, which further augments the paracellular Cl^- permeation (Tatum et al., 2007).

5.4.2.2 Epithelial Cell Adhesion Molecule (EpCAM)

EpCAM, also known as CD326, is found in the lateral membrane including the TJ of the intestinal epithelium (Wu, Mannan, Lu, & Udey, 2013). Knockout of EpCAM in murine models downregulated the claudin-7 protein abundance levels in the intestine (Lei et al., 2012). EpCAM can interact with claudin-7 and regulate its membrane stability and endocytosis (Wu et al., 2013). Loss of

EpCAM in the intestinal epithelium decreases the paracellular permeability to Na^+ but exerts no effect on Cl^- permeability (Lei et al., 2012).

5.4.3 Claudin-8

5.3.3.1 Corticosteroids

The glucocorticoids including dexamethasone and hydrocortisone stimulate the gene transcription of claudin-8 and reduce the paracellular conductance in human primary tracheal epithelial cells (Kielgast et al., 2016). In the colonic epithelium, the mineralocorticoid, aldosterone upregulates the claudin-8 gene transcriptional levels but elicits no change in the TER values (Amasheh et al., 2009).

5.3.3.2 Protein Interaction

Similar to claudin-4, claudin-8 can interact with MUPP1. In the MDCK-II cells, overexpression of MUPP1 recruits the claudin-8 protein to the TJ, resulting in decreased paracellular conductance (Jeansonne, Lu, Goodenough, & Chen, 2003). Interestingly, claudin-8 *cis* interacts with multiple claudins, including claudin-3, -4, and -7 (Hou et al., 2010). In the mouse collecting duct cells, the interaction of claudin-4 with claudin-8 is particularly important for the localization of claudin-4 in the TJ (Fig. 5.4) (Hou et al., 2010). Removal of

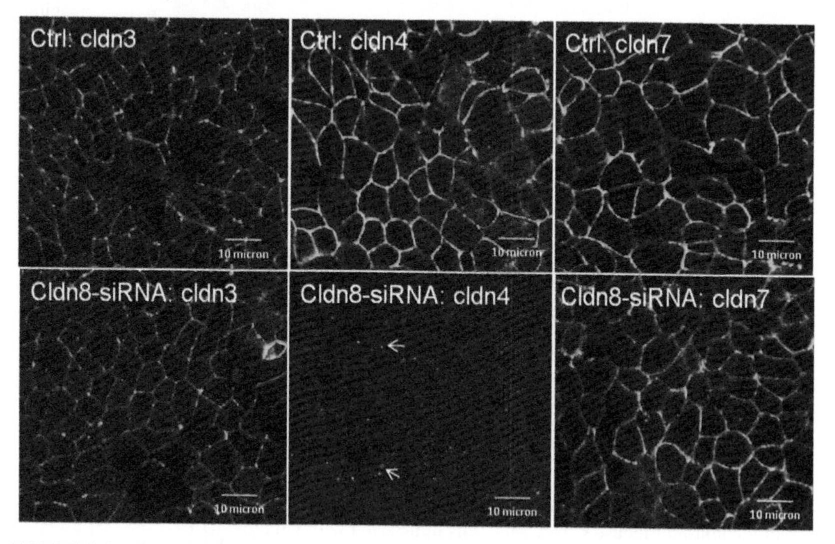

FIGURE 5.4 Claudin-8 recruits claudin-4 to the TJ. In polarized mouse kidney M-1 cells, claudin-4 protein is completely missing from the bicellular TJ but remains in the tricellular TJ (*arrow*) when claudin-8 is depleted from the cells. The localization of claudin-3 or claudin-7 is not affected by the claudin-8 knockdown. Bar: 10 μm. *(Reproduced with permission from Hou, J., Renigunta, A., Yang, J., & Waldegger, S. (2010). Claudin-4 forms paracellular chloride channel in the kidney and requires claudin-8 for tight junction localization. Proceedings of the National Academy of Sciences of the United States of America,* 107, *18010–18015).*

claudin-8 creates a double deletion of both claudins at the level of TJ in the mouse kidney (Gong et al., 2015b). Claudin-8 itself does not appear to form a paracellular anion channel. Instead, it increases the paracellular anion permeability by recruiting claudin-4 to the TJ (Hou et al., 2010). When introduced to the MDCK-II cells, claudin-8 replaces claudin-2 in the TJ, causing decreases in paracellular Na^+ permeability (Angelow, Schneeberger, & Yu, 2007; Yu, Enck, Lencer, & Schneeberger, 2003). Such claudin incompatibility is likely due to the competition between claudin-2 and claudin-8 for their common binding partner, for example, claudin-3, -4 or -7.

5.3.3.3 Ubiquitination

There are two more genes linked to FHHt, PHA-II, or Gordon's syndrome, which encode proteins important in the cullin-really interesting new gene E3 ubiquitin ligase (CRL) complex—Kelch-like 3 (KLHL3) and Cullin 3 (CUL3) (Boyden et al., 2012). KLHL3 directly interacts with claudin-8 and regulates

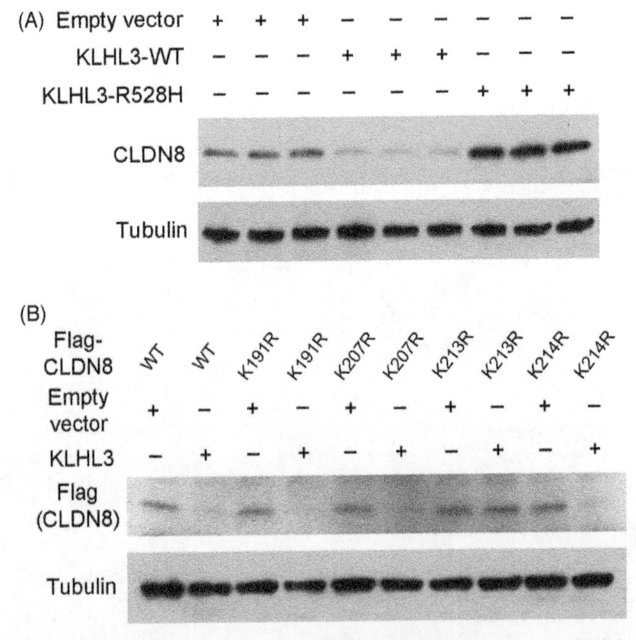

FIGURE 5.5 KLHL3 regulates claudin-8 protein abundance levels. (A) Cellular abundance level of claudin-8 in mouse kidney M-1 cells transfected with wildtype KLHL3, R528H KLHL3, or empty vector. Note that wildtype KLHL3 decreases, whereas R528H mutant increases, the claudin-8 protein levels. (B) Cellular abundance level of claudin-8 in M-1 cells co-transfected with KLHL3 and wildtype or lysine mutated claudin-8. Note that the K213R mutant claudin-8 is resistant to KLHL3-induced degradation, suggesting that K213 is where ubiquitination takes place. *(Reproduced with permission from Gong, Y., Wang, J., Yang, J., Gonzales, E., Perez, R., & Hou, J. (2015b). KLHL3 regulates paracellular chloride transport in the kidney by ubiquitination of claudin-8. Proceedings of the National Academy of Sciences of the United States of America, 112, 4340–4345).*

its ubiquitination and degradation (Gong et al., 2015b). In the mouse collecting duct cells, transfection of the wildtype KLHL3 decreased whereas the FHHt-causing mutant form (R528H) of KLHL3 increased the paracellular permeability to Cl^-. The FHHt-causing mutation in KLHL3 plays a dominant negative role by abolishing its binding, ubiquitination, and degradation of claudin-8 (Fig. 5.5) (Gong et al., 2015b).

REFERENCES

Amasheh, S., Milatz, S., Krug, S. M., Bergs, M., Amasheh, M., Schulzke, J. D., et al. (2009). Na+ absorption defends from paracellular back-leakage by claudin-8 upregulation. *Biochemical and Biophysical Research Communications, 378*, 45–50.

Angelow, S., Schneeberger, E. E., & Yu, A. S. (2007). Claudin-8 expression in renal epithelial cells augments the paracellular barrier by replacing endogenous claudin-2. *The Journal of Membrane Biology, 215*, 147–159.

Boyden, L. M., Choi, M., Choate, K. A., Nelson-Williams, C. J., Farhi, A., Toka, H. R., et al. (2012). Mutations in kelch-like 3 and cullin 3 cause hypertension and electrolyte abnormalities. *Nature, 482*, 98–102.

Claude, P. (1978). Morphological factors influencing transepithelial permeability: a model for the resistance of the zonula occludens. *The Journal of Membrane Biology, 39*, 219–232.

Claude, P., & Goodenough, D. A. (1973). Fracture faces of zonulae occludentes from "tight" and "leaky" epithelia. *The Journal of Cell Biology, 58*, 390–400.

Cong, X., Zhang, Y., Li, J., Mei, M., Ding, C., Xiang, R. L., et al. (2015). Claudin-4 is required for modulation of paracellular permeability by muscarinic acetylcholine receptor in epithelial cells. *Journal of Cell Science, 128*, 2271–2286.

Eisenman, G., & Horn, R. (1983). Ionic selectivity revisited: the role of kinetic and equilibrium processes in ion permeation through channels. *The Journal of Membrane Biology, 76*, 197–225.

Gong, Y., Yu, M., Yang, J., Gonzales, E., Perez, R., Hou, M., et al. (2014). The Cap1-claudin-4 regulatory pathway is important for renal chloride reabsorption and blood pressure regulation. *Proceedings of the National Academy of Sciences of the United States of America, 111*, E3766–E3774.

Gong, Y., Renigunta, V., Zhou, Y., Sunq, A., Wang, J., Yang, J., Renigunta, A., Baker, LA, & Hou, J. (2015a). Biochemical and biophysical analyses of tight junction permeability made of claudin-16 and claudin-19 dimerization. *Mol Biol Cell., 26*(24), 4333–4346.

Gong, Y., Wang, J., Yang, J., Gonzales, E., Perez, R., & Hou, J. (2015b). KLHL3 regulates paracellular chloride transport in the kidney by ubiquitination of claudin-8. *Proceedings of the National Academy of Sciences of the United States of America, 112*, 4340–4345.

Hou, J., Paul, D. L., & Goodenough, D. A. (2005). Paracellin-1 and the modulation of ion selectivity of tight junctions. *Journal of Cell Science, 118*, 5109–5118.

Hou, J., Gomes, A. S., Paul, D. L., & Goodenough, D. A. (2006). Study of claudin function by RNA interference. *The Journal of Biological Chemistry, 281*, 36117–36123.

Hou, J., Renigunta, A., Yang, J., & Waldegger, S. (2010). Claudin-4 forms paracellular chloride channel in the kidney and requires claudin-8 for tight junction localization. *Proceedings of the National Academy of Sciences of the United States of America, 107*, 18010–18015.

Ikari, A., Atomi, K., Yamazaki, Y., Sakai, H., Hayashi, H., Yamaguchi, M., et al. (2013). Hyperosmolarity-induced up-regulation of claudin-4 mediated by NADPH oxidase-dependent H2O2 production and Sp1/c-Jun cooperation. *Biochimica et Biophysica Acta, 1833*, 2617–2627.

Jeansonne, B., Lu, Q., Goodenough, D. A., & Chen, Y. H. (2003). Claudin-8 interacts with multi-PDZ domain protein 1 (MUPP1) and reduces paracellular conductance in epithelial cells. *Cellular and Molecular Biology (Noisy-le-Grand, France)*, *49*, 13–21.

Kahle, K. T., Macgregor, G. G., Wilson, F. H., Van Hoek, A. N., Brown, D., Ardito, T., et al. (2004). Paracellular Cl- permeability is regulated by WNK4 kinase: insight into normal physiology and hypertension. *Proceedings of the National Academy of Sciences of the United States of America*, *101*, 14877–14882.

Kielgast, F., Schmidt, H., Braubach, P., Winkelmann, V. E., Thompson, K. E., Frick, M., et al. (2016). Glucocorticoids regulate tight junction permeability of lung epithelia by modulating claudin 8. *American Journal of Respiratory Cell and Molecular Biology*, *54*, 707–717.

Krug, S. M., Gunzel, D., Conrad, M. P., Rosenthal, R., Fromm, A., Amasheh, S., et al. (2012). Claudin-17 forms tight junction channels with distinct anion selectivity. *Cellular and Molecular Life Sciences*, *69*, 2765–2778.

Lanaspa, M. A., Andres-Hernando, A., Rivard, C. J., Dai, Y., & Berl, T. (2008). Hypertonic stress increases claudin-4 expression and tight junction integrity in association with MUPP1 in IMCD3 cells. *Proceedings of the National Academy of Sciences of the United States of America*, *105*, 15797–15802.

Le Moellic, C., Boulkroun, S., Gonzalez-Nunez, D., Dublineau, I., Cluzeaud, F., Fay, M., et al. (2005). Aldosterone and tight junctions: modulation of claudin-4 phosphorylation in renal collecting duct cells. *American Journal of Physiology Cell Physiology*, *289*, C1513–C1521.

Lei, Z., Maeda, T., Tamura, A., Nakamura, T., Yamazaki, Y., Shiratori, H., et al. (2012). EpCAM contributes to formation of functional tight junction in the intestinal epithelium by recruiting claudin proteins. *Developmental Biology*, *371*, 136–145.

Liu, L., Hering-Smith, K. S., Schiro, F. R., & Hamm, L. L. (2002). Serine protease activity in m-1 cortical collecting duct cells. *Hypertension*, *39*, 860–864.

Shinoda, T., Shinya, N., Ito, K., Ohsawa, N., Terada, T., Hirata, K., et al. (2016). Structural basis for disruption of claudin assembly in tight junctions by an enterotoxin. *Scientific Reports*, *6*, 33632.

Tanaka, H., Yamamoto, Y., Kashihara, H., Yamazaki, Y., Tani, K., Fujiyoshi, Y., et al. (2016). Claudin-21 has a paracellular channel role at tight junctions. *Molecular and Cellular Biology*, *36*, 954–964.

Tatum, R., Zhang, Y., Lu, Q., Kim, K., Jeansonne, B. G., & Chen, Y. H. (2007). WNK4 phosphorylates ser(206) of claudin-7 and promotes paracellular Cl(-) permeability. *FEBS Letters*, *581*, 3887–3891.

Van Itallie, C. M., Fanning, A. S., & Anderson, J. M. (2003). Reversal of charge selectivity in cation or anion-selective epithelial lines by expression of different claudins. *American Journal of Physiology Renal Physiology*, *285*, F1078–F1084.

Van Itallie, C. M., Rogan, S., Yu, A., Vidal, L. S., Holmes, J., & Anderson, J. M. (2006). Two splice variants of claudin-10 in the kidney create paracellular pores with different ion selectivities. *American Journal of Physiology Renal Physiology*, *291*, F1288–F1299.

Wilson, F. H., Disse-Nicodeme, S., Choate, K. A., Ishikawa, K., Nelson-Williams, C., Desitter, I., et al. (2001). Human hypertension caused by mutations in WNK kinases. *Science (New York, NY)*, *293*, 1107–1112.

Wu, C. J., Mannan, P., Lu, M., & Udey, M. C. (2013). Epithelial cell adhesion molecule (EpCAM) regulates claudin dynamics and tight junctions. *The Journal of Biological Chemistry*, *288*, 12253–12268.

Yamauchi, K., Rai, T., Kobayashi, K., Sohara, E., Suzuki, T., Itoh, T., et al. (2004). Disease-causing mutant WNK4 increases paracellular chloride permeability and phosphorylates claudins. *Proceedings of the National Academy of Sciences of the United States of America, 101*, 4690–4694.

Yu, A. S., Enck, A. H., Lencer, W. I., & Schneeberger, E. E. (2003). Claudin-8 expression in Madin-Darby canine kidney cells augments the paracellular barrier to cation permeation. *The Journal of Biological Chemistry*, *278*, 17350–17359.

Yu, A. S., Cheng, M. H., Angelow, S., Gunzel, D., Kanzawa, S. A., Schneeberger, E. E., et al. (2009). Molecular basis for cation selectivity in claudin-2-based paracellular pores: identification of an electrostatic interaction site. *The Journal of General Physiology*, *133*, 111–127.

Chapter 6

Paracellular Water Channel

6.1 CONTROVERSY OVER WATER PERMEABILITY OF TIGHT JUNCTION

6.1.1 Evidence for Paracellular Water Pathway

The paracellular route of water permeation has been described in a wide range of tissues, including the bladder epithelium (Civan & DiBona, 1978), the corneal endothelium (Fischbarg, 2010), the kidney proximal tubule (Carpi-Medina & Whittembury, 1988), and the salivary gland (Murakami, Shachar-Hill, Steward, & Hill, 2001). When aquaporin-1, the only aquaporin expressed in the corneal endothelium, was knocked out by genetic deletion, the transendothelial water permeability was reduced by less than 20%, suggesting that the paracellular pathway accounted for over 80% of the transendothelial water transport (Fischbarg, 2010). In the kidney proximal tubule, the paracellular water reabsorption was estimated to account for as much as half of the total transepithelial water reabsorption (Carpi-Medina & Whittembury, 1988). In a study of the plasma membrane permeability to water in the acinar epithelium of the salivary gland, the value obtained, which is almost entirely due to that of the basolateral membrane, was too small to explain the transepithelial water transport by osmosis (Steward, Seo, Rawlings, & Case, 1990). In the renal epithelium, overexpression of the claudin-2 protein, which is known to make a paracellular cation channel, significantly increased the transepithelial water permeability (Rosenthal et al., 2010). The net transepithelial water reabsorption was found to be reduced by 30% in isolated and perfused proximal tubules from the claudin-2 knockout mouse kidneys (Muto et al., 2010).

6.1.2 Evidence Against Paracellular Water Pathway

Spring and coworkers developed an elegant optical microscopic approach to directly record the paracellular permeability to water (Kovbasnjuk, Leader, Weinstein, & Spring, 1998). It was based upon the hypothesis that water flow across the tight junction would lead to dilution of the high-molecular-weight fluorescent marker trapped in the lateral intercellular space (LIS) (Fig. 3.8). With this optical approach, Spring and coworkers measured the paracellular water permeability in the low-resistance, renal epithelial cell line—MDCK-II, known to express high levels of claudin-2, but found near zero water permeability across the bicellular tight junction (Kovbasnjuk et al., 1998).

The Paracellular Channel. http://dx.doi.org/10.1016/B978-0-12-814635-4.00006-1

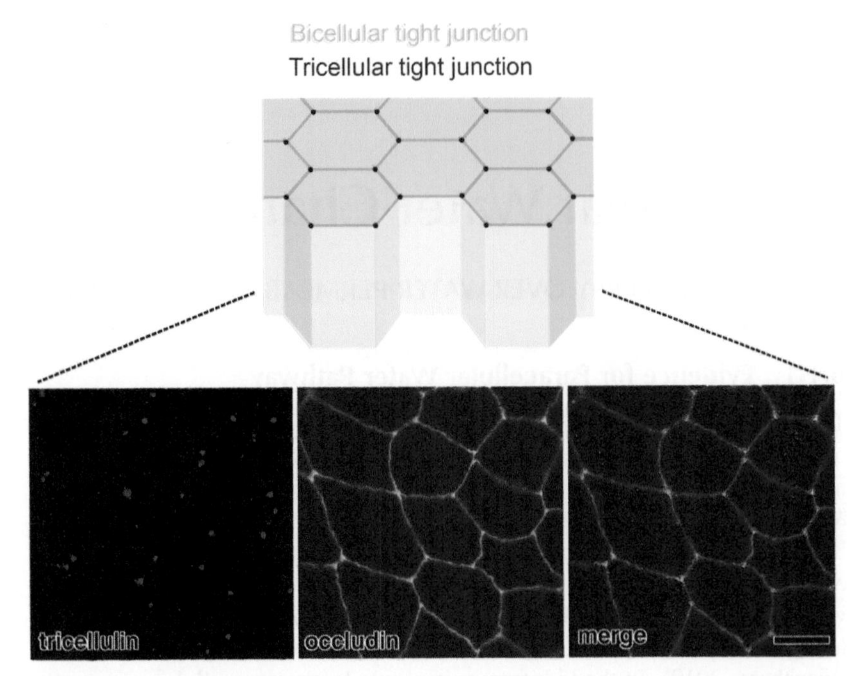

FIGURE 6.1 Tricellular tight junction. The tTJs and bTJs in mouse mammary gland Eph4 epithelial cells are labeled with tricellulin and occludin respectively. Bar: 10 μm. *(Reproduced with permission from Ikenouchi, J., Furuse, M., Furuse, K., Sasaki, H., Tsukita, S., & Tsukita, S. (2005). Tricellulin constitutes a novel barrier at tricellular contacts of epithelial cells. The Journal of Cell Biology,* 171, *939–945).*

6.2 NEW CONCEPT OF TRICELLULAR TIGHT JUNCTION

6.2.1 Ultrastructure of Tricellular Tight Junction

Regular bicellular tight junctions (bTJs) cannot seal some exceptional regions, namely the tricellular tight junctions (tTJs), where the corners of three (or more) polygonal epithelial cells meet (Fig. 6.1). The tTJ is ultrastructurally different from the bTJ. Freeze-fracture replica electron microscopy reveals that the TJ strands are not continuous at the tricellular junction, where they make a 90 degrees turn to extend towards the basal direction (Staehelin, Mukherjee, & Williams, 1969; Wade & Karnovsky, 1974). As a result, the tTJ is composed of three pairs of TJ strands that are arranged vertically and known as the central sealing elements (Fig. 6.2) (Staehelin, 1973). Because the three paired strands of the central sealing elements cannot eliminate the extracellular space among the three cells, it has been speculated that there is a narrow channel of ~1 μm in length and of ~10 nm in diameter, known as the central tube (Staehelin, 1973).

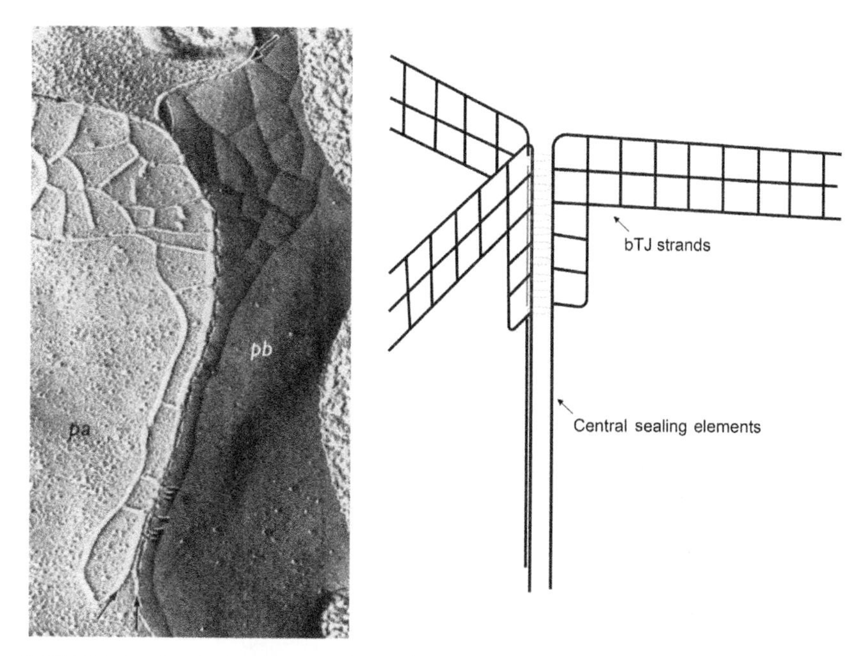

FIGURE 6.2 **Tricellular tight junction architecture.** *Left*: Freeze-fracture replica electron micrograph showing the characteristic extension of the sealing-element meshwork of the tTJ. The two most apical TJ strands (*top arrows*) of the adjoining cells turn from horizontal into vertical orientation in the central region of the tTJ. Due to tilting of the membranes, they appear to fuse (*arrowheads*), but towards the basal end of the extension these two TJ strands can be resolved again (*bottom arrows*). *pa*: plasma membrane of cell *a*; *pb*: plasma membrane of cell *b*. *Right*: Schematic diagram showing the tTJ architecture based upon freeze-fracture replica electron micrograph. In the central area of the tTJ, three parallel and closely spaced (~10 nm apart) vertical strands are illustrated, which arise as extensions of the most apical bTJ strands. *(Reproduced with permission from Staehelin, L.A. (1973). Further observations on the fine structure of freeze-cleaved tight junctions. Journal of Cell Science, 13, 763–786).*

6.2.2 Molecular Components of Tricellular Tight Junction

6.2.2.1 Tricellulin

Tricellulin is the first identified tTJ protein (Ikenouchi et al., 2005). Tricellulin (also known as marvelD2), occludin, and marvelD3 belong to the TJ-associated MARVEL domain containing protein (TAMP) family (Chapter 2, Section 2.6). Tricellulin has four transmembrane domains, two extracellular loops and large amino- and carboxyl-terminal cytoplasmic domains (Fig. 6.3). The carboxyl-terminal domain in tricellulin binds to ZO-1 (Riazuddin et al., 2006). Freeze-fracture replica immunolabeling experiments have revealed the subcellular localization of tricellulin along the central sealing elements (Fig. 6.4) (Ikenouchi et al., 2005). Removal of tricellulin disrupts the characteristic structure of the central sealing elements in the mouse inner ear, which suggests that tricellulin

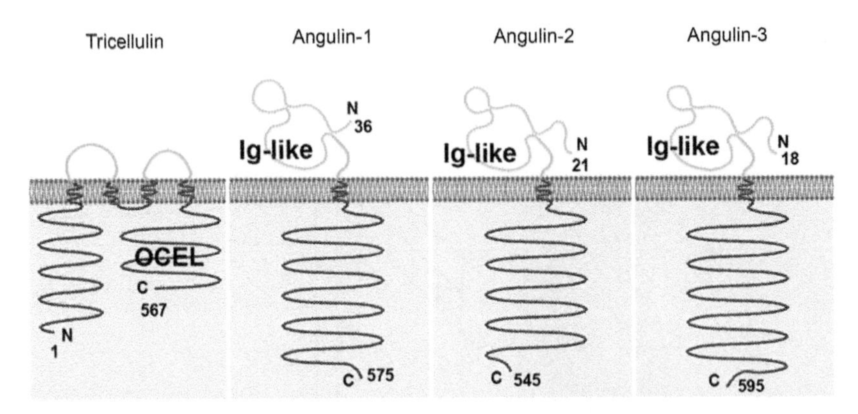

FIGURE 6.3 The tTJ protein family. Human tricellulin and angulin proteins are shown. *Top*, Extracellular; *bottom*, intracellular. *Ig-like*, Immunoglobulin-like domain; *OCEL*, occludin ELL-like domain. *(Reproduced with permission from Higashi, T., & Miller, A.L. (2017). Tricellular junctions: how to build junctions at the TRICkiest points of epithelial cells. Molecular Biology of the Cell, 28, 2023–2034).*

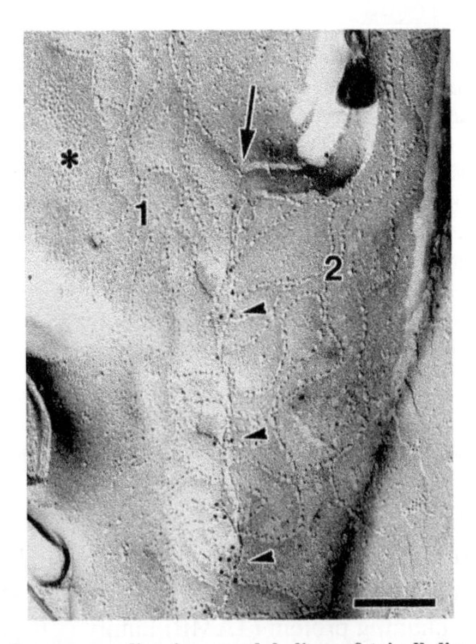

FIGURE 6.4 Freeze-fracture replica immunolabeling of tricellulin. Freeze-fracture replica obtained from the tricellular junction of Eph4 cells is immunolabeled with anti-tricellulin antibody. The central sealing elements *(arrow)*, identified between the bTJ networks of two adjacent cells (1 and 2), are specifically labeled with anti-tricellulin antibody *(arrowheads)*. *Asterisk* indicates adherens junction. Bar: 200 nm. *(Reproduced with permission from Ikenouchi, J., Furuse, M., Furuse, K., Sasaki, H., Tsukita, S., & Tsukita, S. (2005). Tricellulin constitutes a novel barrier at tricellular contacts of epithelial cells. The Journal of Cell Biology, 171, 939–945).*

is important for the formation of tTJ (Nayak et al., 2013). Nevertheless, the tricellulin protein lacks *trans*-interaction when transfected into the HEK293 cells (Cording et al., 2013). It is therefore speculated that tricelllulin may organize the tTJ architecture by its *cis*-interaction with other tTJ proteins such as angulins (Masuda et al., 2011).

6.2.2.2 *Angulins*

The angulin protein family comprises angulin-1 (also known as lipolysis-stimulated lipoprotein receptor [LSR]), angulin-2 (also known as immunoglobulin [Ig]-like domain-containing receptor [ILDR] 1), and angulin-3 (also known as ILDR2, LISCH-like, or C1orf32) (Higashi et al., 2013; Masuda et al., 2011). Angulins have a single transmembrane domain, an amino-terminal extracellular Ig-like domain, and a carboxyl-terminal cytoplasmic domain (Fig. 6.3). Similar to tricellulin, the angulin-1 protein has been localized to the central sealing elements by freeze-fracture replica immunolabeling (Masuda et al., 2011). Angulins recruit tricellulin to the tTJ by *cis*-interaction (Masuda et al., 2011). In angulin-1 knockdown cells, tricellulin is delocalized from the tTJ but relocated to the bTJ (Masuda et al., 2011).

6.2.3 Permeability of Tricellular Tight Junction

6.2.3.1 *Solute Permeability*

It is now well-known that TJ contains two pathways of different size selectivity (Chapter 4, Section 4.5). The high-capacity, size-restrictive pathway allows permeation of small solutes and most ions with diameter of <8 Å, which can be regulated by the claudin protein family in the bTJ (Van Itallie et al., 2008). The low-capacity, size-independent pathway allows permeation of macromolecules with diameter of >8 Å, which may utilize the central tube in the tTJ owing to its large dimension—~10 nm in diameter (Staehelin, 1973). Compatible with this hypothesis, knockdown of angulin-1 in the mouse mammary gland Eph4 epithelial cells significantly increased the permeability of macromolecules ranging in sizes from 0.9 to 9.0 nm in diameter (equivalent to 300 Da–40 kDa in molecular weight) (Higashi et al., 2013). Tricellulin, when expressed at low levels in the MDCK-II cells, impeded the permeation of macromolecules with sizes from 1.3–4.6 nm in diameter (900 Da–10 kDa in molecular weight) but showed no effect on molecules smaller than 1.3 nm in diameter (<900 Da in molecular weight) (Krug et al., 2009). It appears that tricellulin only reduces the diameter of the central tube to a cut-off size of 1.3 nm, while angulins physically block the pathway in the tTJ.

6.2.3.2 *Water Permeability*

Hou and coworkers have made a serendipitous discovery that the angulin-2 knockout mice developed the phenotypes similar to a hereditary human disease—*diabetes insipidus*, which included polydipsia, polyuria and renal

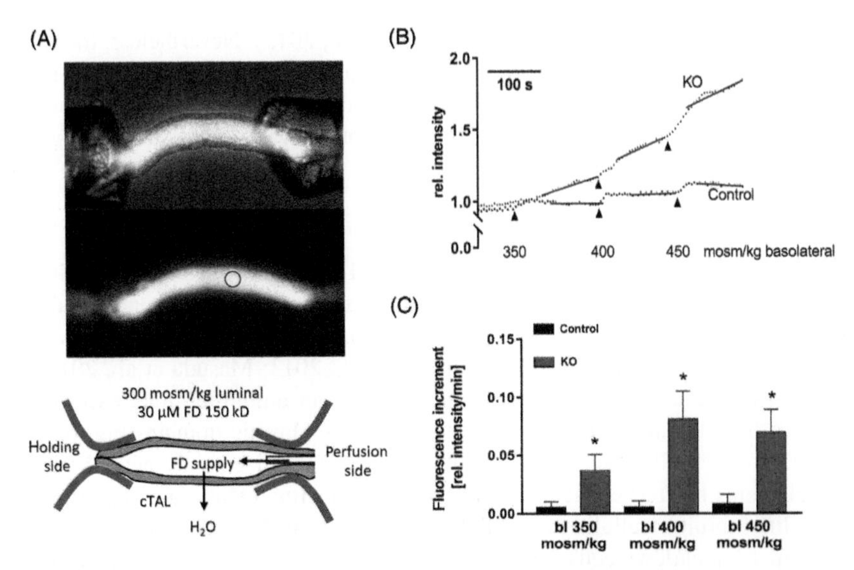

FIGURE 6.5 Water permeability in perfused mouse renal tubule. (A) The thick ascending limbs (TAL) from wildtype (control) and angulin-2 knockout (KO) mouse kidneys were perfused ex vivo. Light microscopy demonstrates the microperfusion system with a live TAL tubule in perfusion, which contains 150 kDa FITC-dextran (FD) in the lumen. The luminal fluorescent intensity was measured continuously at the focal point depicted by the *red circle*. (B) Real-time luminal FITC fluorescence recording after stepwise increases in basolateral osmolality. (C) Fluorescence increment as a direct measurement of water permeability across control and KO TAL tubules. $^*P < 0.05$, $n = 6$ animals; data are given as mean ± standard error of mean of nine and six perfused TAL tubules. *(Reproduced with permission from Gong, Y., Himmerkus, N., Sunq, A., Milatz, S., Merkel, C., Bleich, M. et al. (2017). ILDR1 is important for paracellular water transport and urine concentration mechanism. Proceedings of the National Academy of Sciences of the United States of America, 114, 5271–5276).*

concentrating defect (Gong et al., 2017). The renal tubular epithelium that was impermeable to water in wildtype mice became highly permeable to water in angulin-2 knockout mice (Fig. 6.5). The paracellular permeabilities to Na^+ and Cl^- were not altered in the knockout mouse kidney tubules, indicating that the bTJ function was retained when angulin-2 was removed from the tTJ. Molecular analyses revealed normal gene expression profiles and subcellular localization patterns for the bTJ structural components—ZO-1 and claudins in the knockout mouse kidney (Gong et al., 2017). More importantly, the transcellular water pathway remained intact when the paracellular water pathway was perturbed. In both wildtype and knockout mouse kidneys, vasopressin, also known as antidiuretic hormone, elicited similar increases in transcellular water permeability, indicating that aquaporin function was not affected by angulin-2 deletion (Gong et al., 2017). These results have provided strong evidence that the tTJ contains a paracellular water pathway, which is normally inhibited by the angulin protein and is independent from the transcellular water pathway.

6.2.4 Regulation of Tricellular Tight Junction

6.2.4.1 Phosphorylation

Tricellulin and angulins can be regulated by phosphorylation. The first line of evidence comes from Western immunoblots of epithelial cell lysates, which reveal that the tricellulin protein migrates as multiple bands and that the high-molecular-weight band is sensitive to alkaline phosphatase (Ikenouchi et al., 2005). In human pancreatic HPAC cells, activation of the c-Jun N-terminal kinase (JNK) hyperphosphorylates tricellulin and reduces the paracellular permeabilities to ions and solutes (Kojima et al., 2010). The JNK kinase is also responsible for angulin phosphorylation. In the mammary gland Eph4 epithelial cells, JNK1 and JNK2 directly phosphorylate angulin-1 at the Ser288 site (Nakatsu et al., 2014). The dephosphorylated mutant form (S288A) of angulin-1 fails to localize in the tTJ, indicating that angulin-1 phosphorylation is crucial for its protein trafficking process (Nakatsu et al., 2014).

6.2.4.2 Osmolality

The sea lamprey (*Petromyzon marinus*) must defend the extracellular fluid content in face of surrounding osmolality changes, that is, when migrating from fresh water to sea water. The paracellular permeabilities to solutes and water in the gill epithelium are regulated by the levels of osmolality in the surrounding environments (Chasiotis, Kolosov, Bui, & Kelly, 2012). Low environmental osmolality has been found to increase the mRNA and the protein expression levels of tricellulin in the gill epithelium (Kolosov, Bui, Donini, Wilkie, & Kelly, 2017).

6.2.4.3 Pharmacologic Reagents

6.2.4.3.1 Sodium Caprate

Sodium caprate, the sodium salt of the aliphatic saturated 10-carbon chain fatty acid–capric acid, is the only absorption-enhancing reagent clinically approved as a component of the rectal ampicillin suppository (Lindmark et al., 1997). In a human intestinal cell line—HT-29/B6, sodium caprate transiently increases the paracellular permeabilities to small ions and macromolecules of sizes ranging from 0.9 to 4.6 nm in diameter (300 Da–10 kDa in molecular weight) (Krug et al., 2013). Confocal microscopy reveals marked delocalization of claudin-5 and tricellulin from the bTJ and the tTJ respectively. Biotin labeling experiments show that the tTJ appears to be the predominant leaky pathway in response to sodium caprate treatments (Krug et al., 2013).

6.2.4.3.2 Trictide

The peptidomimetic—trictide, a synthetic peptide derived from the second extracellar loop domain in tricellulin, increases the paracellular permeabilities of ions, as well as of solutes with sizes up to 4.6 nm in diameter

(10 kDa in molecular weight) in human colorectal adenocarcinoma Caco2 cells (Cording et al., 2017). Trictide causes the tTJ proteins, tricellulin and angulin-1, to relocate from tTJs to bTJs, and the bTJ proteins, claudin-1 and occludin, to redistribute from bTJs to the cytoplasm. Interestingly, trictide appears to break the *cis*-interactions between tricellulin and the bTJ proteins such as claudin-1 and occludin, presumably owing to the conformational change in tricellulin induced by its binding to trictide (Cording et al., 2017).

6.2.4.3.3 Angubindin-1

Clostridium perfringens iota-toxin causes antibiotics associated enterotoxemia in rabbits (Sakurai, Nagahama, Oda, Tsuge, & Kobayashi, 2009). It consists of an enzymatic component (Ia) and a receptor-binding component (Ib). Angulin-1 is the receptor for the Ib component of iota-toxin (Papatheodorou et al., 2011). Krug and coworkers have developed an angulin-1 binder, named angubindin-1, which comprises an Ib fragment corresponding to amino acid sequence 421–664 and binds to angulin-1 and -3, but not angulin-2 (Krug et al., 2017). Application of angubindin-1 to the mouse intestines resulted in the removal of angulin-1 and tricellulin from the tTJ, accompanied by increased paracellular permeabilities to molecules with sizes up to 9.0 nm in diameter (40 kDa in molecular weight) (Krug et al., 2017).

REFERENCES

Carpi-Medina, P., & Whittembury, G. (1988). Comparison of transcellular and transepithelial water osmotic permeabilities (Pos) in the isolated proximal straight tubule (PST) of the rabbit kidney. *Pflugers Archiv: European Journal of Physiology*, *412*, 66–74.

Chasiotis, H., Kolosov, D., Bui, P., & Kelly, S. P. (2012). Tight junctions, tight junction proteins and paracellular permeability across the gill epithelium of fishes: A review. *Respiratory Physiology & Neurobiology*, *184*, 269–281.

Civan, M. M., & DiBona, D. R. (1978). Pathways for movement of ions and water across toad urinary bladder. III. Physiologic significance of the paracellular pathway. *The Journal of Membrane Biology*, *38*, 359–386.

Cording, J., Berg, J., Kading, N., Bellmann, C., Tscheik, C., Westphal, J. K., et al. (2013). In tight junctions, claudins regulate the interactions between occludin, tricellulin and marvelD3, which, inversely, modulate claudin oligomerization. *Journal of Cell Science*, *126*, 554–564.

Cording, J., Arslan, B., Staat, C., Dithmer, S., Krug, S. M., Kruger, A., et al. (2017). Trictide, a tricellulin-derived peptide to overcome cellular barriers. *Annals of the New York Academy of Sciences*, 1405(1): 89–101.

Fischbarg, J. (2010). Fluid transport across leaky epithelia: Central role of the tight junction and supporting role of aquaporins. *Physiological Reviews*, *90*, 1271–1290.

Gong, Y., Himmerkus, N., Sunq, A., Milatz, S., Merkel, C., Bleich, M., et al. (2017). ILDR1 is important for paracellular water transport and urine concentration mechanism. *Proceedings of the National Academy of Sciences of the United States of America*, *114*, 5271–5276.

Higashi, T., & Miller, A. L. (2017). Tricellular junctions: How to build junctions at the TRICkiest points of epithelial cells. *Molecular Biology of the Cell*, *28*, 2023–2034.

Higashi, T., Tokuda, S., Kitajiri, S., Masuda, S., Nakamura, H., Oda, Y., et al. (2013). Analysis of the 'angulin' proteins LSR, ILDR1 and ILDR2--tricellulin recruitment, epithelial barrier function and implication in deafness pathogenesis. *Journal of Cell Science, 126*, 966–977.

Ikenouchi, J., Furuse, M., Furuse, K., Sasaki, H., Tsukita, S., & Tsukita, S. (2005). Tricellulin constitutes a novel barrier at tricellular contacts of epithelial cells. *The Journal of Cell Biology, 171*, 939–945.

Kojima, T., Fuchimoto, J., Yamaguchi, H., Ito, T., Takasawa, A., Ninomiya, T., et al. (2010). c-Jun N-terminal kinase is largely involved in the regulation of tricellular tight junctions via tricellulin in human pancreatic duct epithelial cells. *Journal of Cellular Physiology, 225*, 720–733.

Kolosov, D., Bui, P., Donini, A., Wilkie, M. P., & Kelly, S. P. (2017). A role for tight junction-associated MARVEL proteins in larval sea lamprey (Petromyzon marinus) osmoregulation. *The Journal of Experimental Biology, 220*(Pt 20), 3657–3670.

Kovbasnjuk, O., Leader, J. P., Weinstein, A. M., & Spring, K. R. (1998). Water does not flow across the tight junctions of MDCK cell epithelium. *Proceedings of the National Academy of Sciences of the United States of America, 95*, 6526–6530.

Krug, S. M., Amasheh, S., Richter, J. F., Milatz, S., Gunzel, D., Westphal, J. K., et al. (2009). Tricellulin forms a barrier to macromolecules in tricellular tight junctions without affecting ion permeability. *Molecular Biology of the Cell, 20*, 3713–3724.

Krug, S. M., Amasheh, M., Dittmann, I., Christoffel, I., Fromm, M., & Amasheh, S. (2013). Sodium caprate as an enhancer of macromolecule permeation across tricellular tight junctions of intestinal cells. *Biomaterials, 34*, 275–282.

Krug, S. M., Hayaishi, T., Iguchi, D., Watari, A., Takahashi, A., Fromm, M., et al. (2017). Angubindin-1, a novel paracellular absorption enhancer acting at the tricellular tight junction. *Journal of Controlled Release: Official Journal of the Controlled Release Society, 260*, 1–11.

Lindmark, T., Soderholm, J. D., Olaison, G., Alvan, G., Ocklind, G., & Artursson, P. (1997). Mechanism of absorption enhancement in humans after rectal administration of ampicillin in suppositories containing sodium caprate. *Pharmaceutical Research, 14*, 930–935.

Masuda, S., Oda, Y., Sasaki, H., Ikenouchi, J., Higashi, T., Akashi, M., et al. (2011). LSR defines cell corners for tricellular tight junction formation in epithelial cells. *Journal of Cell Science, 124*, 548–555.

Murakami, M., Shachar-Hill, B., Steward, M. C., & Hill, A. E. (2001). The paracellular component of water flow in the rat submandibular salivary gland. *The Journal of Physiology, 537*, 899–906.

Muto, S., Hata, M., Taniguchi, J., Tsuruoka, S., Moriwaki, K., Saitou, M., et al. (2010). Claudin-2-deficient mice are defective in the leaky and cation-selective paracellular permeability properties of renal proximal tubules. *Proceedings of the National Academy of Sciences of the United States of America, 107*, 8011–8016.

Nakatsu, D., Kano, F., Taguchi, Y., Sugawara, T., Nishizono, T., Nishikawa, K., et al. (2014). JNK1/2-dependent phosphorylation of angulin-1/LSR is required for the exclusive localization of angulin-1/LSR and tricellulin at tricellular contacts in EpH4 epithelial sheet. *Genes to Cells: Devoted to Molecular and Cellular Mechanisms, 19*, 565–581.

Nayak, G., Lee, S. I., Yousaf, R., Edelmann, S. E., Trincot, C., Van Itallie, C. M., et al. (2013). Tricellulin deficiency affects tight junction architecture and cochlear hair cells. *The Journal of Clinical Investigation, 123*, 4036–4049.

Papatheodorou, P., Carette, J. E., Bell, G. W., Schwan, C., Guttenberg, G., Brummelkamp, T. R., et al. (2011). Lipolysis-stimulated lipoprotein receptor (LSR) is the host receptor for the binary toxin Clostridium difficile transferase (CDT). *Proceedings of the National Academy of Sciences of the United States of America, 108*, 16422–16427.

Riazuddin, S., Ahmed, Z. M., Fanning, A. S., Lagziel, A., Kitajiri, S., Ramzan, K., et al. (2006). Tricellulin is a tight-junction protein necessary for hearing. *American Journal of Human Genetics*, *79*, 1040–1051.

Rosenthal, R., Milatz, S., Krug, S. M., Oelrich, B., Schulzke, J. D., Amasheh, S., et al. (2010). Claudin-2, a component of the tight junction, forms a paracellular water channel. *Journal of Cell Science*, *123*, 1913–1921.

Sakurai, J., Nagahama, M., Oda, M., Tsuge, H., & Kobayashi, K. (2009). Clostridium perfringens iota-toxin: Structure and function. *Toxins*, *1*, 208–228.

Staehelin, L. A. (1973). Further observations on the fine structure of freeze-cleaved tight junctions. *Journal of Cell Science*, *13*, 763–786.

Staehelin, L. A., Mukherjee, T. M., & Williams, A. W. (1969). Fine structure of frozen-etched tight junctions. *Die Naturwissenschaften*, *56*, 142.

Steward, M. C., Seo, Y., Rawlings, J. M., & Case, R. M. (1990). Water permeability of acinar cell membranes in the isolated perfused rabbit mandibular salivary gland. *The Journal of Physiology*, *431*, 571–583.

Van Itallie, C. M., Holmes, J., Bridges, A., Gookin, J. L., Coccaro, M. R., Proctor, W., et al. (2008). The density of small tight junction pores varies among cell types and is increased by expression of claudin-2. *Journal of Cell Science*, *121*, 298–305.

Wade, J. B., & Karnovsky, M. J. (1974). The structure of the zonula occludens: A single fibril model based on freeze-fracture. *The Journal of Cell Biology*, *60*, 168–180.

Chapter 7

Paracellular Channel in Organ System

7.1 EPITHELIAL SYSTEM

7.1.1 Skin

7.1.1.1 Epidermal Tight Junction

The epidermis is a multilayered stratified epithelium made of several cell layers (Fig. 7.1). The innermost basal layer, stratum basale (SB), consists in undifferentiated keratinocytes, stem cells, melanocytes, and Merkel cells. On top of this layer resides the spinous layer, stratum spinosum (SS). The subsequent granular layer, stratum granulosum (SG), consists of 3–5 cell layers. Tight junctions (TJs) are found in the second layer of stratum granulosum (SG2) (Yoshida et al., 2013). The outermost layer, stratum corneum (SC), consists in corneocytes, that are, dead cells, and intercellular lipids. The primary role of the epidermis is a tissue barrier against pathogen invasion from the external environment. While SC was previously considered to be the only epidermal barrier, it is until recently that the importance of TJ in epidermal barrier function has been recognized. The expression profiles of the claudin proteins reflect the complexity of the epidermis (Table 7.1). Among the claudins analyzed so far, claudin-1, -7, and -12 are found in all living layers from SB to SG. Because TJ is made only by the SG2 cells, the claudin expression in other cell layers is found in the plasma membrane (Brandner, McIntyre, Kief, Wladykowski, & Moll, 2003; Furuse et al., 2002; Troy, Rahbar, Arabzadeh, Cheung, & Turksen, 2005). A recent study elegantly shows how TJs are maintained during cell migration from the SG3 to the SG2 layer. When an epidermal cell moves up from a lower layer to SG2, it forms new TJs that are situated beneath the existing TJs in SG2. This transient twin TJ structure resembles a double-edged polygon, termed Kelvin's tetrakaidecahedron. As the newly formed TJs mature, the existing TJs disappear over time (Fig. 7.2) (Yokouchi et al., 2016).

7.1.1.2 Paracellular Barrier Function

7.1.1.2.1 Claudin-1

Claudin-1 knockout mice died within 1 day of birth with wrinkled skin and epidermal water loss (Fig. 7.3A) (Furuse et al., 2002). The epidermal tissue

The Paracellular Channel. http://dx.doi.org/10.1016/B978-0-12-814635-4.00007-3

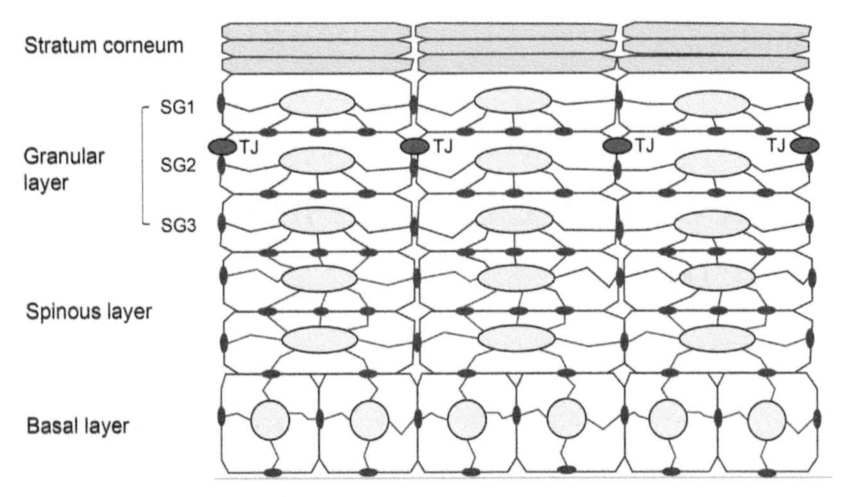

FIGURE 7.1 Schematic diagram of the epidermis. The different strata of the skin epidermis are indicated on the left. The granular layer is composed of three epithelial cell layers (SG1–SG3). TJs are found in the second layer (SG2). The outer layer of the skin, termed as stratum corneum, was previously considered to be the only epidermal barrier. *(Reproduced with permission from Tsuruta, D., Green, K.J., Getsios, S., & Jones, J.C. (2002). The barrier function of skin: how to keep a tight lid on water loss. Trends in Cell Biology, 12, 355–357)*

morphology and the TJ ultrastructure were not affected in the knockout animals. The paracellular permeabilities to molecules with sizes up to 557 Da became significantly increased in the claudin-1 knockout mouse skin (Fig. 7.3B) (Furuse et al., 2002). Although claudin-1 is not present in the SC layer, the SC structure and function are altered in the claudin-1 knockout mouse (Sugawara et al., 2013). Because the water-holding capacity of SC is reduced in the knockout animal, which in part explains the increased epidermal water loss, whether the TJ in SG2 regulates the paracellular water permeability remains to be determined (Sugawara et al., 2013).

7.1.1.2.2 Claudin-6

Transgenic overexpression of claudin-6 driven by the involucrin promoter in the suprabasal cell layers of the mouse epidermis resulted in epidermal barrier defects associated with alterations in TJ structure and function (Turksen & Troy, 2002). The paracellular permeability to β-galactosidase, a 464-kDa protein, was increased in the transgenic mouse epidermis. The transepidermal water permeability was also higher in the transgenic animal, which led to dry skin and weight loss (Turksen & Troy, 2002). Reduction in transgene dosage partially corrected the epidermal barrier defects, but the epidermal differentiation defects persisted in the transgenic mouse (Troy et al., 2005). The carboxyl-terminal domain in claudin is required for its TJ localization (Chapter 2, Section 2.3). Transgenic overexpression of two carboxyl-domain

TABLE 7.1 Claudin Gene Expression Profiles in the Epidermis

Genes	SB	Lower SS	Upper SS	SG	References
Claudin-1	+	+	+	+	Furuse et al. (2002); Haftek et al. (2011); Igawa et al. (2011)
Claudin-2	–	–	–	+	Telgenhoff, Ramsay, Hilz, Slusarewicz, & Shroot (2008)
Claudin-3	–	–	–	+	Watson et al. (2007)
Claudin-4	–	–	+	+	Brandner et al. (2003); Furuse et al. (2002); Morita, Tsukita, & Miyachi, (2004)
Claudin-5	–	–	–	+	Peltonen, Riehokainen, Pummi, & Peltonen (2007)
Claudin-6	–	–	+	+	Turksen & Troy (2002)
Claudin-7	+	+	+	+	Brandner, Kief, Wladykowski, Houdek, & Moll, (2006); Kirschner et al. (2009)
Claudin-11	–	–	–	+	Troy et al. (2005)
Claudin-12	+	+	+	+	Troy et al. (2005)
Claudin-17	–	–	–	+	Brandner et al. (2003)
Claudin-18	–	–	+	+	Troy et al. (2005)

SB, Stratum basale; *SG*, stratum granulosum; *SS*, stratum spinosum.

deletion mutations in claudin-6 (Δ187 and Δ196) retained normal epidermal barrier function, which suggests that the TJ localization of claudin-6 is crucial for causing the abnormality in paracellular permeability (Arabzadeh, Troy, & Turksen, 2006; Troy, Arabzadeh, Lariviere, Enikanolaiye, & Turksen, 2009).

7.1.1.3 Tricellular Tight Junction

The skin is a major immunologic barrier against microbial pathogens and allergens. Langerhans cells are resident dendritic cells (antigen-presenting immune cells) located in the epidermis and elicit immune responses against various foreign antigens (Merad, Ginhoux, & Collin, 2008). Kubo and coworkers have revealed that Langerhans cells elongate their dendrites to penetrate the epidermal TJs in the SG2 layer and survey the environment for antigens located outside the TJs. Interestingly, when dendrites are inserted between the epidermal cells, they often penetrate the tricellular tight junctions (tTJs) and recruit tricellulin to the newly formed contact sites of the

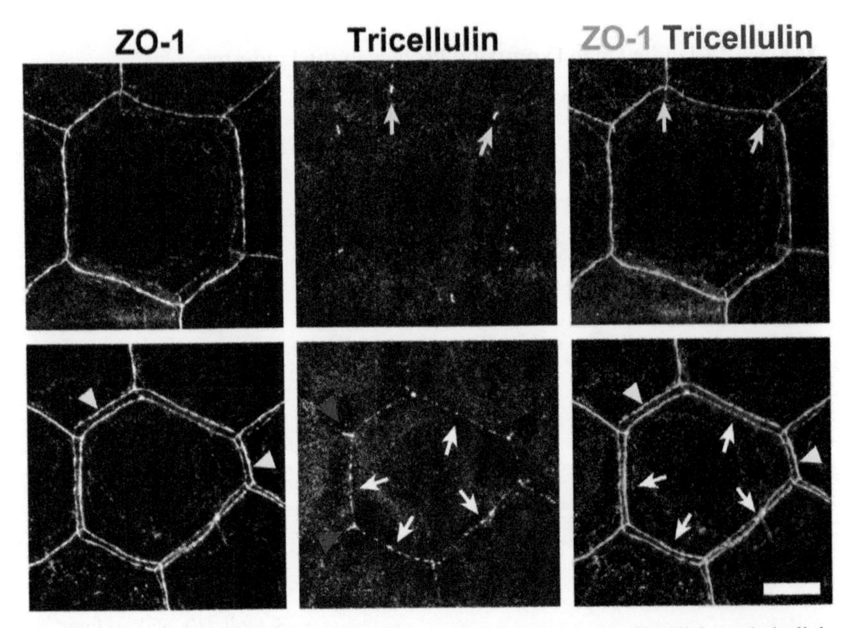

FIGURE 7.2 TJ biogenesis in the epidermis. Subcellular localization of bicellular and tricellular TJ components (ZO-1 and tricellulin) on single- and double-edged polygons. *Yellow arrowheads*, edges of the exterior, exisiting polygon; *white arrows*, edges of the interior, newly formed polygon; *red arrowheads*, vertical edges connecting the vertices of double-edged polygons; *yellow arrows*, vertices of single-edged polygons. Bar: 10 μm. *(Reproduced with permission from Yokouchi, M., Atsugi, T., Logtestijn, M.V., Tanaka, R.J., Kajimura, M., Suematsu, M., Furuse, M., Amagai, M., & Kubo, A. (2016). Epidermal cell turnover across tight junctions based on Kelvin's tetrakaidecahedron cell shape. eLife, 5, e19593)*

dendrites with epidermal cell membrane, where new tTJs are established to maintain the epidermal barrier function (Fig. 7.4) (Kubo, Nagao, Yokouchi, Sasaki, & Amagai, 2009).

7.1.2 Lung

7.1.2.1 Alveolar Claudin Expression

The distal airway epithelium is composed primarily of the alveolar epithelium that covers 99% of the airspace surface area in the lung, and contains thin, squamous type I cells, and thick, cuboidal type II cells (Fig. 7.5). Active fluid reabsorption across the alveolar epithelium is vital to the ability of the lung to remove alveolar fluid at the time of birth, as well as when pathologic conditions cause pulmonary edema (Matthay, Folkesson, & Clerici, 2002). Transcellular channels and paracellular channels both play important roles in this process. The claudins expressed by the alveolar epithelium are claudin-3, -4, -5, -7, and -18 (Daugherty et al., 2004; Wang et al., 2003). The mRNA levels of alveolar

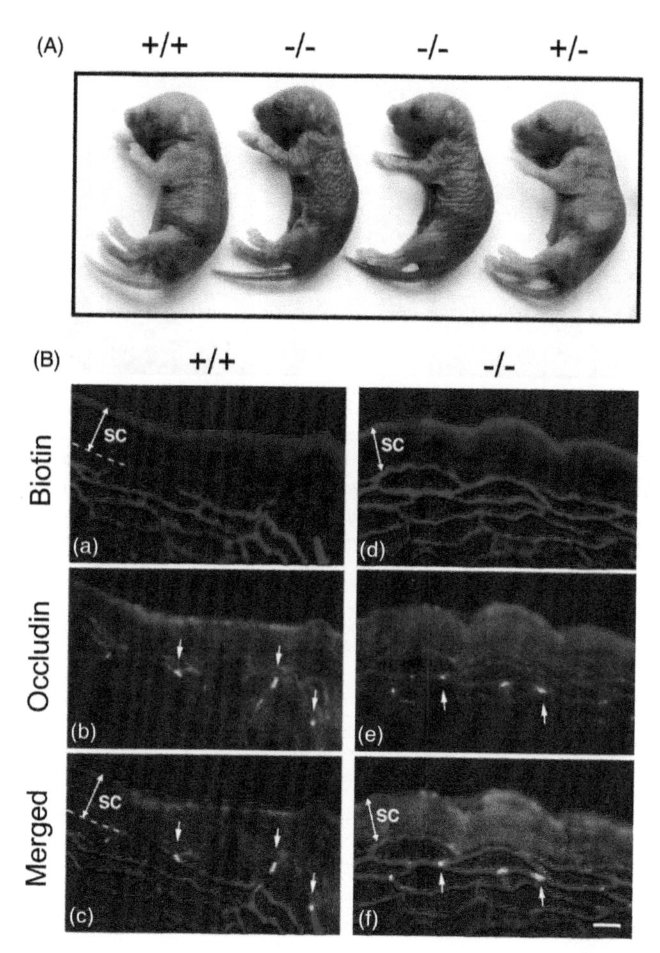

FIGURE 7.3 Impaired epidermal barrier function in claudin-1 knockout mouse. (A) Images of 12-h-old claudin-1$^{+/+}$, claudin-1$^{+/-}$, and claudin-1$^{-/-}$ mice. Claudin-1$^{-/-}$ mice are characterized by wrinkled skin and die within 1 day of birth. (B) In the claudin-1$^{+/+}$ mouse epidermis (a–c), the injected biotin tracer (557 Da) diffuses through the paracellular space from stratum basale to the second layer of stratum granulosum, but this diffusion is stopped at the occludin-positive TJs (*arrows*). In the claudin-1$^{-/-}$ mouse epidermis (d–f), the diffusion of injected biotin tracer is not prevented by the occludin-positive TJs (*arrows*). Instead, the biotin tracer passes through these TJs to reach the border between stratum granulosum and stratum corneum. Bar: 10 μm. SC, stratum corneum. *(Reproduced with permission from Furuse, M., Hata, M., Furuse, K., Yoshida, Y., Haratake, A., Sugitani, Y., Noda, T., Kubo, A., & Tsukita, S. (2002). Claudin-based tight junctions are crucial for the mammalian epidermal barrier: a lesson from claudin-1-deficient mice. The Journal of Cell Biology, 156, 1099–1111)*

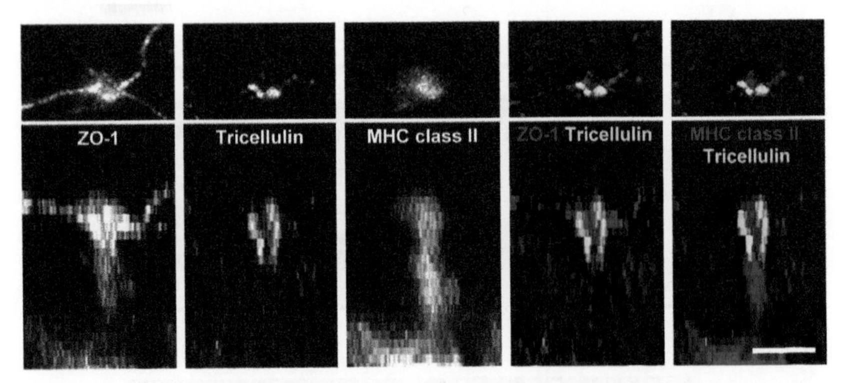

FIGURE 7.4 Langerhans cell extends dendrite to penetrate the epidermal tTJ. Top row shows *en face* images and bottom row presents 90 degree rotated images of bicellular TJ (labeled with ZO-1), tricellular TJ (labeled with tricellulin), and Langerhans cell dendrite (labeled with MHC class II). Bar: 5 μm. *(Reproduced with permission from Kubo, A., Nagao, K., Yokouchi, M., Sasaki, H., & Amagai, M. (2009). External antigen uptake by Langerhans cells with reorganization of epidermal tight junction barriers. The Journal of Experimental Medicine, 206, 2937–2946)*

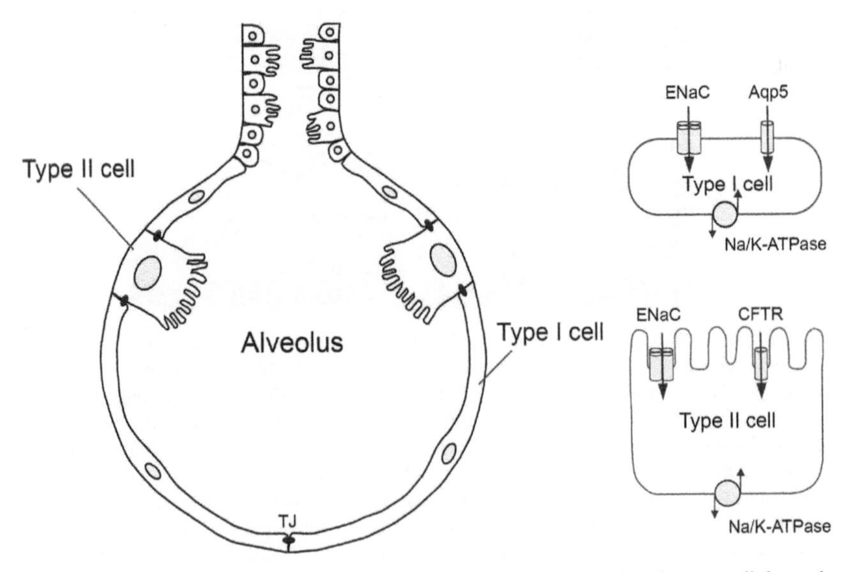

FIGURE 7.5 Schematic diagram of the distal pulmonary epithelium. The transcellular pathways made of ion and water channels, such as epithelial Na^+ channel (ENaC), cystic fibrosis transmembrane conductance regulator (CFTR), aquaporin-5 (Aqp5), and so on, were previously considered to be the only pathway regulating salt and water transport in the alveolus. The TJ constitutes a novel paracellular pathway important for alveolar fluid clearance. *(Reproduced with permission from Matthay, M.A., Folkesson, H.G., & Clerici, C. (2002). Lung epithelial fluid transport and the resolution of pulmonary edema. Physiological reviews, 82, 569–600)*

claudin-5 and -7 are 10–100-fold lower than claudin-3, -4, and -18 mRNA levels (LaFemina et al., 2010). There are heterogeneities in claudin gene expression in the alveolar epithelium. For example, the type II alveolar cells express over 15-fold more claudin-3 proteins than the type I alveolar cells (LaFemina et al., 2010). The heterotypic TJs made by type I–type II alveolar cells may differ in structure and function from the homotypic TJs found between type I–type I or type II–type II cells due to claudin expression differences. Even though these heterotypic TJs are rare, they may possess unique permeability characteristics important for paracellular fluid reabsorption.

7.1.2.2 Paracellular Permeability and Alveolar Fluid Clearance

7.1.2.2.1 Claudin-4

Claudin-4 is highly expressed by both type I and type II alveolar cells (Wang et al., 2003). Knockout mouse studies have revealed that claudin-4 is important for regulating paracellular permeability and alveolar fluid clearance in the distal lung (Kage et al., 2014). Deletion of claudin-4 from the alveolar epithelium reduced the paracellular permeability of 5-carboxyfluorescein (376 Da) without affecting the paracellular permeability of Na^+ and Cl^-. The transcellular ionic conductance and alveolar fluid clearance rate were also decreased in the knockout mouse alveolus. Compared with wildtype animals, the claudin-4 knockout mice were more susceptible to acute lung injury and developed significant pulmonary edema after exposure to hyperoxia (Kage et al., 2014). The protective role of claudin-4 against acute lung injury is further demonstrated by a study using *Clostridium perfringens* enterotoxin (CPE) to disrupt *trans* claudin-4 interactions between the alveolar cells (Wray et al., 2009). In mice, injected CPE decreased the alveolar fluid clearance rate and stimulated pulmonary edema in response to ventilator-induced mechanical stress, under conditions known to be safe to animals without the CPE injection (Wray et al., 2009).

7.1.2.2.2 Claudin-18

There are two claudin-18 splice variants: claudin-18.1, which is found primarily in the lung, and claudin-18.2, which is expressed specifically in the stomach (Niimi et al., 2001). To investigate the role of claudin-18.1 in pulmonary function, two groups have generated independent lines of claudin-18 knockout mice, one of which carries deletion of exon 2–3 and the other carries deletion of all exons in the claudin-18 gene (LaFemina et al., 2014; Li et al., 2014). Claudin-18 plays an important role in alveolar development. The alveolar surface area in claudin-18 knockout mouse lung is markedly smaller than that in wildtype mouse lung by 4-week-old age (LaFemina et al., 2014). On the other hand, deletion of claudin-18 protected the knockout animals from ventilator-induced lung injury by increasing the alveolar fluid clearance rate (Li et al., 2014). The paracellular permeabilities to ions and solutes (5-carboxyfluorescein, 376 Da, FITC-albumin, 67 kDa, and TRITC-dextran, 155 kDa) were significantly increased in

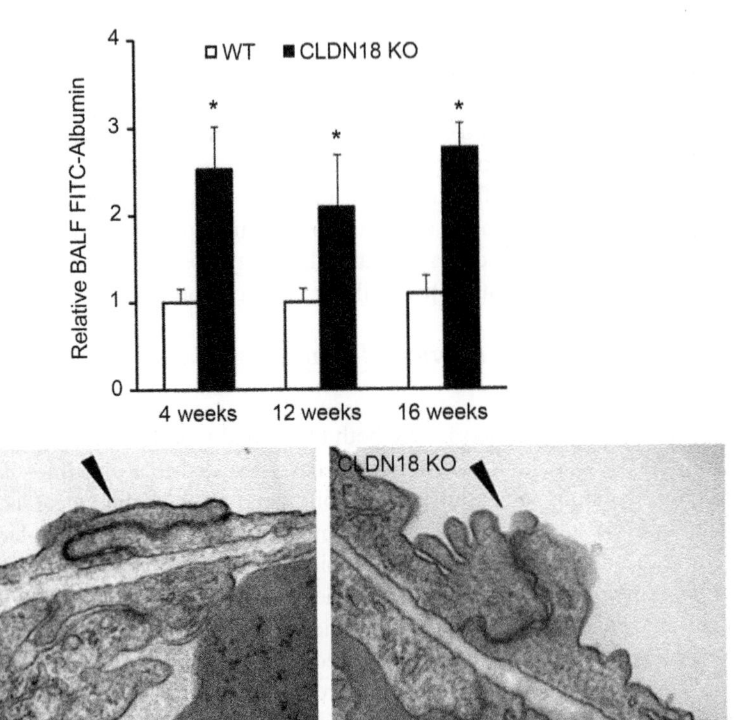

FIGURE 7.6 Alveolar epithelial barrier dysfunction in claudin-18 knockout mouse. (Upper panel) Bidirectional alveolar fluid (BALF) permeability in wildtype (WT) and CLDN18 KO mice of 4, 12, and 16 weeks of age as measured by the accumulation rate of FITC-albumin in lavage fluid 4 h after intraperitoneal injection of tracer. *Asterisk*, $P < 0.05$. (Lower panel) Transmission electron micrographs showing the TJs (*arrowheads*) between alveolar type 1 epithelial cells in WT and CLDN18 KO mice at 8 weeks of age. Note the membrane ruffling and splaying near the TJ of the KO mouse epithelium, suggesting altered TJ morphology in the absence of claudin-18. *(Reproduced with permission from LaFemina, M.J., Sutherland, K.M., Bentley, T., Gonzales, L.W., Allen, L., Chapin, C.J., Rokkam, D., Sweerus, K.A., Dobbs, L.G., Ballard, P.L., et al. (2014). Claudin-18 deficiency results in alveolar barrier dysfunction and impaired alveologenesis in mice. American Journal of Respiratory Cell and Molecular Biology, 51, 550–558)*

claudin-18 knockout mouse alveolus, but the transcellular ionic conductance was not affected by the ablation of claudin-18 (Fig. 7.6) (LaFemina et al., 2014; Li et al., 2014).

7.1.3 Liver

7.1.3.1 Atypical Polarization of Hepatocytes

Most epithelial cells, such as the intestinal and renal cells, are polarized into two membrane domains: apical and basolateral domains along the plane of the tissue. In contrast, hepatocytes have a unique polarization arrangement in which

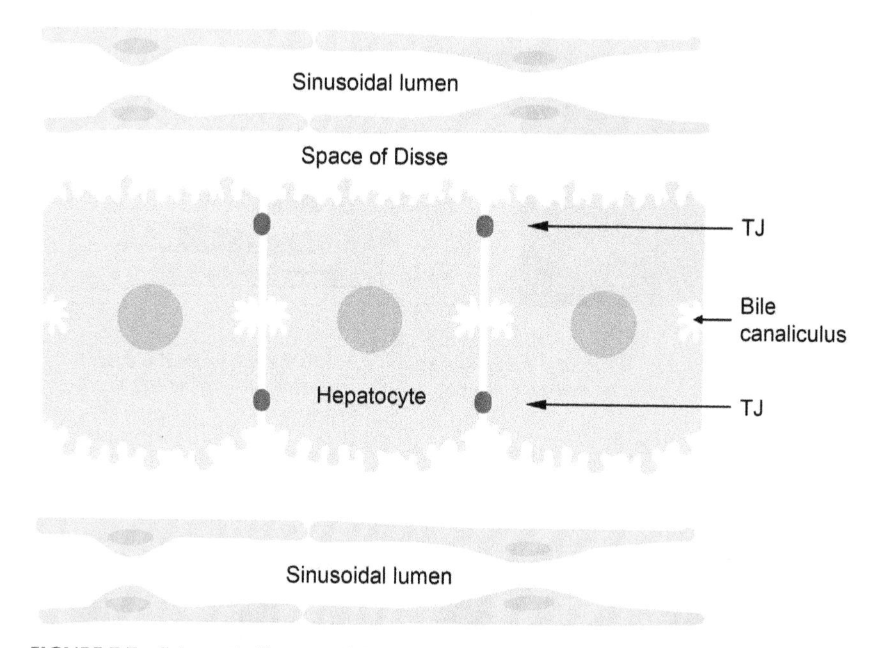

FIGURE 7.7 **Schematic diagram of the hepatic lobule.** Hepatocytes are polarized in a different way from simple squamous or cuboidal epithelia. Adjacent hepatocytes make two TJs to create a tubular lumen in the lateral extracellular space known as the bile canaliculus (BC). The TJs regulate electrolyte and water balances between the BCs and the sinusoids via the space of Disse.

two TJs are made between adjacent cells to partition the plasma membrane into three domains: a luminal domain delimiting the bile canaliculus (BC), and two basolateral domains surrounded by the perisinusoidal space (also known as the space of Disse) (Fig. 7.7) (Treyer & Musch, 2013). Bile production is an isosmotic process. After the hepatocyte secretes bile acids into the BC, an osmotic gradient develops from the perisinusoidal space to the BC lumen, which drives water, electrolytes, and solutes into bile, via either the transcellular pathway or the paracellular pathway (Boyer, 2013). The gene expression of claudin-1, -2, -3, -4, -5, -7, and -18 has been detected in the TJ of normal or carcinomatous hepatocytes (Rao & Samak, 2013). In fact, the first identified claudins: claudin-1 and claudin-2, were purified from the TJ fraction of chicken bile canaliculi (Furuse, Fujita, Hiiragi, Fujimoto, & Tsukita, 1998).

7.1.3.2 Paracellular Permeability and Bile Volume Control

The claudin-2 gene expression demonstrates a remarkable lobular gradient with progressive increases from periportal to perivenous regions (Rahner, Mitic, & Anderson, 2001). Knockout mouse studies have revealed an important role for claudin-2 in bile electrolyte and water secretion (Matsumoto et al., 2014). The bile volume in the claudin-2 knockout mouse is about half of the level in the

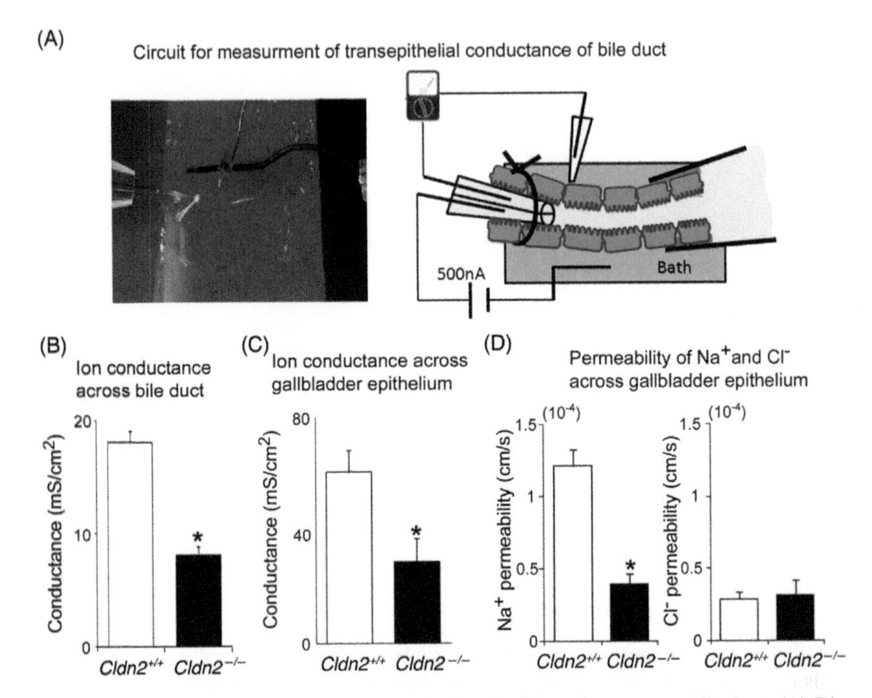

FIGURE 7.8 Transepithelial conductance in claudin-2 knockout mouse bile duct. (A) Diagrams illustrating the Ussing method for measuring the transepithelial conductance of isolated bile ducts ex vivo. (B–C) Conductance across the $Cldn2^{+/+}$ and $Cldn2^{-/-}$ mouse bile duct (B) or gallbladder epithelium (C). (D) Paracellular permeability of the $Cldn2^{+/+}$ and $Cldn2^{-/-}$ mouse gallbladder epithelium to Na^+ and Cl^-. Note that loss of claudin-2 reduces the paracellular permeability to Na^+ but not to Cl^-. *Asterisk, $P < 0.05$. (Reproduced with permission from Matsumoto, K., Imasato, M., Yamazaki, Y., Tanaka, H., Watanabe, M., Eguchi, H., Nagano, H., Hikita, H., Tatsumi, T., Takehara, T., et al. (2014). Claudin 2 deficiency reduces bile flow and increases susceptibility to cholesterol gallstone disease in mice. Gastroenterology, 147, 1134–1145.e1110)*

wildtype animal. While the circulating levels of bile acids remain the same, the concentration of bile acids in the knockout mouse gallbladder is approximately twice that in the wildtype animal, indicating that bile water secretion is reduced in the knockout animal. The ionic concentrations in the bile of the knockout mouse are almost identical to those in the wildtype mouse, indicating that bile electrolyte secretion is also decreased. Electrophysiological recordings in the Ussing chamber have confirmed a significant decrease in paracellular Na^+ permeability across the bile duct of the claudin-2 knockout mouse (Fig. 7.8) (Matsumoto et al., 2014), compatible with the concept of claudin-2 making a paracellular cation channel (Chapter 4). Reduction in bile volume secretion predisposed the claudin-2 knockout mouse to cholesterol gallstone formation owing to increased bile solute concentration (Matsumoto et al., 2014).

7.1.4 Gastrointestinal Tract

7.1.4.1 Anatomy and Physiology

The primary function of the gastrointestinal epithelium is by twofold. First, the epithelium forms a tissue barrier to prevent the entry of harmful substances, such as pathogens and toxins. Second, the epithelium absorbs nutrients, ions, and water to sustain cellular metabolism and systemic homeostasis. The TJ is vital to the barrier and transport functions of the epithelium. The three-dimensional structure of epithelium exhibits significant variation along the gastrointestinal tract. For example, the stomach is organized into pit, neck, and base; the small intestine into villus and crypt; and the colon into surface and crypt (Fig. 7.9). The composition of TJ proteins, in particular claudins, not only varies along the length of the gastrointestinal tract but also along the crypt-villus axis, as a part of a sophisticated program of epithelial differentiation (Noah, Donahue, & Shroyer, 2011).

7.1.4.2 Stomach

7.1.4.2.1 Gastric Claudin Expression

Claudin-3, -4, -5, and -18 have been detected in the stomach (Niimi et al., 2001; Rahner et al., 2001). Among them, claudin-3 and -18 are highly expressed by the cells in the pit and the neck; whereas claudin-4 expression is restricted to the cells in the base (Niimi et al., 2001; Rahner et al., 2001). Claudin-5 is expressed throughout the gastric regions (Rahner et al., 2001). The gastric epithelium is highly regenerative. When the gastric stem cells migrate from the isthmus, an anatomic definition of the region connecting the pit to the neck, toward the surface or the base of the gastric unit, they adopt a claudin expression profile belonging to the destination (Kim & Shivdasani, 2016).

7.1.4.2.2 Paracellular H^+ Permeability

The claudin-18 gene has two splice variants: claudin-18.1, which is found primarily in the lung, and claudin-18.2, which is expressed exclusively in the stomach (Niimi et al., 2001). In the gastric epithelium, the claudin-18.2 protein is localized not only in the TJ but also along the basolateral membrane (Niimi et al., 2001). The claudin-18 knockout mouse developed atrophic gastritis due to neutrophil infiltration and progressive losses of the parietal and chief cells in the stomach (Fig. 7.10) (Hayashi et al., 2012). The paracellular permeabilities of the gastric epithelium to H^+ and Na^+ were increased in the knockout animal, which resulted in H^+ backleak from the gastric lumen to the submucosal space as a potential insult to trigger tissue damage and chronic inflammation. The paracellular permeability to biotin (443 Da) or dextran (4 kDa) was not affected when claudin-18 was removed from the gastric epithelium (Hayashi et al., 2012).

7.1.4.2.3 Peculiar Role of Occludin

Occludin, a MARVEL domain containing protein, plays an important role in the stomach. The occludin knockout mice developed marked losses of gastric

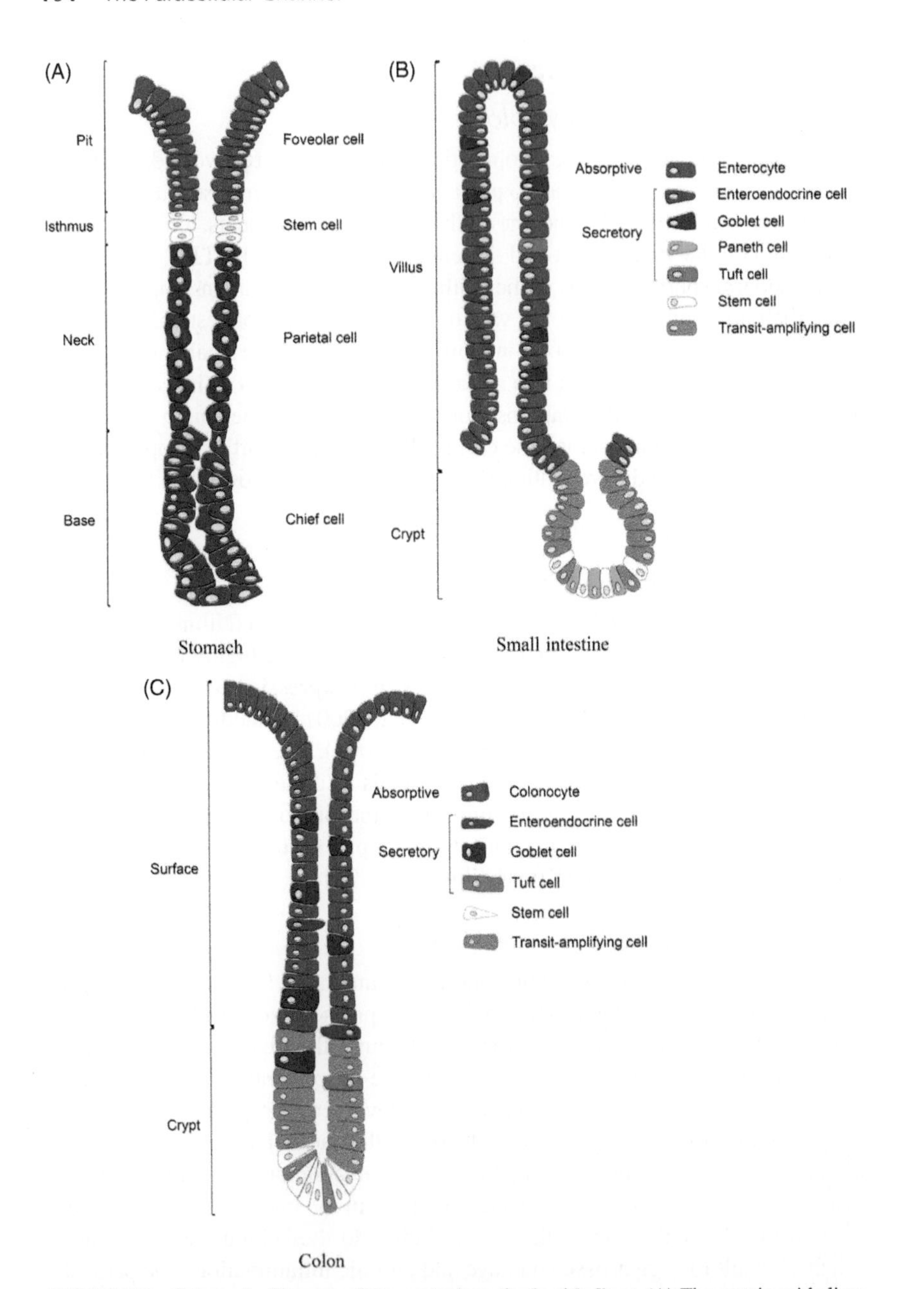

FIGURE 7.9 Schematic diagram of the gastrointestinal epithelium. (A) The gastric epithelium is organized into pit, neck and base regions, and composed of foveolar cell, parietal cell, chief cell, and stem cell. (B) The small intestinal epithelium is organized into villus and crypt regions, and composed of enterocyte, enteroendocrine cell, goblet cell, Paneth cell, tuft cell, transit-amplifying cell and stem cell. (C) The colonic epithelium is organized into surface and crypt regions, and composed of colonocyte, enteroendocrine cell, goblet cell, tuft cell, transit-amplifying cell, and stem cell.

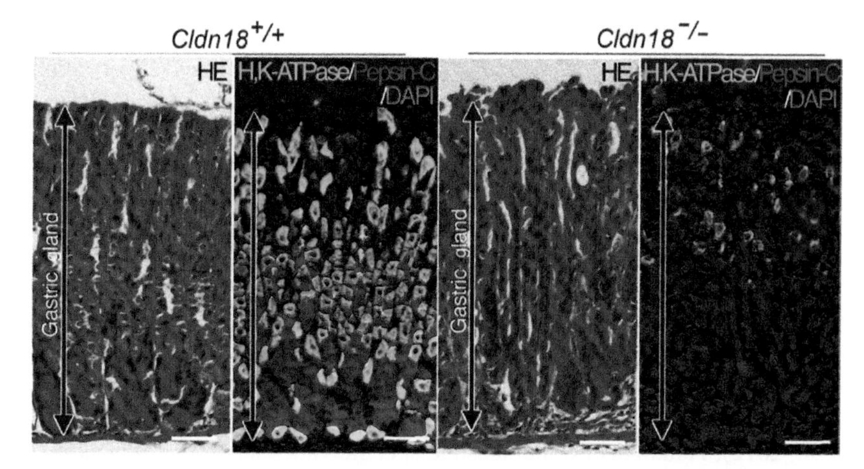

FIGURE 7.10 **Gastritis in claudin-18 knockout mouse.** Light microscopic images of H&E-stained paraffin sections and immunofluorescence images of sections stained with antibodies against H,K-ATPase (a marker for parietal cell) and pepsin C (a marker for chief cell). Bar: 50 μm. *(Reproduced with permission from Hayashi, D., Tamura, A., Tanaka, H., Yamazaki, Y., Watanabe, S., Suzuki, K., Suzuki, K., Sentani, K., Yasui, W., Rakugi, H., et al. (2012). Deficiency of claudin-18 causes paracellular H^+ leakage, up-regulation of interleukin-1beta, and atrophic gastritis in mice. Gastroenterology, 142, 292–304)*

parietal and chief cells with concomitant foveolar cell hyperplasia at 3−6 weeks of age (Saitou et al., 2000). Gastric acid secretion was almost completely abolished in the knockout mouse stomach (Schulzke et al., 2005). The knockout mice developed chronic gastritis even though the paracellular barrier was intact in the stomach (Saitou et al., 2000; Schulzke et al., 2005).

7.1.4.3 Small Intestine

7.1.4.3.1 Claudins in Enterocytes

Claudin-1, -2, -3, -4, -5, -7, -12, and -15 are expressed by the enterocytes of the small intestine (Fujita et al., 2006; Fujita et al., 2008; Holmes, Van Itallie, Rasmussen, & Anderson, 2006; Lameris et al., 2013; Rahner et al., 2001). Similar to the gastric epithelium, the claudin gene expression in the enterocyte varies significantly along the crypt-villus axis. For example, claudin-2 is restricted to the crypt, whereas claudin-7 is concentrated in the apex of the villus (Fujita et al., 2006; Rahner et al., 2001). Claudin-1, -5, and -15 are expressed throughout the small intestine (Fujita et al., 2006; Holmes et al., 2006). The paracellular permeability also varies along the crypt-villus axis. The proliferative crypt compartment where the intestinal stem cells reside contains leakier TJs than the quiescent villus compartment in part due to altered TJ ultrastructures (Marcial, Carlson, & Madara, 1984).

7.1.4.3.2 Paracellular Na^+ Permeability

7.1.4.3.2.1 Claudin-2 Claudin-2 is a well-characterized paracellular cation channel. The knockout mouse of claudin-2 is without apparent morphologic or cytotoxic phenotype in the small intestine (Tamura et al., 2011). The luminal Na^+ content or the intestinal absorption rate of glucose was not affected by the deletion of claudin-2, despite significantly reduced paracellular Na^+ permeability in ex vivo perfused small intestines from both infant and adult claudin-2 knockout mice (Tamura et al., 2011).

7.1.4.3.2.2 Claudin-7 The global claudin-7 knockout mice died within 9 days after birth (Ding et al., 2012). The small intestine in the knockout animal showed mucosal ulceration, epithelial cell sloughing, and inflammation. While the TJ ultrastructure remained intact, cell adhesion and cell-matrix interaction were both weakened in the knockout mouse small intestine (Ding et al., 2012). Paradoxically, the intestine-specific knockout mouse of claudin-7 showed grossly normal phenotypes in the small intestine (Tanaka et al., 2015). The paracellular barrier function was retained in the small intestines of both knockout mouse models (Ding et al., 2012; Tanaka et al., 2015).

7.1.4.3.2.3 Claudin-15 The claudin-15 knockout mouse developed an interesting phenotype of enlarged small intestine, also known as megaintestine (Tamura et al., 2008). The small intestine was approximately twice longer in length and twice wider in diameter in the knockout mouse than in the wildtype mouse (Fig. 7.11A). The number of transit-amplifying cells in the crypt was increased by at least two-fold in the knockout mouse small intestine (Fig. 7.11B) (Tamura et al., 2008). The paracellular Na^+ permeability in the small intestine was significantly reduced in the knockout animal (Fig. 7.12) (Tamura et al., 2011). The paracellular Na^+ permeation from the submucosa to the intestinal lumen supplies the Na^+ ions needed for the Na^+-dependent absorption of glucose via the Na^+/glucose cotransporter 1 (SGLT1). The luminal Na^+ concentration in the small intestine was significantly lower in the claudin-15 knockout mouse than in the wildtype mouse. The low luminal Na^+ level inhibited SGLT1 and caused glucose malabsorption (Tamura et al., 2011). It appears that claudin-15 cooperates with claudin-2 to regulate the paracellular Na^+ permeability in the small intestine. Dual deletion of claudin-15 and claudin-2 created a more lethal phenotype than single knockouts. The double knockout mice manifested malnutrition, growth retardation, and death by 4-week of age (Wada, Tamura, Takahashi, & Tsukita, 2013). The TJs in these animals were almost completely impermeable to Na^+, which caused luminal Na^+ paucity and glucose malabsorption in the small intestine. As a result, hypoglycemia became prevalent in the double knockout animal (Wada et al., 2013).

7.1.4.4 Colon

7.1.4.4.1 Colonic Claudin Expression

The architecture of the colonic epithelium is markedly different from that of the small intestine. The structure of villus is absent, and the colonic epithelium

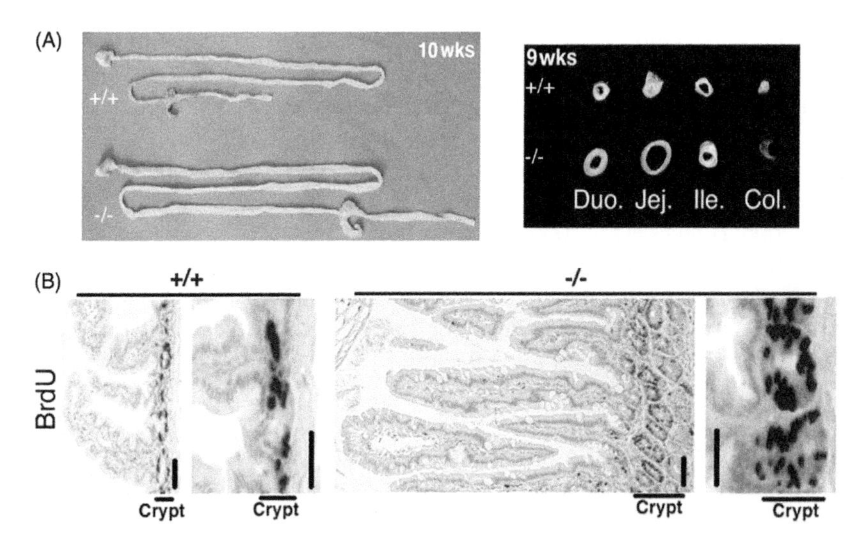

FIGURE 7.11 **Megaintestine in claudin-15 knockout mouse.** (A) Longitudinal and transverse views of small and large intestines in adult Cldn15$^{+/+}$ and Cldn15$^{-/-}$ mice. The length and diameter of Cldn15$^{-/-}$ mouse small intestines—duodenum and jejunum are increased by twofold compared with those of Cldn15$^{+/+}$ small intestines at the age of 9–10 weeks. (B) Cell proliferation assay as examined using a cell proliferation marker, BrdU, in the crypts of the small intestines in Cldn15$^{+/+}$ and Cldn15$^{-/-}$ mice. Note that the proliferation zone in the crypt is extended by 2 times in width in Cldn15$^{-/-}$ mouse compared with that in Cldn15$^{+/+}$ mouse. Bar: 50 μm. *(Reproduced with permission from Tamura, A., Kitano, Y., Hata, M., Katsuno, T., Moriwaki, K., Sasaki, H., Hayashi, H., Suzuki, Y., Noda, T., Furuse, M., et al. (2008). Megaintestine in claudin-15-deficient mice. Gastroenterology, 134, 523–534)*

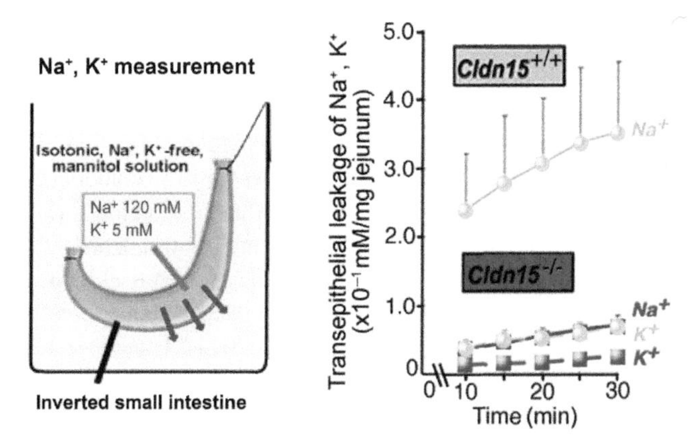

FIGURE 7.12 **Paracellular cation permeability in claudin-15 knockout mouse small intestine.** (Left panel) Drawing of the experimental setup for measuring transepithelial Na^+ and K^+ permeabilities in an inverted mouse small intestine. (Right panel) Time course of paracellular Na^+ and K^+ leakage from the basolateral side of the small intestine in adult Cldn15$^{+/+}$ and Cldn15$^{-/-}$ mice. Note that the paracellular Na^+ leakage is markedly reduced in the knockout mouse intestine. *(Reproduced with permission from Tamura, A., Hayashi, H., Imasato, M., Yamazaki, Y., Hagiwara, A., Wada, M., Noda, T., Watanabe, M., Suzuki, Y., and Tsukita, S. (2011). Loss of claudin-15, but not claudin-2, causes Na^+ deficiency and glucose malabsorption in mouse small intestine. Gastroenterology, 140, 913–923)*

is organized into surface and crypt. The claudin gene expression profile in the colon, however, resembles that in the small intestine. Claudin-1, -2, -3, -4, -5, -7, -8, -12, and -15 have been found in the colonic epithelium (Fujita et al., 2006; Fujita et al., 2008; Holmes et al., 2006; Rahner et al., 2001). The colonic claudin gene expression shows axial variation from the surface to the crypt. For example, claudin-3, -4, and -7 accumulate in the surface cell population (Fujita et al., 2006; Holmes et al., 2006; Rahner et al., 2001). The cells in the crypt preferentially express claudin-2, -5, and -15 (Fujita et al., 2006; Rahner et al., 2001). Claudin-12 is found throughout the colon (Fujita et al., 2008). When the colonic stem cell migrates from the crypt to the surface, the claudin gene expression alters according to the fate of the cell (Barker, 2014).

7.1.4.4.2 Paracellular Solute Barrier

7.1.4.4.2.1 Claudin-1 A transgenic mouse model has recently been generated to overexpress the claudin-1 gene in the colon with the villin promoter (Pope et al., 2014). The epithelial lineage in the colon is reprogrammed by the claudin-1 gene. There were marked losses of the goblet cells and compensated proliferation of the colonocytes in the transgenic mouse colon. While the paracellular permeability to dextran (4 kDa) remained unaffected across the colonic epithelium, the colonic mucus barrier was disrupted owing to the losses of the goblet cell and the key mucus protein: mucin-2, which the goblet cell secretes. As a result, the transgenic mice were more susceptible to dextran sulfate sodium (DSS)-induced colitis (Pope et al., 2014).

7.1.4.4.2.2 Claudin-2 Transgenic overexpression of claudin-2 in the mouse colon with the villin promoter generated different results from the claudin-1 overexpression model. The paracellular permeabilities to Na^+ and solute (dextran, 4 kDa) were both increased across the colonic epithelium in the claudin-2 transgenic mouse (Ahmad et al., 2014). Apart from being a paracellular channel, claudin-2 plays an important role in colonocyte proliferation. There were markedly more BrdU (a marker for proliferation) positive colonocytes in claudin-2 transgenic mice than in wildtype mice (Ahmad et al., 2014). The overexpressed claudin-2 gene protected the transgenic mouse from DSS-induced colitis. The claudin-2 overexpression stimulated proliferation-associated gene expression while suppressed inflammation-associated gene expression in the colon of the DSS-treated animal (Ahmad et al., 2014). Claudin-2 also plays a vital role in colonic clearance of *Citrobacter rodentium*. Compared with wildtype mice, claudin-2 knockout mice experienced increased mucosal colonization by *C. rodentium*, prolonged pathogen shedding, amplified cytokine responses, and greater tissue injury, whereas claudin-2 overexpression mice became resistant to *C. rodentium* infection (Tsai et al., 2017). The claudin-2 dependent paracellular permeabilities to Na^+ and water facilitated the development of diarrhea, which in turn reduced *C. rodentium* burden and mucosal inflammation in the colon (Tsai et al., 2017).

7.1.4.4.2.3 Claudin-7 The villin promoter-driven, intestine-specific deletion of claudin-7 in mice caused colonic inflammation (Fig. 7.13A) (Tanaka et al., 2015). The paracellular permeabilities to small solutes, such as Lucifer Yellow (457 Da) and *N*-formylmethionyl-leucyl-phenylalanine (FMLP) (438 Da), were increased across the colonic epithelium in the knockout mouse (Fig. 7.13B). FMLP is a major bacterial metabolite that can diffuse from the intestinal lumen to the lamina propria to trigger inflammatory responses (Marasco et al., 1984). When

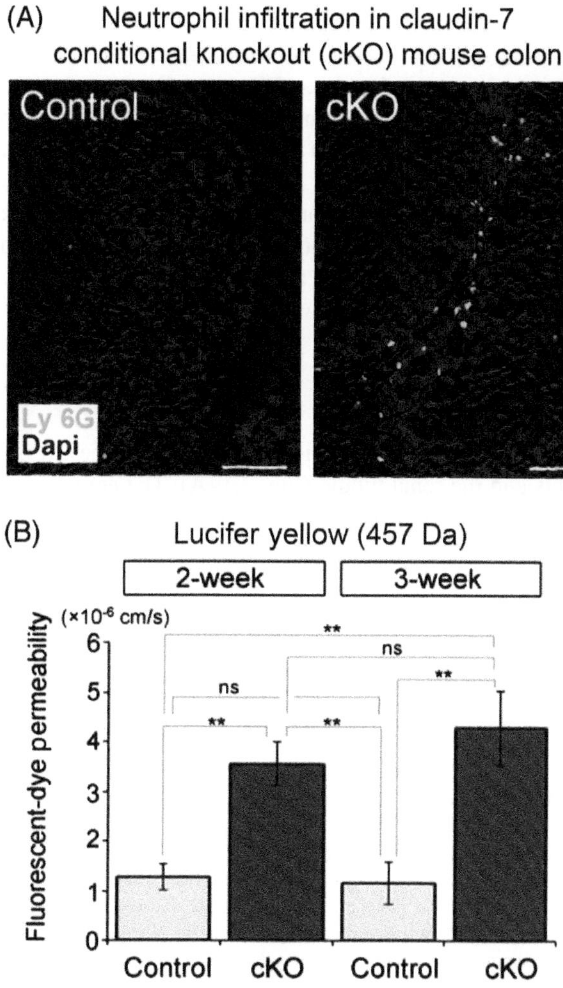

FIGURE 7.13 Colonic abnormality in claudin-7 conditional knockout (cKO) mouse. (A) Immunofluorescence for Ly-6G (a neutrophil marker) with counter-staining for the nucleus in the colons of 3-week-old control and Cldn7 cKO mice. Bar: 100 μm. (B) Paracellular permeability of 457-Da Lucifer Yellow across the colons of 2-week-old or 3-week-old control and Cldn7 cKO mice. ns, not significant; *Double Asterisk*, $P < 0.01$. *(Reproduced with permission from Tanaka, H., Takechi, M., Kiyonari, H., Shioi, G., Tamura, A., and Tsukita, S. (2015). Intestinal deletion of Claudin-7 enhances paracellular organic solute flux and initiates colonic inflammation in mice. Gut, 64, 1529–1538)*

injected into the lumen of colon, FMLP stimulated inflammation-associated gene expression in knockout mice but not in wildtype mice, which suggests that claudin-7 prevents the entry of chemotactic bacterial metabolites into the submucosa by impeding the paracellular permeation (Tanaka et al., 2015).

7.1.4.4.2.4 JAM-A The junctional adhesion molecule (JAM) proteins are immunoglobulin (Ig)-like molecules found in the TJs of epithelial and endothelial cells (Chapter 2, Section 2.7). The JAM-A knockout mouse developed colonic inflammation characterized by polymorphonuclear leukocyte infiltration and lymphoid aggregates (Laukoetter et al., 2007). The paracellular permeabilities to ions and solutes (dextran, 4 kDa) were both increased across the colonic epithelium of the knockout animal, which might have allowed luminal antigens to leak into the submucosa and to elicit immune responses. In addition to its permeant role, JAM-A regulates the regenerative ability of the colonic epithelium. In the JAM-A knockout mouse colon, the Ki-67 (a marker for proliferation) positive cell population expanded from the crypt to the lumen along the normally quiescent crypt-to-surface axis (Laukoetter et al., 2007).

7.1.5 Kidney

7.1.5.1 Anatomy and Physiology

The nephron is the structural and functional unit of the kidney. It is composed of the glomerulus and the renal tubule (Fig. 7.14A). The epithelial cells in the glomerulus and the renal tubule play vital roles in determining the permeability of these structures to ion, solute, and water, and hence fulfilling the glomerular filtration function and the tubular reabsorption function, respectively. Transepithelial transport occurs via the transcellular and the paracellular pathways. The paracellular permeability in different nephron segments varies as a result of the composition of the claudin molecules in the segment. The segment-specific claudin expression profiles are summarized in Table 7.2.

FIGURE 7.14 Schematic diagram of the renal epithelium. (A) The nephron consists in the glomerulus (G) and the renal tubule, which includes the proximal convoluted tubule (PCT), the proximal straight tubule (PST), the thin descending limb (tDL), the thin ascending limb (tAL), the thick ascending limb (TAL), the distal convoluted tubule (DCT), the connecting tubule (CNT), and the collecting duct (CD). (B) In the proximal tubule, both Na^+ and HCO_3^- are reabsorbed via the Na^+/H^+ exchanger 3 (NHE3) on the apical membrane and the Na^+-HCO_3^- cotransporter (NBC) on the basolateral membrane. This process increases the Cl^- concentration in the proximal tubule luminal fluid, which drives paracellular Cl^- reabsorption down the chemical gradient of Cl^-. Paracellular Cl^- reabsorption creates a lumen-positive voltage, which then drives paracellular reabsorption of Na^+. The reabsorption of Na^+ and glucose via the Na^+-glucose cotransporter (SGLT) in the proximal tubule generates a significant osmotic gradient that drives the reabsorption of water. (C) In thick ascending limb, Na^+, K^+, and Cl^- are reabsorbed through the luminal membrane Na^+-K^+-$2Cl^-$ cotransporter (NKCC2). Na^+ is secreted into the basolateral interstitium via the Na^+/K^+–ATPase. Cl^- is secreted into the basolateral interstitium via the chloride channel (ClCkb/barttin). K^+ is recycled into the lumen through the renal outer medullary potassium channel (ROMK). Due to ▶

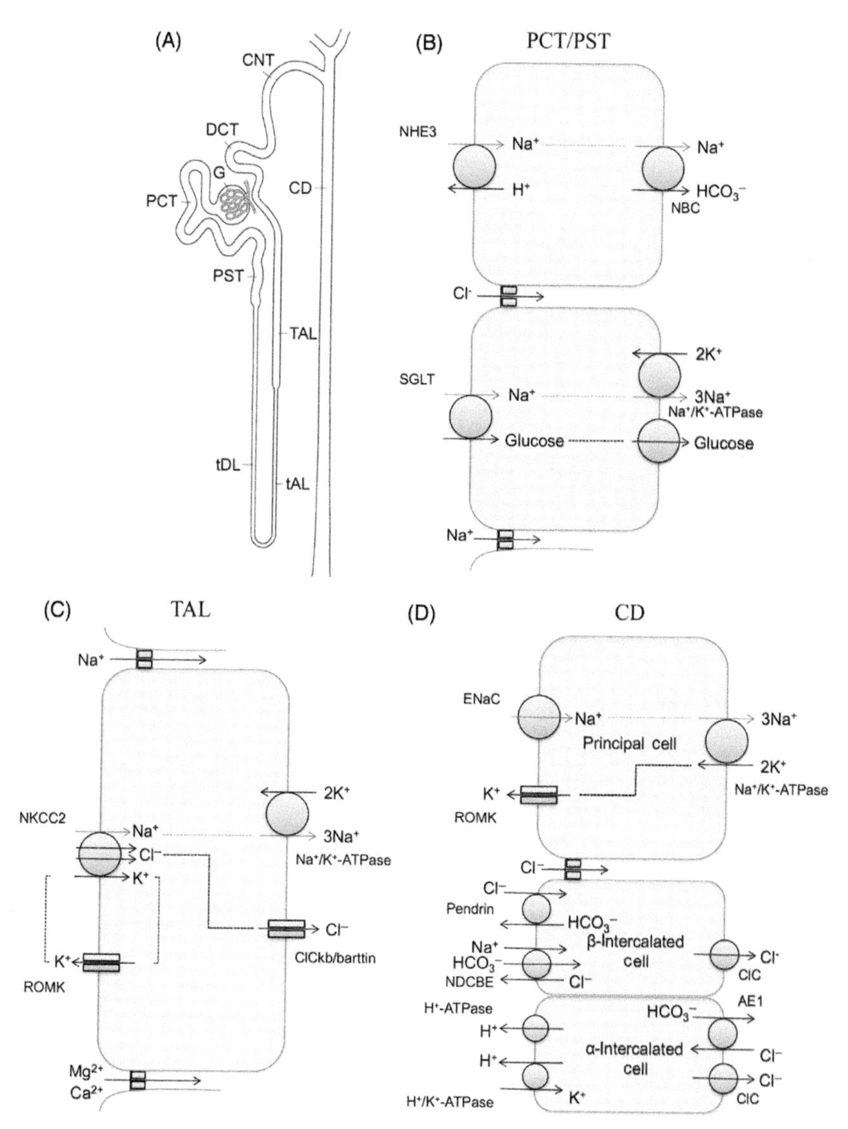

continuous reabsorption of NaCl, a NaCl gradient develops from interstitial space to lumen, which generates a lumen-positive diffusion voltage to drive paracellular Mg^{++} and Ca^{++} reabsorption. (D) In the collecting duct, Na^+ is reabsorbed through the epithelial sodium channel (ENaC). Na^+ is secreted into the basolateral interstitium via the Na^+/K^+–ATPase. K^+ is secreted into the lumen via ROMK. Because of the unidirectional Na^+ reabsorption, a lumen–negative voltage develops, which drives paracellular Cl^- absorption. A parallel electroneutral transport pathway for Cl^- uses the Cl^-/HCO_3^- exchanger (pendrin) and the Na^+-driven Cl^-/HCO_3^- exchanger (NDCBE) on the apical membrane and the chloride channel (ClC) on the basolateral membrane of the β-intercalated cell. The α-intercalated cell primarily handles H^+ secretion via the H^+-ATPase and the H^+/K^+-ATPase on the luminal membrane. *(Reproduced with permission from Hou, J., Rajagopal, M., and Yu, A.S. (2013). Claudins and the kidney. Annual Review of Physiology, 75, 479–501)*

TABLE 7.2 Claudin Gene Expression Profile in the Kidney

Genes	Podocyte	PEC	PT	tDL	tAL	TAL	DCT	CNT/CD	References
Claudin-1	–	+	–	–	–	–	–	–	Kiuchi-Saishin et al. (2002)
Claudin-2	–	+	+	+	–	–	–	–	Enck, Berger, & Yu, (2001); Kiuchi-Saishin et al. (2002)
Claudin-3	–	–	–	–	+	+	+	+	Kiuchi-Saishin et al. (2002)
Claudin-4	–	–	–	+	–	–	+	+	Gong et al. (2014); Kiuchi-Saishin et al. (2002)
Claudin-5	+	–	–	–	–	–	–	–	Koda et al. (2011)
Claudin-6	+	–	–	–	–	–	–	–	Zhao et al. (2008)
Claudin-7	–	–	–	+	–	–	+	+	Li, Huey, & Yu (2004)
Claudin-8	–	–	–	+	–	–	+	+	Gong et al. (2015); Kiuchi-Saishin et al. (2002); Li et al. (2004)
Claudin-10a	–	–	+	–	–	–	–	–	Breiderhoff et al. (2012); Kiuchi-Saishin et al. (2002)
Claudin-10b	–	–	–	–	–	+	–	–	Breiderhoff et al. (2012); Kiuchi-Saishin et al. (2002)
Claudin-14	–	–	–	–	–	+	–	–	Gong et al. (2012)
Claudin-16	–	–	–	–	–	+	–	–	Hou et al. (2007); Simon et al. (1999)
Claudin-17	–	–	+	–	–	–	–	–	Krug et al. (2012)
Claudin-19	–	–	–	–	–	+	–	–	Hou et al. (2009); Konrad et al. (2006)

CD, collecting duct; *CNT*, connecting tubule; *DCT*, distal convoluted tubule; *PEC*, parietal epithelial cell; *PT*, proximal tubule; *tDL*, thin descending limb of Henle's loop; *tAL*, thin ascending limb of Henle's loop; *TAL*, thick ascending limb of Henle's loop.

7.1.5.2 Glomerulus

7.1.5.2.1 Slit Diaphragm

The glomerulus consists in Bowman's capsule and the glomerular tuft. Bowman's capsule is covered by a thin layer of squamous epithelium known as the parietal epithelial cell (PEC). The glomerular tuft is a microvascular bed that contains three types of cells, including the glomerular endothelial cell, the visceral epithelial cell, also known as podocyte, and the mesangial cell. The glomerular filtration barrier comprises of the endothelial cell, the podocyte, and the glomerular basement membrane. During the glomerular development, the presumptive podocytes are connected by the TJ, which is found near the apical membrane of the podocytes (Quaggin & Kreidberg, 2008). Mature podocytes are devoid of the TJ but form a special cell junction known as slit diaphragm (SD) (Grahammer, Schell, & Huber, 2013). The SD is made of two types of membrane proteins, nephrin and podocin, and acts as a major impediment to albumin permeation across the glomerular filtration barrier (Kestila et al., 1998; Schwarz et al., 2001).

7.1.5.2.2 Claudin-1

The normal podocyte expresses little or no claudin. Gong and coworkers have generated a transgenic overexpression mouse model for claudin-1 in the glomerular podocyte (Gong, Sunq, Roth, & Hou, 2016). Induction of claudin-1 gene expression in mature podocytes caused microalbuminuria and changes in the SD on the ultrastructural level, including foot process effacement, loss of slit space, and reappearance of TJ. Mechanistically, claudin-1 interacts with nephrin and podocin to destabilize the SD in the podocyte (Gong et al., 2016).

7.1.5.3 Proximal Tubule

7.1.5.3.1 Paracellular NaCl and Water Reabsorption

The primary role of the proximal tubule is bulk reabsorption of NaCl and water from the glomerular ultrafiltrate (Fig. 7.14B). Approximately, half of the filtered NaCl is reabsorbed through the paracellular pathway in the proximal tubule (Rector, 1983). Whether the proximal tubule retains paracellular water permeability is highly controversial. Genetic knockout of aquaporin-1 in mice showed reduction in the proximal tubular water reabsorption only by 80%, suggesting that an alternative water permeation pathway may exist (Schnermann et al., 1998). Nevertheless, when Preisig and Berry measured the paracellular permeability of the proximal tubule to mannitol or sucrose as a surrogate for water, they concluded that the paracellular water permeability was $< 2\%$ of overall transepithelial water permeability (Preisig & Berry, 1985).

7.1.5.3.2 Claudin-2

Claudin-2 functions in vitro as a paracellular cation channel (Yu et al., 2009). Genetic ablation of claudin-2 in mice caused defects specific to the proximal tubule

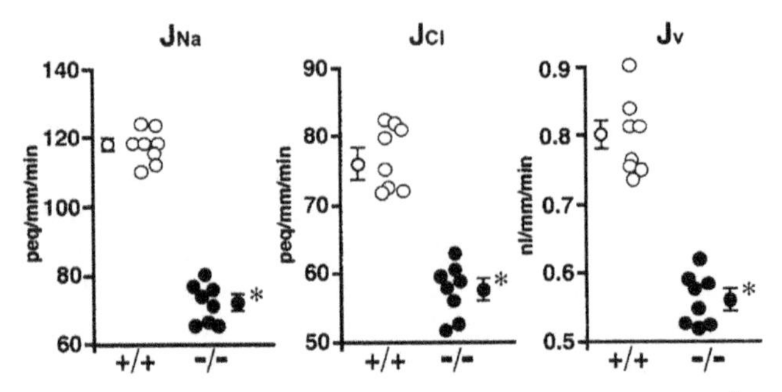

FIGURE 7.15 Proximal tubular reabsorption of Na^+, Cl^-, and volume in $Cldn2^{+/+}$ and $Cldn2^{-/-}$ mouse. Net transepithelial reabsorption rates of Na^+, Cl^-, and volume (J_{Na}, J_{Cl}, and J_v, respectively) in isolated and perfused S2 segments of the proximal tubules are shown. *Asterisk, $P < 0.001$. (Reproduced with permission from Muto, S., Hata, M., Taniguchi, J., Tsuruoka, S., Moriwaki, K., Saitou, M., Furuse, K., Sasaki, H., Fujimura, A., Imai, M., et al. (2010). Claudin-2-deficient mice are defective in the leaky and cation-selective paracellular permeability properties of renal proximal tubules. Proceedings of the National Academy of Sciences of the United States of America, 107, 8011–8016)*

of the kidney, owing to decreases in the proximal tubular reabsorption of Na^+, Cl^-, and volume (Fig. 7.15) (Muto et al., 2010). Ex vivo perfusion of the proximal tubule from the claudin-2 knockout mouse kidney revealed a marked decrease in paracellular cation selectivity (P_{Na}/P_{Cl}). The proximal tubular water reabsorption was also reduced in the knockout mouse kidney (Muto et al., 2010). Yu and coworkers have uncovered an additional role for claudin-2 in rationalizing renal energy efficiency (Pei et al., 2016). Reduction of the paracellular permeability in the proximal tubule led to increased transcellular permeabilities in the downstream tubules at the expense of oxygen consumption. Such an energy burden predisposed the renal medulla to ischemic injuries in the claudin-2 knockout mouse (Pei et al., 2016).

7.1.5.4 Thick Ascending Limb of Henle's Loop

7.1.5.4.1 Paracellular Ca^{++} and Mg^{++} Reabsorption

The thick ascending limb of Henle's loop is a nephron segment important for reabsorbing Ca^{++} and Mg^{++} via the paracellular pathway (Fig. 7.14C). The driving force for divalent cation reabsorption is the lumen-positive voltage difference, which is generated by two additive mechanisms: (1) transcellular reabsorption of NaCl through the apical Na-K-2Cl cotransporter, coupled with K^+ secretion through the apical K channel, gives rise to a spontaneous lumen-positive potential (~ +8 mV) (Burg & Green, 1973); (2) dilution of the luminal fluid through constant NaCl reabsorption creates a diffusion potential through the cation-selective TJ, which adds up to +30 mV to the total transepithelial voltage difference (Greger, 1985).

7.1.5.4.2 Claudin-16

Genetic mutations in the claudin-16 gene, formerly known as paracellin-1, cause an autosomal recessive renal disorder in man: familial hypomagnesemia

TABLE 7.3 Plasma and Urine Electrolyte Levels in WT and CLDN16 Knockdown (KD) Mice

Group	WT	KD	Significance
Weight (g)	25.78 ± 1.27	24.00 ± 1.00	n.s.
UV (μl/min)	2.11 ± 0.04	2.28 ± 0.03	n.s.
Osm (mOsm/L)	1504.2 ± 123.0	1334.0 ± 125.9	n.s.
GFR (ml/min/100g)	0.89 ± 0.09	0.65 ± 0.09	n.s.
Mean BP (mmHg)	91.91 ± 0.94	71.56 ± 2.24	$P < 0.05$
P_{Na} (mmol/L)	138.76 ± 2.24	141.59 ± 2.60	n.s.
P_K (mmol/L)	4.45 ± 0.20	3.68 ± 0.20	$P < 0.05$
P_{Mg} (mmol/L)	0.71 ± 0.06	0.51 ± 0.03	$P < 0.05$
P_{Ca} (mmol/L)	2.01 ± 0.08	2.16 ± 0.04	n.s.
FE_{Na} (%)	0.74 ± 0.22	0.98 ± 0.25	n.s.
E_{Na} (μEq/min/100g)	0.77 ± 0.24	0.85 ± 0.24	n.s.
FE_K (%)	26.05 ± 4.24	53.49 ± 11.81	$P < 0.05$
E_K (μEq/min/100g)	1.02 ± 0.23	1.11 ± 0.20	n.s.
FE_{Mg} (%)	18.01 ± 3.22	73.28 ± 10.34	$P < 0.05$
E_{Mg} (μEq/min/100g)	0.106 ± 0.025	0.214 ± 0.022	$P < 0.05$
FE_{Ca} (%)	0.75 ± 0.09	2.80 ± 0.54	$P < 0.05$
E_{Ca} (nEq/min/100g)	12.99 ± 2.34	38.62 ± 8.44	$P < 0.05$

UV: urine volume; *GFR:* glomerular filtration rate; *Osm:* urine osmolality; *BP:* blood pressure; P_{Na}, P_K, P_{Mg}, P_{Ca}: plasma Na^+, K^+, Mg^{2+} and Ca^{2+}concentrations measured in anesthetized animals during renal clearance study; E_{Na}, E_K, E_{Mg}, E_{Ca}: absolute Na^+, K^+, Mg^{2+} and Ca^{2+}excretion; FE_{Na}, FE_K, FE_{Mg}, FE_{Ca}: fractional excretion of Na^+, K^+, Mg^{2+} and Ca^{2+}. $N = 9$ animals.
Reproduced with permission from Hou, J., Shan, Q., Wang, T., Gomes, A.S., Yan, Q., Paul, D.L., Bleich, M., and Goodenough, D.A. (2007). Transgenic RNAi depletion of claudin-16 and the renal handling of magnesium. *The Journal of Biological Chemistry, 282*, 17114–17122.

and hypercalciuria with nephrocalcinosis (FHHNC, OMIM #248250) (Simon et al., 1999). FHHNC is characterized by severe renal losses of Ca^{++} and Mg^{++}. Hou and coworkers have generated a claudin-16 deficient mouse model using RNA interference technology (Hou et al., 2007). The claudin-16 knockdown mouse developed the phenotypes recapitulating the main features of human FHHNC, including significantly reduced plasma Mg^{++} levels and excessive urinary excretion of Ca^{++} and Mg^{++} (Table 7.3). Calcium deposits were observed in the medullary interstitium of the knockdown mouse kidney, indicating signs of nephrocalcinosis. Ex vivo tubule perfusion experiments revealed that when claudin-16 was removed from the thick ascending limb, the paracellular cation selectivity (P_{Na}/P_{Cl}) markedly diminished (Table 7.4) (Hou et al., 2007). The fact that the lumen-positive NaCl diffusion potential depends upon paracellular cation selectivity suggests that claudin-16 facilitates divalent cation reabsorption

TABLE 7.4 Properties of the Thick Ascending Limbs of Loops of Henle in WT and CLDN16 Knockdown (KD) Mice

	WT	KD	*P*
Length constant (μm)	85.6 ± 5.1	80.3 ± 4.3	n.s.[a]
V_{te} (mV)	9.0 ± 0.7	7.4 ± 0.8	n.s.[a]
R_{te} (Ω.cm²)	13.6 ± 1.3	12.3 ± 1.2	n.s.[a]
I_{sc} (μA/cm²)	703 ± 72	645 ± 96	n.s.[a]
PD (mV)	−18.0 ± 1.4	−6.6 ± 0.9	<0.01[a]
P_{Na}/P_{Cl}	3.1 ± 0.3	1.5 ± 0.1	<0.01[a]
P_{Na}/P_{Li}	1.6 ± 0.1	1.7 ± 0.1	n.s.[b]
P_{Na}/P_{Mg}	1.9 ± 0.1	2.3 ± 0.2	n.s.[b]

I_{sc}, equivalent short-circuit current; *n.s.*, not significant; *PD*, diffusion potential; R_{te}, transepithelial resistance; V_{te}, transepithelial voltage; P_{Na}, Na^+ permeability; P_{Cl}, Cl^- permeability; P_{Li}, Li^+ permeability; P_{Mg}, Mg^{++} permeability. Data are mean ± SE.
[a]*N = 9 for WT, N = 8 for KD.*
[b]*N = 5 for WT, N = 6 for KD.*
Reproduced with permission from Hou, J., Shan, Q., Wang, T., Gomes, A.S., Yan, Q., Paul, D.L., Bleich, M., & Goodenough, D.A. (2007). Transgenic RNAi depletion of claudin-16 and the renal handling of magnesium. *The Journal of Biological Chemistry, 282*, 17114–17122.

by establishing the electrical driving force. Interestingly, the thick ascending limb from a knockout mouse model for claudin-16 showed more selective decreases in paracellular divalent cation permeabilities relative to monovalent cation permeability, which suggests that claudin-16 may confer divalent cation selectivity to the paracellular pathway (Will et al., 2010).

7.1.5.4.3 Claudin-19

Genetic mutations in the claudin-19 gene cause FHHNC with ocular involvement (OMIM #248190). The claudin-19 knockdown mouse phenocopied the claudin-16 knockdown mouse and developed the FHHNC-like phenotypes of reduced plasma Mg^{++} levels and excessive renal losses of Ca^{++} and Mg^{++} (Hou et al., 2009). Mechanistically, claudin-16 and claudin-19 *cis*-interact and co-polymerize into the TJ (Hou et al., 2008). The interaction between claudin-16 and claudin-19 is speculated to stabilize the claudins in the TJ membrane. Without the binding partner, neither claudin was localized in the TJ of thick ascending limb cells, which might be a result of trafficking defect or membrane instability (Fig. 7.16) (Hou et al., 2009).

7.1.5.4.4 Claudin-14

Claudin-14 is an intriguing gene. Its mRNA and protein expression levels in the kidney are extremely low, but its promoter activities are particularly high in the thick ascending limb (Gong et al., 2012). Genome-wide association studies (GWAS) have identified claudin-14 as a major risk gene of hypercalciuric

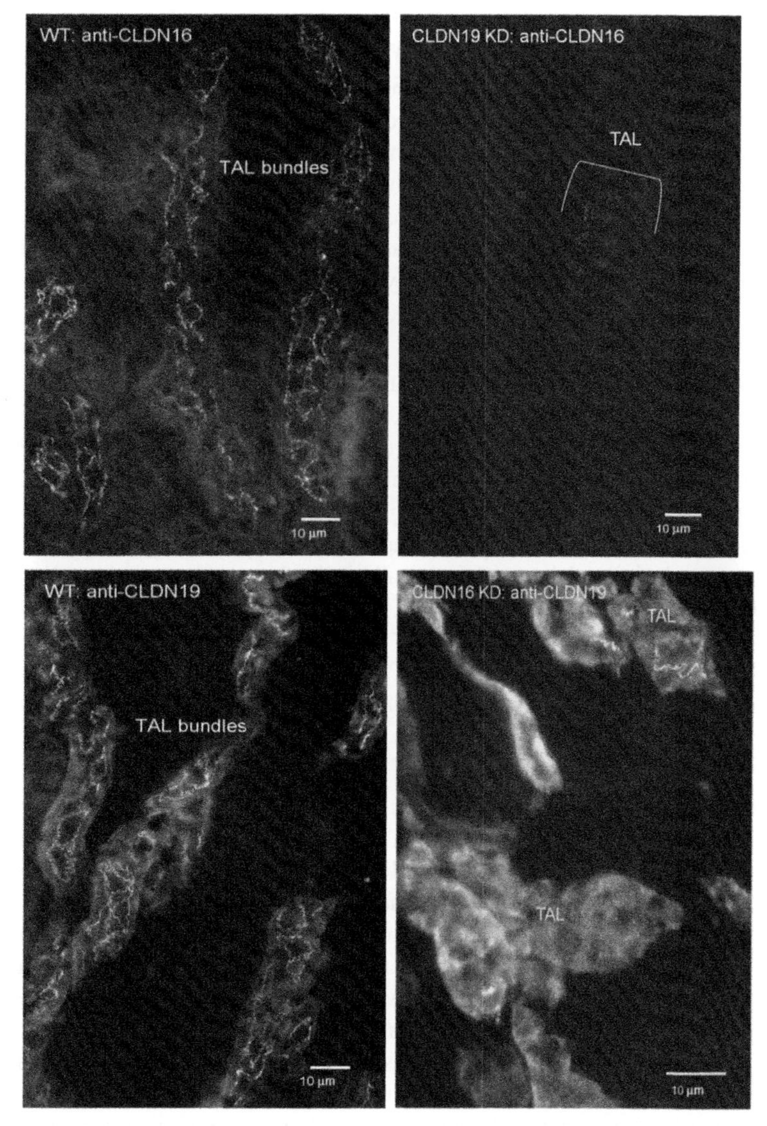

FIGURE 7.16 **Claudin-16 and claudin-19 co-assembly into TJ.** Cryostat sagittal sections (10 µm thick) from wildtype (WT), claudin-16 knockdown (CLDN16 KD), and claudin-19 knockdown (CLDN19 KD) mouse kidneys were immunolabeled with anti-claudin-16 or anti-claudin-19 antibody to show claudin-16 and claudin-19 localization in the TJs of thick ascending limbs. Bar: 10 µm. *(Reproduced with permission from Hou, J., Reniguntа, A., Gomes, A.S., Hou, M., Paul, D.L., Waldegger, S., and Goodenough, D.A. (2009). Claudin-16 and claudin-19 interaction is required for their assembly into tight junctions and for renal reabsorption of magnesium. Proceedings of the National Academy of Sciences of the United States of America, 106, 15350–15355)*

nephrolithiasis and reduced bone mineral density (Thorleifsson et al., 2009). Transgenic overexpression of claudin-14 in mouse thick ascending limbs generated renal tubular defects featured by increased urinary excretion of Ca^{++} and Mg^{++} (Gong & Hou, 2014). Hypercalcemia has been shown to upregulate the claudin-14 gene expression on both mRNA and protein levels in the thick ascending limb (Gong et al., 2012). When fed a high Ca^{++} diet, the claudin-14 knockout mouse developed hypermagnesemia, hypomagnesiuria, and hypocalciuria, indicating that claudin-14 acts as a negative regulator of renal divalent cation reabsorption to defend against hypercalcemia (Gong et al., 2012).

7.1.5.4.5 Claudin-10b

The claudin-10 gene has two splice variants: claudin-10a, which is expressed by the proximal tubule, and claudin-10b, which is found in the thick ascending limb (Van Itallie et al., 2006). The distal tubule-specific knockout of claudin-10 in mice selectively deleted the claudin-10b gene in the thick ascending limb without affecting the claudin-10a gene in the proximal tubule (Breiderhoff et al., 2012). The claudin-10b knockout mouse developed hypermagnesemia, hypomagnesiuria, and hypocalciuria, allegedly due to hyper-reabsorption of divalent cations in the thick ascending limb. Ex vivo tubule perfusion experiments revealed that the paracellular Na^+ permeability (P_{Na}) was reduced but the divalent cation selectivity (P_{Ca}/P_{Na} or P_{Mg}/P_{Na}) was increased in the knockout mouse thick ascending limb (Breiderhoff et al., 2012).

7.1.5.4.6 Tricellular Tight Junction

Angulin-2 [also known as immunoglobulin (Ig)-like domain-containing receptor (ILDR) 1] is a key component of the tTJ (Chapter 6, Section 6.2.2). Angulin-2 is expressed by the distal tubules including the thick ascending limb in the kidney (Gong et al., 2017). The thick ascending limb is generally considered water impermeable due to the lack of aquaporin expression, which allows effective dilution of the luminal fluid when NaCl is actively reabsorbed via the transcellular pathway. The separation of NaCl from water reabsorption in the thick ascending limb fuels the countercurrent multiplication mechanism to establish the osmotic gradient along the cortico-medullary axis in the interstitium, which in turn drives water reabsorption in the collecting duct through aquaporin-2 (Smith, 1959). When angulin-2 was removed from the mouse kidney, the thick ascending limb became highly permeable to water (Fig. 6.5) (Gong et al., 2017). As a result, the cortico-medullary osmotic gradient dissipated in the interstitium, causing urinary concentration defect. Generation of the lumen-positive NaCl diffusion potential in the thick ascending limb also depends upon the NaCl concentration difference between the lumen and the interstitium (Greger, 1981). The elevated luminal tonicity in the knockout mouse thick ascending limb is expected to diminish the diffusion potential, that is, the driving force for divalent cation reabsorption. Consistent with this theory, the tubular reabsorption rate for Ca^{++} was markedly reduced in the angulin-2 knockout mouse kidney (Gong et al., 2017).

7.1.5.5 Aldosterone Sensitive Distal Tubule

7.1.5.5.1 Paracellular Cl^- Reabsorption

The aldosterone sensitive distal tubules comprise the distal convoluted tubule, the connecting tubule and the collecting duct. While the TJ function in the distal convoluted tubule is not well understood, the paracellular pathways in the connecting tubule and the collecting duct play important roles in tubular Cl^- reabsorption (Fig. 7.14D) (O'Neil & Sansom, 1984). The paracellular Cl^- permeability is essential to establish electrical coupling with the electrogenic Na^+ reabsorption that takes place via the epithelial Na^+ channel (ENaC) on the luminal membrane (Sansom, Weinman, & O'Neil, 1984).

7.1.5.5.2 Claudin-4

The global claudin-4 knockout mouse developed a wide spectrum of defects in the kidney and the urinary tract (Fujita, Hamazaki, Noda, Oshima, & Minato, 2012). The renal defects included increases in urinary excretion of Ca^{++} and Cl^-, and decreases in urinary osmolality. Urothelial hyperplasia was pronounced in the knockout mouse, which frequently led to urinary tract obstruction and hydronephrosis. In order to delineate the kidney-specific role, Gong and coworkers have generated a collecting duct-specific knockout mouse model for claudin-4 (Gong et al., 2014). The conditional knockout mouse developed hypotension, hypochloremia, and metabolic alkalosis due to excessive urinary excretion of Cl^- (Table 7.5). Accumulation of luminal Cl^- in the knockout mouse collecting duct is expected to depolarize the apical membrane, and consequently inhibit ENaC, causing renal loss of Na^+. The renin–angiotensin–aldosterone system upregulates claudin-4 to defend the blood pressure in the face of volume depletion. When fed a low NaCl diet, the conditional knockout mouse continued to lose Cl^-, Na^+ and volume through the kidney, thereby presenting exacerbated hypotension (Gong et al., 2014).

7.1.5.5.3 Claudin-7

The claudin-7 knockout mouse died of severe salt wasting, chronic dehydration, and growth retardation within 12 days after birth (Tatum et al., 2010). The urinary excretion rates of Na^+, K^+, and Cl^- were invariably higher in the knockout mouse. The renal losses of electrolytes and volume activated the renin–angiotensin–aldosterone system in the knockout mouse to provide a feedback control mechanism (Tatum et al., 2010).

7.1.5.5.4 Claudin-8

The collecting duct-specific knockout mouse model for claudin-8 has been generated with the same approach as claudin-4 (Gong et al., 2015). The conditional claudin-8 knockout mouse phenocopied the conditional claudin-4 knockout mouse in many ways including the renal losses of Cl^-, Na^+ and volume but developed more severe hypotension. The phenotypic similarity of claudin-8

TABLE 7.5 Plasma and Urine Electrolyte Levels in Claudin-4 KO and Control Mice

Groups	$CLDN4^{flox/flox}$	$CLDN4^{flox/flox}/Aqp2^{Cre}$	Significance
Weight (g)	20.94 ± 0.64	20.08 ± 0.40	n.s.
UV (μL/24 h.g)	46.01 ± 3.21	60.32 ± 4.21	$P = 0.012$
Osm (Osm/kg H_2O)	1.653 ± 0.066	1.815 ± 0.101	n.s.
Urine (pH)[a]	6.06 ± 0.10	6.13 ± 0.24	n.s.
GFR (mL/24 h.g)	3.07 ± 0.35	3.29 ± 0.30	n.s.
Hematocrit (%)[b]	47.3 ± 0.8	48.9 ± 1.0	n.s.
Mean BP (mmHg)	95.2 ± 2.5	88.7 ± 1.4	$P = 0.036$
Serum (pH)	7.43 ± 0.01	7.44 ± 0.01	n.s.
P_{HCO3} (mmol/L)	22.5 ± 0.8	25.0 ± 0.9	$P = 0.045$
P_{Cl} (mmol/L)	105.0 ± 0.9	101.1 ± 1.4	$P = 0.022$
P_{Na} (mmol/L)	142.3 ± 0.9	145.8 ± 2.2	n.s.
P_K (mmol/L)	3.85 ± 0.16	4.28 ± 0.14	$P = 0.060$
FE_{Cl} (%)	1.30 ± 0.18	2.33 ± 0.26	$P = 0.003$
E_{Cl} (μmol/24 h.g)	3.71 ± 0.42	7.27 ± 0.84	$P = 0.0007$
FE_{Na} (%)	0.81 ± 0.11	1.23 ± 0.13	$P = 0.022$
E_{Na} (μmol/24 h.g)	3.10 ± 0.33	5.15 ± 0.55	$P = 0.004$
FE_K (%)	18.74 ± 1.74	19.60 ± 3.09	n.s.
E_K (μmol/24 h.g)	1.93 ± 0.15	2.39 ± 0.23	n.s.

BP, blood pressure; E_{Cl}, absolute excretion level of Cl^-; E_{Na}, absolute excretion level of Na^+; E_K, absolute excretion level of K^+; FE_{Cl}, fractional excretion level of Cl^-; FE_{Na}, fractional excretion level of Na^+; FE_K, fractional excretion level of K^+; *GFR*, glomerular filtration rate; *Osm*, urinary osmolality; P_{HCO3}, plasma HCO_3^- level; P_{Cl}, plasma Cl^- level; P_{Na}, plasma Na^+ level; P_K, plasma K^+ level; *UV*, urinary volume. $N = 15$ animals.
[a]*Spot urine collection (n = 8).*
[b]*In sex-matched (male) animals.*
Reproduced with permission from Gong, Y., Yu, M., Yang, J., Gonzales, E., Perez, R., Hou, M., Tripathi, P., Hering-Smith, K.S., Hamm, L.L., & Hou, J. (2014). The Cap1-claudin-4 regulatory pathway is important for renal chloride reabsorption and blood pressure regulation. *Proceedings of the National Academy of Sciences of the United States of America, 111*, E3766–3774.

to claudin-4 knockout mouse might be a result of the molecular interaction between these two claudins (Hou, Renigunta, Yang, & Waldegger, 2010). Deletion of claudin-8 from the collecting duct disrupted the TJ localization of claudin-4, which resulted in a double knockout on the TJ level and a more complete closure of the paracellular Cl^- pathway (Gong et al., 2015). When excessive Cl^- accumulates in the lumen, K^+ is secreted by the collecting duct to maintain the charge neutrality of the luminal fluid. As a result, the claudin-8 knockout mouse developed hypokalemia due to renal K^+ loss, a phenotype not shared by the claudin-4 knockout mouse (Gong et al., 2015).

7.1.6 Testis

7.1.6.1 Blood-Testis Barrier (BTB)

The seminiferous tubule is lined by a complex stratified epithelium containing two distinct populations of cells, the spermatogenic cell that develops into spermatozoon, and the Sertoli cell that plays a supportive role. The blood-testis barrier (BTB) consists of the TJ between the Sertoli cells that restricts the paracellular permeation of ions and solutes, thereby creating a microenvironment within the seminiferous tubule and providing immune privilege to meiotic spermatocytes and postmeiotic spermatids (Fig. 7.17). Claudin-1, -3, -5, -11, -12, and -13 are expressed by the Sertoli cell (Chakraborty et al., 2014; Dietze, Shihan, Stammler, Konrad, & Scheiner-Bobis, 2015; Haverfield et al., 2014; Meng, Holdcraft, Shima, Griswold, & Braun, 2005; Morita, Sasaki, Fujimoto, Furuse, & Tsukita, 1999a; Morrow et al., 2009). How spermatocyte traverses the Sertoli cell TJ without compromising its integrity is a major mystery. Smith and Braun have provided critical insights of the molecular mechanism underlying this process. They revealed that an intermediate compartment enclosing the migrating spermatocytes was formed by "new" and "old" TJs above and below the spermatocytes, respectively (Fig. 7.18) (Smith & Braun, 2012). Claudin-3 was transiently incorporated into the new TJs and then replaced by claudin-11.

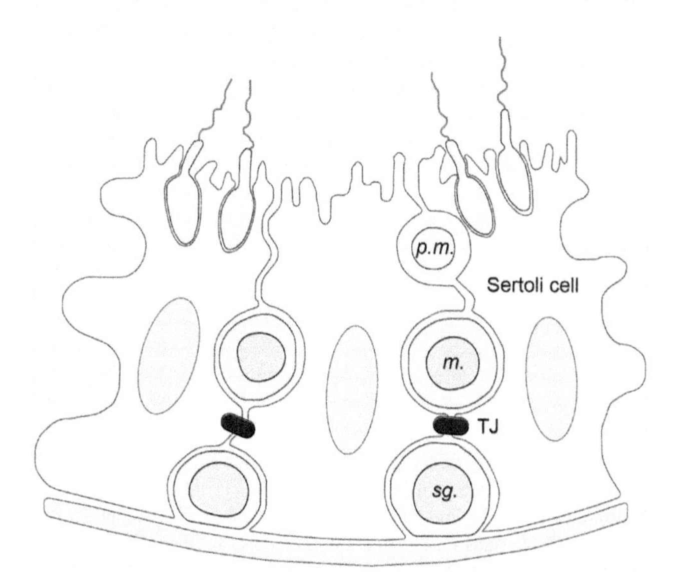

FIGURE 7.17 Schematic diagram of the seminiferous epithelium. The TJ seals adjacent Sertoli cells to create the adluminal compartment in which meiotic (m.) and postmeiotic (p.m.) germ cells reside, whereas their progenitor cells [spermatogonia (sg.)] reside in the basal compartment with free access to circulating factors.

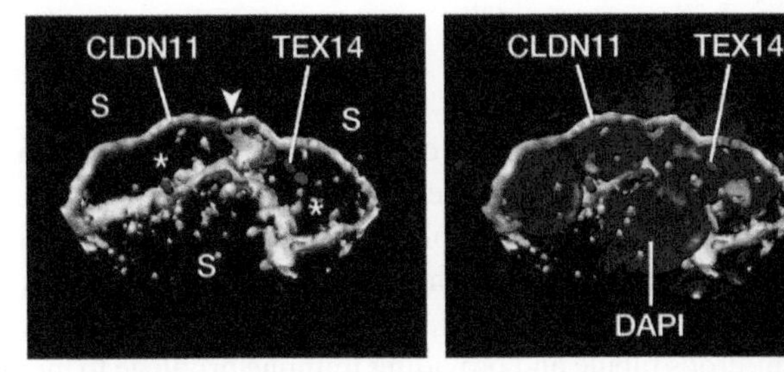

FIGURE 7.18 **Preleptotene spermatocyte migration across Sertoli cell TJ.** During germ cell migration across the Sertoli cell TJ, the leptotene spermatocyte is enclosed within a transient compartment (*Asterisk*) enclosed by a double-layer TJ that sandwiches the leptotene spermatocyte. The *arrowhead* shows the connection between TJ layers formed at the tricellular junction of three Sertoli cells (S). TEX14 *(red)* marks the cytoplasmic bridge between the leptotene spermatocytes. TEX14 localization to the tricellular junction (*arrowhead*) reveals that the cytoplasmic bridge between spermatocytes passes through the tricellular junction. *(Reproduced with permission from Smith, B.E., & Braun, R.E. (2012). Germ cell migration across Sertoli cell tight junctions. Science, 338, 798–802)*

Dissolution of the old TJs released the spermatocytes into the adluminal compartment (Smith & Braun, 2012).

7.1.6.2 Claudin-11

The male claudin-11 knockout mouse was infertile due to spermatocyte degeneration (Gow et al., 1999). The overall integrity of the BTB was retained in the knockout mouse because no autoantibody was present in the adluminal space. The TJ, however, completely disappeared from the Sertoli cell membrane (Fig. 7.19) (Gow et al., 1999). The Sertoli cells themselves lost polarity, detached from the basement membrane, and underwent sloughing in the knockout mouse testis (Mazaud-Guittot et al., 2010).

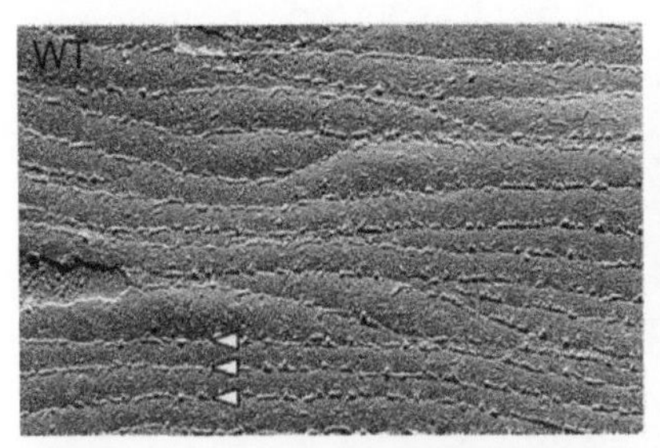

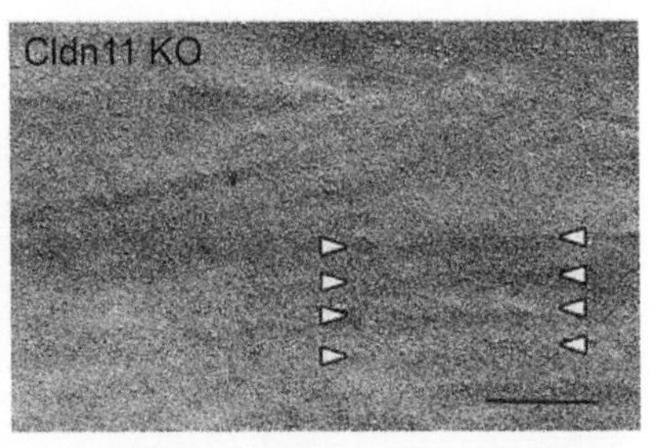

FIGURE 7.19 **TJ ultrastructural abnormality in claudin-11 knockout mouse testis.** Freeze-fracture electron micrographs showing TJ organization in wildtype (WT) and Cldn11 KO mouse testes. Note the parallel TJ strands in WT mouse Sertoli cells (left panel: *arrowheads*). TJ strands are absent from KO mouse Sertoli cells. Ripples in the fractured membrane indicate remnants of TJ (right panel: *arrowheads*). Bar: 200 nm. *(Reproduced with permission from Gow, A., Southwood, C.M., Li, J.S., Pariali, M., Riordan, G.P., Brodie, S.E., Danias, J., Bronstein, J.M., Kachar, B., & Lazzarini, R.A. (1999). CNS myelin and sertoli cell tight junction strands are absent in Osp/claudin-11 null mice. Cell, 99, 649–659)*

7.2 ENDOTHELIAL SYSTEM

7.2.1 Blood-Brain Barrier

The endothelium in the brain vasculature is different from the endothelium in the non-neural tissues by three ways: (1) the highly specialized TJ restricts the paracellular pathway; (2) the rate of transcytosis is extremely low; and (3) the lack of membrane fenestration limits the direct exchange of solutes and water. The term, blood-brain barrier (BBB) refers to the low permeability of the brain vasculature. The ensheathing pericytes and astrocytes exert regulatory control over the BBB via various secretory factors, and with the neurons form the neurovascular unit (Fig. 7.20) (Obermeier, Daneman, & Ransohoff, 2013). Claudin-1, -3, -5, and -12 are expressed by the endothelium of the brain vasculature (Liebner, Kniesel, Kalbacher, & Wolburg, 2000; Morita, Sasaki, Furuse, & Tsukita, 1999b; Ohtsuki, Yamaguchi, Katsukura, Asashima, & Terasaki, 2008; Wolburg et al., 2003).

7.2.2 Claudin-5

The claudin-5 knockout mouse died from unknown causes within hours after birth (Nitta et al., 2003). The microvessels in the knockout mouse brain were morphologically normal with intact TJs. The vascular permeabilities to small molecules (< 800 Da) were, however, markedly increased in the knockout mouse brain, indicating that claudin-5 regulates the size-dependent paracellular pathway in the BBB (Fig. 7.21) (Nitta et al., 2003). Depletion of claudin-5 in the

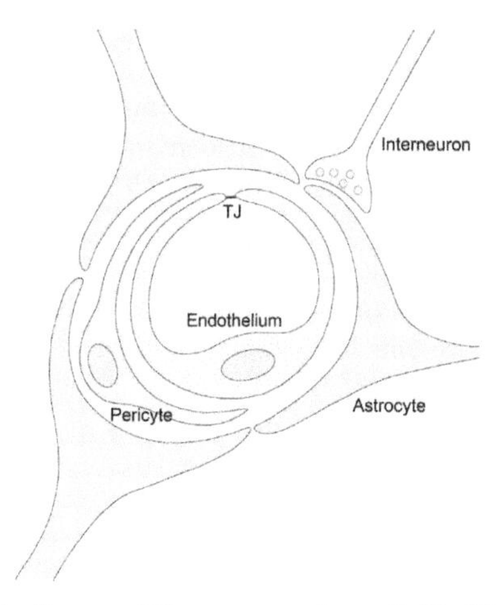

FIGURE 7.20 Schematic diagram of the neurovascular unit. The TJ seals the neighboring endothelial cells. The endothelium's abluminal membrane is covered by the basement membrane in which the pericyte is embedded. The astrocytes extend their foot processes to encircle the endothelial cell and the pericyte. The neuron interacts with the astrocyte to regulate the endothelial permeability. *(Reproduced with permission from Abbott, N.J., Ronnback, L., & Hansson, E. (2006). Astrocyte-endothelial interactions at the blood-brain barrier. Nature Reviews Neuroscience, 7, 41–53)*

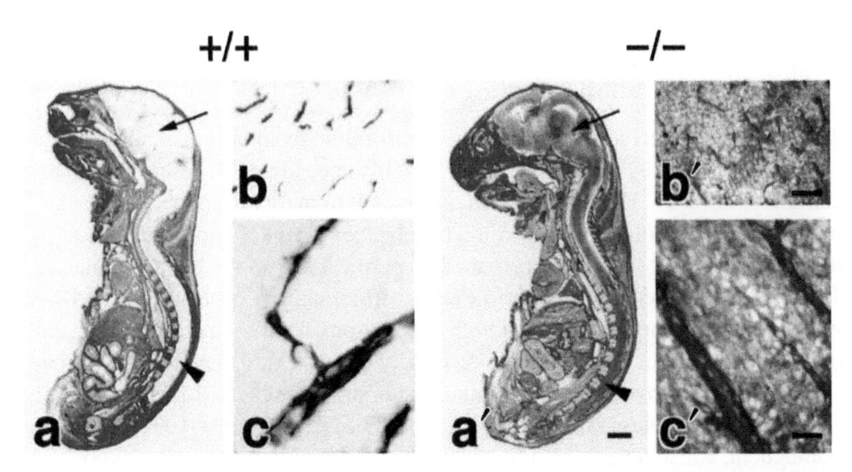

FIGURE 7.21 **Increased BBB permeability in claudin-5 knockout mouse brain.** Primary amine-reactive biotinylation reagent (443 Da) was perfused from the heart of resuscitated E18.5 embryos. After 5-min incubation, sagittal frozen sections from the whole body (a and a′) or the brain (b, b′, c, and c′) were incubated with HRP-conjugated streptavidin to visualize the biotin signal by peroxidase activity. At low magnification in the wildtype mouse (a), biotin was completely excluded from the brain (*arrow*) and the spinal cord (*arrowhead*); in sharp contrast, the $Cldn5^{-/-}$ mouse brain (a′, *arrow*) and spinal cord (a′, *arrowhead*) showed intense peroxidase activity. At high magnification, in the $Cldn5^{-/-}$ mouse (b′ and c′) but not in the wildtype mouse brain (b and c), biotin infiltrated deep into the parenchyma. Bar: (a and a′) 2 mm; (b and b′) 40 μm; (c and c′) 10 μm. *(Reproduced with permission from Nitta, T., Hata, M., Gotoh, S., Seo, Y., Sasaki, H., Hashimoto, N., Furuse, M., and Tsukita, S. (2003). Size-selective loosening of the blood-brain barrier in claudin-5-deficient mice. The Journal of Cell Biology, 161, 653–660)*

mouse brain by intravenous injection of siRNA molecules increased the vascular permeability to water and ameliorated the cerebral edema caused by traumatic brain injury (Campbell et al., 2012). Greene and coworkers have generated a transgenic inducible knockdown mouse model for claudin-5 using RNA interference to circumvent the perinatal lethality due to constitutive deletion (Greene et al., 2017). The knockdown mice developed characteristic features of schizophrenia, including impairments in learning and memory, anxiety-like behaviors, and sensorimotor gating defects. The extravasation rate of a gadolinium tracer (742 Da) was significantly higher in the knockdown mouse brain, indicating increased BBB permeability (Greene et al., 2017).

7.2.3 Tricellular Tight Junction

Angulin-1 [also known as lipolysis-stimulated lipoprotein receptor (LSR)] is enriched in the tTJ of the brain vasculature (Iwamoto, Higashi, & Furuse, 2014). The angulin-1 knockout mouse died in uterus by the developmental stage of E15.5 (Sohet et al., 2015). At E14.5, the vascular permeability to biotin (446 Da), but not to albumin (67 kDa), was significantly increased in the knockout mouse brain. The bicellular TJ components, including claudin-5 and occludin, were not affected by the deletion of angulin-1 (Sohet et al., 2015).

7.3 NERVOUS SYSTEM

7.3.1 Autotypic Tight Junction

The ensheathment of neurons with the myelin enables rapid saltatory conduction of action potentials in the nervous system. To facilitate this process, the intramyelinic space must be sealed by TJs to prevent electric current leakage. Peters first observed a radial component in the intraperiod line that appeared as rod-like thickening under transmission electron microscopy and hypothesized it to be the TJ (Peters, 1961; Peters, 1964). Freeze fracture replica electron microscopy revealed linear intramembranous strands of ~10 nm in diameter between the myelin lamellae and resembling the TJ structure found in epithelial or endothelial cells (Dermietzel, Leibstein, & Schunke, 1980; Reale, Luciano, & Spitznas, 1975). This type of TJ is formed by one cell, for example, the oligodendrocyte in the central nervous system (CNS) or the Schwann cell in the peripheral nervous system (PNS), and termed as autotypic TJ. The autotypic TJ seals the intramyelinic space and provides electrical insulation to the myelinated axon (Fig. 7.22).

7.3.2 Claudin-11

Claudin-11 is expressed by the oligodendrocyte and located in the autotypic TJs between the paranodal loops, in the outer loop, and in the Schmidt-Lantermann incisures of the CNS myelin (Gow et al., 1999; Morita et al., 1999a). The claudin-11 knockout mouse developed the neurological defects including hindlimb weakness and reduced nerve conduction velocity due to reduced insulation of the myelinated axons in the CNS (Gow et al., 1999). Freeze fracture replica electron microscopy revealed that the interlaminar TJs were absent from the CNS myelin of the knockout mouse (Fig. 7.23) (Gow et al., 1999). The knockout mouse exhibited increased action potential thresholds and activated internodal K^+ channels in the nerves of the CNS (Devaux & Gow, 2008). Notably, these changes were the most pronounced for the small axons in the optic nerve (<1 μm in diameter), which are normally ensheathed by relatively thin myelin sheaths. In contrast, the large axons in the ventral spinal cord with thicker myelin sheaths were unaffected by the deletion of claudin-11 (Devaux & Gow, 2008).

7.3.3 Claudin-19

Claudin-19 is expressed by the Schwann cell and located in the autotypic TJs between the paranodal loops, in the inner and outer mesaxons, and in the Schmidt-Lantermann incisures of the PNS myelin (Miyamoto et al., 2005). Knockout of claudin-19 in mice caused a complete loss of TJ from the inner

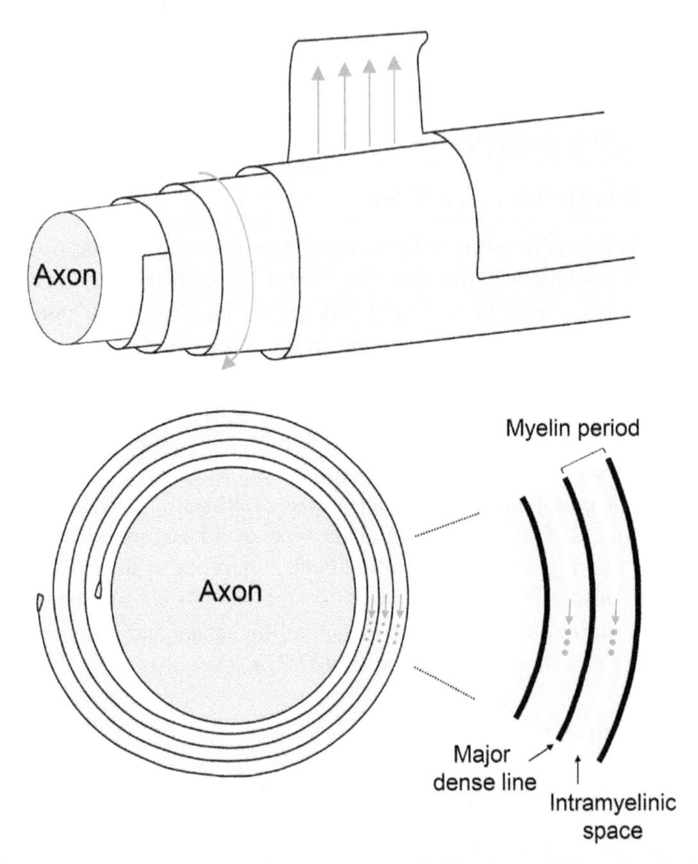

FIGURE 7.22 **Schematic diagram of an axon ensheathed by the myelin lamellae.** A simplified model unravels the myelin as membrane sheets spiraling around the axon (top). The cross-section view reveals the key features of the myelin including the major dense line and the intramyelinic space, which are organized into the myelin period (bottom). Current flow (*green arrow*) into the intramyelinic space is normally blocked by the autotypic TJ (*green dot*). *(Reproduced with permission from Gow, A., & Devaux, J. (2008). A model of tight junction function in central nervous system myelinated axons. Neuron Glia Biology, 4, 307–317)*

and outer mesaxons, in spite of grossly normal morphology of the PNS myelin. Similar to claudin-11, claudin-19 is required to electrically seal the intramyelinic space to prevent current leakage from the myelin. When claudin-19 was knocked out from the PNS, the conduction velocity of the sciatic nerve became markedly slower than that in normal animals (Miyamoto et al., 2005).

7.4 AUDITORY SYSTEM

7.4.1 Stria Vascularis

The cochlea of the inner ear has two compartments with different ionic composition (Fig. 7.24A). The perilymph of the scala vestibuli and the scala tympani

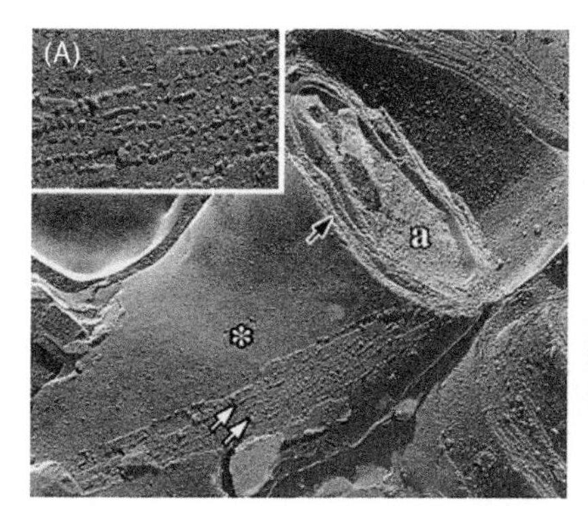

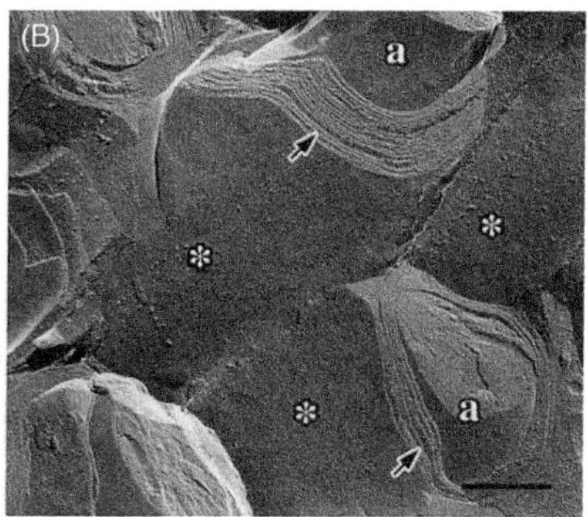

FIGURE 7.23 CNS TJ ultrastructural abnormality in claudin-11 knockout mouse. Freeze-fracture electron micrographs showing TJ organization in wildtype (A) and Cldn11 KO (B) mouse optic nerve. (A) *Black arrow*, internodal myelin; *white arrows*, TJ strands; *asterisk*, outer wrapping of the myelin sheath; *a*, axon. Inset: the TJ strands comprise linear arrays of intramembranous particles. (B) TJ strands are absent. *Black arrow*, internodal myelin; *asterisk*, outer wrapping of the myelin sheath; *a*, axon. Bar: 200 nm. *(Reproduced with permission from Gow, A., Southwood, C.M., Li, J.S., Pariali, M., Riordan, G.P., Brodie, S.E., Danias, J., Bronstein, J.M., Kachar, B., & Lazzarini, R.A. (1999). CNS myelin and sertoli cell tight junction strands are absent in Osp/claudin-11 null mice. Cell, 99, 649–659)*

is low in K^+ concentration but high in Na^+ concentration, similar to the cerebrospinal fluid (Ryan, Wickham, & Bone, 1979). The endolymph of the scala media is unusually rich in K^+ (~150 mM) and held at a positive endocochlear potential (EP) of ~ +90 mV relative to the perilymph (Sterkers, Ferrary, & Amiel, 1988). The endocochlear K^+ concentration and the EP are both indispensable for depolarizing the cochlear hair cells to transduce acoustic stimuli to electrical signals (Hudspeth, 1985). The stria vascularis is considered the primary site for the K^+ secretion from perilymph into endolymph and the development of positive EP (Sterkers et al., 1988; Tasaki & Spyropoulos, 1959). The stria vascularis consists of four epithelial cell layers: the fibrocyte, the basal cell, the intermediate cell, and the marginal cell. K^+ is actively taken up from the intrastrial space into the marginal cell through the basolateral Na/K-ATPase and Na-K-2Cl cotransporter and then secreted into the endolymph through the apical KCNQ1/KCNE1 K^+ channels. This process creates a low K^+ concentration in the intrastrial space, which facilitates the K^+ diffusion across the apical membrane of the intermediate cell via the K^+ channel Kir4.1 (Fig. 7.24B) (Ikeda & Morizono, 1989). The EP is generated by unidirectional K^+ diffusion across the apical membranes of the intermediate cell and the marginal cell as a result of two diffusion potentials (Nin et al., 2008). The TJ functions as a paracellular diffusion barrier to K^+ and prevents the dissipating of the K^+ concentration gradient through the paracellular shunt. There are two types of TJs in the stria vascularis, one made between the marginal cells and the other made between the basal cells (Fig. 7.24B). Claudin-1, -2, -3, -8, -9, -10, -12, -14, and -18 are expressed in the marginal cells, whereas the basal cells only express claudin-11 (Kitajiri et al., 2004a).

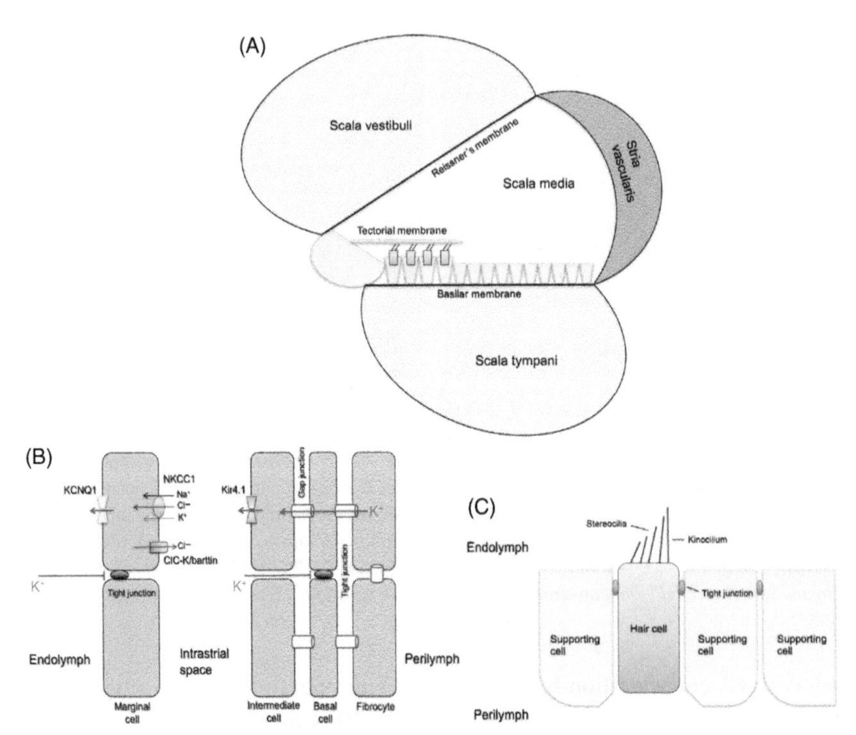

FIGURE 7.24 Schematic diagram of the cochlea. (A) The cochlea consists in two extracellular compartments with different ionic composition—the Na^+ enriched perilymph in the scala vestibuli and the scala tympani, and the K^+ enriched endolymph in the scala media. (B) The stria vascularis is a multilayered epithelium in which the TJs between the marginal cells, which face the endolymph, and the TJs between the basal cells, which face the perilymph, create an electrically isolated intrastrial space. The endocochlear potential is generated by unidirectional K^+ diffusion from the perilymph to the endolymph. (C) The TJs between the hair cell and the supporting cell or between the supporting cells form a paracellular barrier to impede K^+ permeation from the endolymph to the perilymph. *(Reproduced with permission from Hou, J. (2013). A connected tale of claudins from the renal duct to the sensory system. Tissue Barriers, 1, e24968)*

7.4.1.1 Claudin-11

Claudin-11 is located in the TJs between the basal cells in the stria vascularis (Gow et al., 2004; Kitajiri et al., 2004b). Two independent lines of claudin-11 knockout mice revealed the same phenotypes in the inner ear (Gow et al., 2004; Kitajiri et al., 2004b). The hearing thresholds of the knockout mice were elevated by 50 dB sound pressure across a wide spectrum of frequencies. While the endocochlear K^+ concentration was maintained at the normal level, the EP was markedly reduced below +30 mV in the knockout mice. Freeze fracture replica electron microscopy showed that the TJ completely disappeared from the basal cells in the knockout mouse cochlea. As a result, the paracellular permeability was markedly increased across the basal cell layer (Fig. 7.25). The paracellular

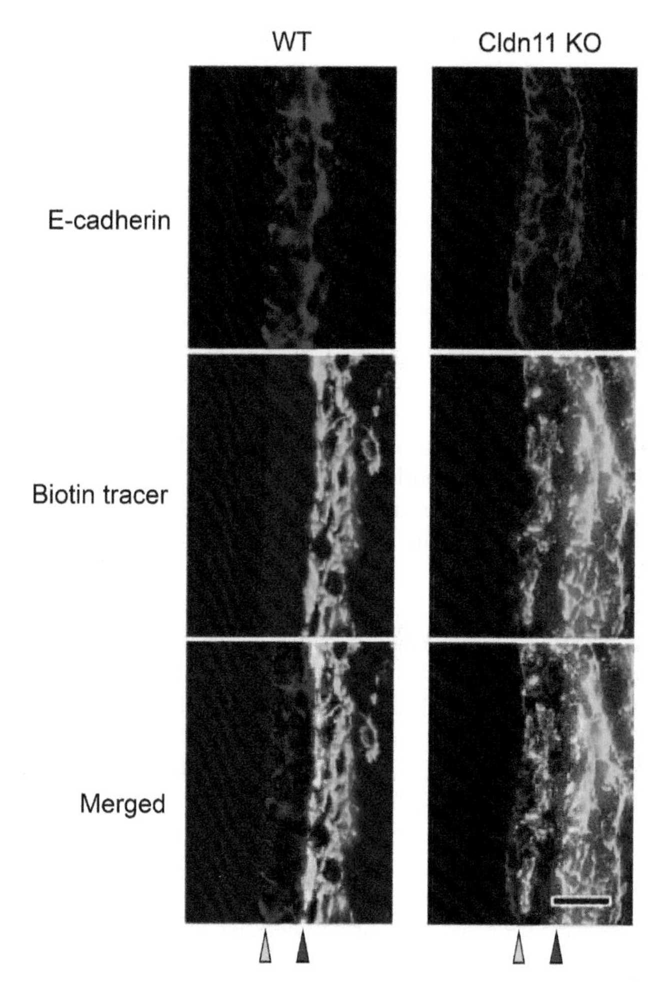

FIGURE 7.25 Increased paracellular permeability in claudin-11 knockout mouse stria vascularis. An isotonic solution containing a biotin tracer of 557 Da was injected into the perilymph from the round window of the cochlea, and after 5 min incubation, the cochlea was dissected out, fixed, sectioned, and stained with anti-E-cadherin antibody *(red)* and streptavidin *(green)* to detect the cell junction and the bound biotin, respectively. In the wildtype (WT) mouse stria vascularis, the biotin tracer diffused freely through the connective tissue until it was stopped by the basal cell layer. In the Cldn11 KO mouse cochlea, however, the biotin tracer was not stopped by the basal cell layer but reached the intrastrial space and was then stopped by the marginal cell layer. *Yellow* and *red arrowheads* indicate the marginal and basal cell layers, respectively. Bar: 20 μm. *(Reproduced with permission from Kitajiri, S.I., Furuse, M., Morita, K., Saishin-Kiuchi, Y., Kido, H., Ito, J., & Tsukita, S. (2004b). Expression patterns of claudins, tight junction adhesion molecules, in the inner ear. Hearing Research, 187, 25–34)*

barrier in the marginal cell was, however, not affected by the deletion of claudin-11. There is no evidence of cochlear pathologic abnormality, including hair cell degeneration in either knockout mouse model.

7.4.2 Hair Cells

Hearing depends upon two types of mechanosensitive hair cells: the inner hair cell (IHC) and the outer hair cell (OHC) in the organ of Corti. Both types of hair cells respond to movements of their apical stereocilia by modulation of cation influx through the mechanosensitive channels. The depolarizing currents across the apical membrane of the hair cells are carried by K^+, instead of Na^+ (Wangemann, 2006). K^+ is then taken up by the apposing supporting cells, secreted into the perilymph in the scala tympani and recycled back to the endolymph in the scala media across the stria vascularis. TJs are made between the sensory hair cell and the neighboring supporting cell or between the supporting cells (Fig. 7.24C). The hair cells and the supporting cells express claudin-1, -2, -3, -9, -10, -12, -14, and -18 (Kitajiri et al., 2004a).

7.4.2.1 Claudin-9

Nakano and coworkers performed ethylnitrosourea (ENU) induced mutagenesis on experimental mice and serendipitously found a mutant mouse model (nmf329) that developed nonsyndromic deafness due to hair cell degeneration (Nakano et al., 2009). The hearing thresholds of the mutant mice were elevated by 60 dB sound pressure at both low and high frequencies. Neither the endocochlear K^+ concentration nor the EP was affected in the mutant mouse. The perilymphatic K^+ concentration was, however, increased in the mutant mouse (Nakano et al., 2009). The hair cell degeneration was rescued when the organ of Corti from the mutant mouse was cultured in low K^+ milieu, suggesting that high perilymphatic K^+ concentration may have triggered the hair cell degeneration (Nakano et al., 2009). Positional cloning reveals that the nmf329 strain carries a missense mutation in the claudin-9 gene, which causes a phenylalanine (F) to leucine (L) substitution at the amino acid locus 35 in the first extracellular loop domain of the claudin-9 protein (Nakano et al., 2009). The first extracellular loop domain in claudin is important for making the paracellular channel pore (Chapter 4, Section 4.3). While the wildtype claudin-9 protein impeded the paracellular permeation of cations including K^+, the F35L mutation in claudin-9 abrogated its paracellular barrier function (Nakano et al., 2009). The paracellular diffusion of K^+ via the OHC TJ from the endolymph to the perilymph may increase the K^+ concentration in the perilymph, which is known to be toxic to the hair cells (Konishi & Salt, 1983).

7.4.2.2 Claudin-14

Genetic mutations in the claudin-14 gene cause autosomal recessive deafness DFNB29 in man (Wilcox et al., 2001). Similar to the claudin-9 mutant mouse (nmf329), the claudin-14 knockout mouse was deaf owing to the degeneration

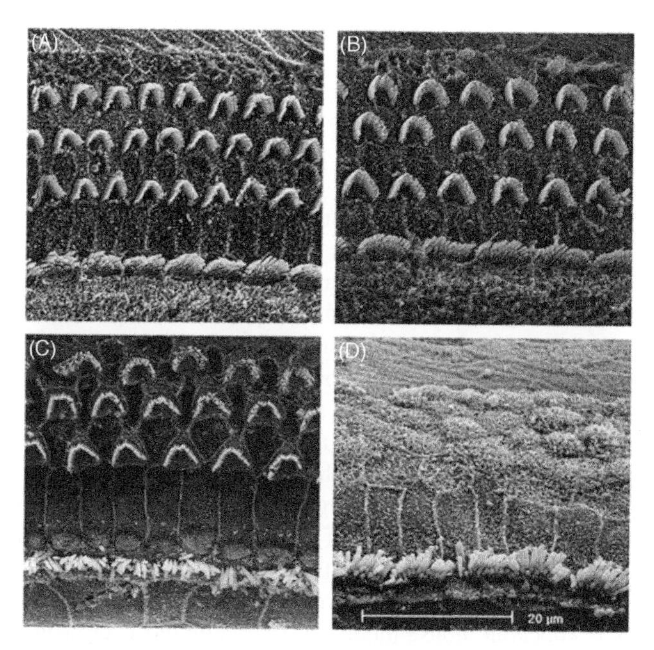

FIGURE 7.26 Hair cell degeneration in the organ of Corti of claudin-14 knockout mouse. Scanning electron micrographs of the organ of Corti reveal normally organized IHCs and OHCs at P7 in wildtype (A) and Cldn14 knockout (B) mice. By P18, IHCs and OHCs appear normal in wildtype mouse (C) but the stereocilia of OHCs are completely missing in Cldn14 knockout mouse (D). Bar: 20 μm. *(Reproduced with permission from Ben-Yosef, T., Belyantseva, I.A., Saunders, T.L., Hughes, E.D., Kawamoto, K., Van Itallie, C.M., Beyer, L.A., Halsey, K., Gardner, D.J., Wilcox, E.R., et al. (2003). Claudin 14 knockout mice, a model for autosomal recessive deafness DFNB29, are deaf due to cochlear hair cell degeneration. Human Molecular Genetics, 12, 2049–2061)*

of hair cells (Fig. 7.26) (Ben-Yosef et al., 2003). The hearing thresholds of the knockout mice were elevated by 50 dB sound pressure across all frequencies analyzed. The EP was normal in the knockout mouse cochlea. Although neither endolymphatic nor perilymphatic K^+ level has been determined for the knockout mouse, the hair cell TJs are predicted to be leakier to K^+ without claudin-14, because claudin-14 is known to reduce the paracellular cation permeability (Ben-Yosef et al., 2003).

7.4.2.3 Occludin

Whether occludin contributes to the paracellular channel is highly debatable. The occludin knockout mouse was deaf and manifested the cochlear phenotypes similar to those found in the claudin-9 and claudin-14 knockout mice (Kitajiri et al., 2014). While neither claudin-9 nor claudin-14 gene expression nor protein localization was affected by the deletion of occludin, the localization of tricellulin, a tTJ protein, was overtly altered in occludin knockout mouse hair cells, which suggests that occludin plays an essential role in the auditory function by maintaining the proper localization of tricellulin in the tTJ (Kitajiri et al., 2014).

7.4.2.4 Tricellular Tight Junction

7.4.2.4.1 Tricellulin

Tricellulin is concentrated at the tTJs of the marginal cells in the stria vascularis, and the hair cells and the supporting cells in the organ of Corti (Riazuddin et al., 2006). Genetic mutations in the tricellulin gene cause autosomal recessive deafness DFNB49 in man (Riazuddin et al., 2006). Nayak and coworkers have generated a transgenic knockin mouse model that carries a nonsense mutation orthologous to the R500X mutation found in human DFNB49 patients, resulting in the truncation of the carboxyl-terminal domain of the tricellulin protein (Nayak et al., 2013). The knockin mouse was deaf due to the degeneration of hair cells. The EP and the paracellular permeability in the stria vascularis were normal in the knockin mouse. Freeze fracture replica electron microscopy revealed conspicuous disruption of the tTJ structure in the organ of Corti from the knockin mouse, which suggests that ions, in particular K^+, may leak through the tTJ and create a toxic microenvironment for the hair cells (Nayak et al., 2013). A knockout mouse model for tricellulin has later been generated and fully recapitulated the cochlear phenotypes in the knockin mouse model (Kamitani et al., 2015).

7.4.2.4.2 Angulin-2

Angulin-2 (also known as ILDR1) is another tTJ protein found in the cochlea. Its expression pattern is very similar to that of tricellulin, including the epithelial cells in the stria vascularis and the organ of Corti (Borck et al., 2011; Morozko et al., 2015). Human genetic mutations in the angulin-2 gene cause autosomal recessive deafness DFNB42 (Borck et al., 2011). The angulin-2 knockout mouse phenocopied the tricellulin knockout mouse and developed early-onset deafness associated with rapid degeneration of the hair cells (Morozko et al., 2015). Mechanistically, the loss of angulin-2 disorganized the tTJ ultrastructure and delocalized tricellulin from the tTJ in the hair cells (Morozko et al., 2015).

REFERENCES

Ahmad, R., Chaturvedi, R., Olivares-Villagomez, D., Habib, T., Asim, M., Shivesh, P., Polk, D. B., Wilson, K. T., Washington, M. K., Van Kaer, L., et al. (2014). Targeted colonic claudin-2 expression renders resistance to epithelial injury, induces immune suppression, and protects from colitis. *Mucosal Immunology*, *7*, 1340–1353.

Arabzadeh, A., Troy, T. C., & Turksen, K. (2006). Role of the Cldn6 cytoplasmic tail domain in membrane targeting and epidermal differentiation in vivo. *Molecular and Cellular Biology*, *26*, 5876–5887.

Barker, N. (2014). Adult intestinal stem cells: critical drivers of epithelial homeostasis and regeneration. *Nature Reviews Molecular Cell Biology*, *15*, 19–33.

Ben-Yosef, T., Belyantseva, I. A., Saunders, T. L., Hughes, E. D., Kawamoto, K., Van Itallie, C. M., Beyer, L. A., Halsey, K., Gardner, D. J., Wilcox, E. R., et al. (2003). Claudin 14 knockout mice, a model for autosomal recessive deafness DFNB29, are deaf due to cochlear hair cell degeneration. *Human Molecular Genetics*, *12*, 2049–2061.

Borck, G., Ur Rehman, A., Lee, K., Pogoda, H. M., Kakar, N., von Ameln, S., Grillet, N., Hildebrand, M. S., Ahmed, Z. M., Nurnberg, G., et al. (2011). Loss-of-function mutations of ILDR1 cause autosomal-recessive hearing impairment DFNB42. *American Journal of Human Genetics*, *88*, 127–137.

Boyer, J. L. (2013). Bile formation and secretion. *Comprehensive Physiology*, *3*, 1035–1078.

Brandner, J. M., McIntyre, M., Kief, S., Wladykowski, E., & Moll, I. (2003). Expression and localization of tight junction-associated proteins in human hair follicles. *Archives of Dermatological Research*, *295*, 211–221.

Brandner, J. M., Kief, S., Wladykowski, E., Houdek, P., & Moll, I. (2006). Tight junction proteins in the skin. *Skin Pharmacology and Physiology*, *19*, 71–77.

Breiderhoff, T., Himmerkus, N., Stuiver, M., Mutig, K., Will, C., Meij, I. C., Bachmann, S., Bleich, M., Willnow, T. E., & Muller, D. (2012). Deletion of claudin-10 (Cldn10) in the thick ascending limb impairs paracellular sodium permeability and leads to hypermagnesemia and nephrocalcinosis. *Proceedings of the National Academy of Sciences of the United States of America*, *109*, 14241–14246.

Burg, M. B., & Green, N. (1973). Function of the thick ascending limb of Henle's loop. *The American Journal of Physiology*, *224*, 659–668.

Campbell, M., Hanrahan, F., Gobbo, O. L., Kelly, M. E., Kiang, A. S., Humphries, M. M., Nguyen, A. T., Ozaki, E., Keaney, J., Blau, C. W., et al. (2012). Targeted suppression of claudin-5 decreases cerebral oedema and improves cognitive outcome following traumatic brain injury. *Nature Communications*, *3*, 849.

Chakraborty, P., William Buaas, F., Sharma, M., Smith, B. E., Greenlee, A. R., Eacker, S. M., & Braun, R. E. (2014). Androgen-dependent sertoli cell tight junction remodeling is mediated by multiple tight junction components. *Molecular Endocrinology*, *28*, 1055–1072.

Daugherty, B. L., Mateescu, M., Patel, A. S., Wade, K., Kimura, S., Gonzales, L. W., Guttentag, S., Ballard, P. L., & Koval, M. (2004). Developmental regulation of claudin localization by fetal alveolar epithelial cells. *American Journal of Physiology Lung Cellular and Molecular Physiology*, *287*, L1266–1273.

Dermietzel, R., Leibstein, A. G., & Schunke, D. (1980). Interlamellar tight junctions of central myelin. II. A freeze fracture and cytochemical study on their arrangement and composition. *Cell and Tissue Research*, *213*, 95–108.

Devaux, J., & Gow, A. (2008). Tight junctions potentiate the insulative properties of small CNS myelinated axons. *The Journal of Cell Biology*, *183*, 909–921.

Dietze, R., Shihan, M., Stammler, A., Konrad, L., & Scheiner-Bobis, G. (2015). Cardiotonic steroid ouabain stimulates expression of blood-testis barrier proteins claudin-1 and -11 and formation of tight junctions in Sertoli cells. *Molecular and Cellular Endocrinology*, *405*, 1–13.

Ding, L., Lu, Z., Foreman, O., Tatum, R., Lu, Q., Renegar, R., Cao, J., & Chen, Y. H. (2012). Inflammation and disruption of the mucosal architecture in claudin-7-deficient mice. *Gastroenterology*, *142*, 305–315.

Enck, A. H., Berger, U. V., & Yu, A. S. (2001). Claudin-2 is selectively expressed in proximal nephron in mouse kidney. *American Journal of Physiology Renal Physiology*, *281*, F966–974.

Fujita, H., Chiba, H., Yokozaki, H., Sakai, N., Sugimoto, K., Wada, T., Kojima, T., Yamashita, T., & Sawada, N. (2006). Differential expression and subcellular localization of claudin-7, -8, -12,

-13, and -15 along the mouse intestine. *The Journal of Histochemistry and Cytochemistry: Official Journal of the Histochemistry Society*, *54*, 933–944.

Fujita, H., Sugimoto, K., Inatomi, S., Maeda, T., Osanai, M., Uchiyama, Y., Yamamoto, Y., Wada, T., Kojima, T., Yokozaki, H., et al. (2008). Tight junction proteins claudin-2 and -12 are critical for vitamin D-dependent Ca^{2+} absorption between enterocytes. *Molecular Biology of the Cell*, *19*, 1912–1921.

Fujita, H., Hamazaki, Y., Noda, Y., Oshima, M., & Minato, N. (2012). Claudin-4 deficiency results in urothelial hyperplasia and lethal hydronephrosis. *PLoS One*, *7*, e52272.

Furuse, M., Fujita, K., Hiiragi, T., Fujimoto, K., & Tsukita, S. (1998). Claudin-1 and -2: novel integral membrane proteins localizing at tight junctions with no sequence similarity to occludin. *The Journal of Cell Biology*, *141*, 1539–1550.

Furuse, M., Hata, M., Furuse, K., Yoshida, Y., Haratake, A., Sugitani, Y., Noda, T., Kubo, A., & Tsukita, S. (2002). Claudin-based tight junctions are crucial for the mammalian epidermal barrier: a lesson from claudin-1-deficient mice. *The Journal of Cell Biology*, *156*, 1099–1111.

Gong, Y., Himmerkus, N., Sunq, A., Milatz, S., Merkel, C., Bleich, M., & Hou, J. (2017). ILDR1 is important for paracellular water transport and urine concentration mechanism. *Proceedings of the National Academy of Sciences of the United States of America*, *114*, 5271–5276.

Gong, Y., & Hou, J. (2014). Claudin-14 underlies Ca^{++}-sensing receptor-mediated Ca^{++} metabolism via NFAT-microRNA-based mechanisms. *Journal of the American Society of Nephrology*, *25*, 745–760.

Gong, Y., Renigunta, V., Himmerkus, N., Zhang, J., Renigunta, A., Bleich, M., & Hou, J. (2012). Claudin-14 regulates renal Ca(+)(+) transport in response to CaSR signalling via a novel microRNA pathway. *The EMBO Journal*, *31*, 1999–2012.

Gong, Y., Sunq, A., Roth, R. A., & Hou, J. (2017). Inducible Expression of Claudin-1 in Glomerular Podocytes Generates Aberrant Tight Junctions and Proteinuria Through Slit Diaphragm Destabilization. *Journal of the American Society of Nephrology*, *28*(1), 106–117.

Gong, Y., Wang, J., Yang, J., Gonzales, E., Perez, R., & Hou, J. (2015). KLHL3 regulates paracellular chloride transport in the kidney by ubiquitination of claudin-8. *Proceedings of the National Academy of Sciences of the United States of America*, *112*, 4340–4345.

Gong, Y., Yu, M., Yang, J., Gonzales, E., Perez, R., Hou, M., Tripathi, P., Hering-Smith, K. S., Hamm, L. L., & Hou, J. (2014). The Cap1-claudin-4 regulatory pathway is important for renal chloride reabsorption and blood pressure regulation. *Proceedings of the National Academy of Sciences of the United States of America*, *111*, E3766–3774.

Gow, A., Davies, C., Southwood, C. M., Frolenkov, G., Chrustowski, M., Ng, L., Yamauchi, D., Marcus, D. C., & Kachar, B. (2004). Deafness in Claudin 11-null mice reveals the critical contribution of basal cell tight junctions to stria vascularis function. *The Journal of Neuroscience*, *24*, 7051–7062.

Gow, A., Southwood, C. M., Li, J. S., Pariali, M., Riordan, G. P., Brodie, S. E., Danias, J., Bronstein, J. M., Kachar, B., & Lazzarini, R. A. (1999). CNS myelin and sertoli cell tight junction strands are absent in Osp/claudin-11 null mice. *Cell*, *99*, 649–659.

Grahammer, F., Schell, C., & Huber, T. B. (2013). The podocyte slit diaphragm--from a thin grey line to a complex signalling hub. *Nature Reviews Nephrology*, *9*, 587–598.

Greene, C., Kealy, J., Humphries, M. M., Gong, Y., Hou, J., Hudson, N., Cassidy, L. M., Martiniano, R., Shashi, V., Hooper, S. R., et al. (2017). Dose-dependent expression of claudin-5 is a modifying factor in schizophrenia. *Molecular Psychiatry*, Oct 10. doi: 10.1038/mp.2017.156. [Epub ahead of print].

Greger, R. (1981). Cation selectivity of the isolated perfused cortical thick ascending limb of Henle's loop of rabbit kidney. *Pflugers Archiv*, *390*, 30–37.

Greger, R. (1985). Ion transport mechanisms in thick ascending limb of Henle's loop of mammalian nephron. *Physiological Reviews*, *65*, 760–797.

Haftek, M., Callejon, S., Sandjeu, Y., Padois, K., Falson, F., Pirot, F., Portes, P., Demarne, F., & Jannin, V. (2011). Compartmentalization of the human stratum corneum by persistent tight junction-like structures. *Experimental Dermatology*, *20*, 617–621.

Haverfield, J. T., Meachem, S. J., Nicholls, P. K., Rainczuk, K. E., Simpson, E. R., & Stanton, P. G. (2014). Differential permeability of the blood-testis barrier during reinitiation of spermatogenesis in adult male rats. *Endocrinology*, *155*, 1131–1144.

Hayashi, D., Tamura, A., Tanaka, H., Yamazaki, Y., Watanabe, S., Suzuki, K., Suzuki, K., Sentani, K., Yasui, W., Rakugi, H., et al. (2012). Deficiency of claudin-18 causes paracellular H+ leakage, upregulation of interleukin-1beta, and atrophic gastritis in mice. *Gastroenterology*, *142*, 292–304.

Holmes, J. L., Van Itallie, C. M., Rasmussen, J. E., & Anderson, J. M. (2006). Claudin profiling in the mouse during postnatal intestinal development and along the gastrointestinal tract reveals complex expression patterns. *Gene Expression Patterns*, *6*, 581–588.

Hou, J., Renigunta, A., Gomes, A. S., Hou, M., Paul, D. L., Waldegger, S., & Goodenough, D. A. (2009). Claudin-16 and claudin-19 interaction is required for their assembly into tight junctions and for renal reabsorption of magnesium. *Proceedings of the National Academy of Sciences of the United States of America*, *106*, 15350–15355.

Hou, J., Renigunta, A., Konrad, M., Gomes, A. S., Schneeberger, E. E., Paul, D. L., Waldegger, S., & Goodenough, D. A. (2008). Claudin-16 and claudin-19 interact and form a cation-selective tight junction complex. *The Journal of Clinical Investigation*, *118*, 619–628.

Hou, J., Renigunta, A., Yang, J., & Waldegger, S. (2010). Claudin-4 forms paracellular chloride channel in the kidney and requires claudin-8 for tight junction localization. *Proceedings of the National Academy of Sciences of the United States of America*, *107*, 18010–18015.

Hou, J., Shan, Q., Wang, T., Gomes, A. S., Yan, Q., Paul, D. L., Bleich, M., & Goodenough, D. A. (2007). Transgenic RNAi depletion of claudin-16 and the renal handling of magnesium. *The Journal of Biological Chemistry*, *282*, 17114–17122.

Hudspeth, A. J. (1985). The cellular basis of hearing: the biophysics of hair cells. *Science*, *230*, 745–752.

Igawa, S., Kishibe, M., Murakami, M., Honma, M., Takahashi, H., Iizuka, H., & Ishida-Yamamoto, A. (2011). Tight junctions in the stratum corneum explain spatial differences in corneodesmosome degradation. *Experimental Dermatology*, *20*, 53–57.

Ikeda, K., & Morizono, T. (1989). Electrochemical profiles for monovalent ions in the stria vascularis: cellular model of ion transport mechanisms. *Hearing Research*, *39*, 279–286.

Iwamoto, N., Higashi, T., & Furuse, M. (2014). Localization of angulin-1/LSR and tricellulin at tricellular contacts of brain and retinal endothelial cells in vivo. *Cell Structure and Function*, *39*, 1–8.

Kage, H., Flodby, P., Gao, D., Kim, Y. H., Marconett, C. N., DeMaio, L., Kim, K. J., Crandall, E. D., & Borok, Z. (2014). Claudin 4 knockout mice: normal physiological phenotype with increased susceptibility to lung injury. *American Journal of Physiology Lung Cellular and Molecular Physiology*, *307*, L524–536.

Kamitani, T., Sakaguchi, H., Tamura, A., Miyashita, T., Yamazaki, Y., Tokumasu, R., Inamoto, R., Matsubara, A., Mori, N., Hisa, Y., et al. (2015). Deletion of tricellulin causes progressive hearing loss associated with degeneration of cochlear hair cells. *Scientific Reports*, *5*, 18402.

Kestila, M., Lenkkeri, U., Mannikko, M., Lamerdin, J., McCready, P., Putaala, H., Ruotsalainen, V., Morita, T., Nissinen, M., Herva, R., et al. (1998). Positionally cloned gene for a novel glomerular protein--nephrin--is mutated in congenital nephrotic syndrome. *Molecular Cell*, *1*, 575–582.

Kim, T. H., & Shivdasani, R. A. (2016). Stomach development, stem cells and disease. *Development*, *143*, 554–565.

Kirschner, N., Poetzl, C., von den Driesch, P., Wladykowski, E., Moll, I., Behne, M. J., & Brandner, J. M. (2009). Alteration of tight junction proteins is an early event in psoriasis: putative involvement of proinflammatory cytokines. *The American Journal of Pathology*, *175*, 1095–1106.

Kitajiri, S. I., Furuse, M., Morita, K., Saishin-Kiuchi, Y., Kido, H., Ito, J., & Tsukita, S. (2004a). Expression patterns of claudins, tight junction adhesion molecules, in the inner ear. *Hearing Research*, *187*, 25–34.

Kitajiri, S., Katsuno, T., Sasaki, H., Ito, J., Furuse, M., & Tsukita, S. (2014). Deafness in occludin-deficient mice with dislocation of tricellulin and progressive apoptosis of the hair cells. *Biology Open*, *3*, 759–766.

Kitajiri, S., Miyamoto, T., Mineharu, A., Sonoda, N., Furuse, K., Hata, M., Sasaki, H., Mori, Y., Kubota, T., Ito, J., et al. (2004b). Compartmentalization established by claudin-11-based tight junctions in stria vascularis is required for hearing through generation of endocochlear potential. *Journal of Cell Science*, *117*, 5087–5096.

Kiuchi-Saishin, Y., Gotoh, S., Furuse, M., Takasuga, A., Tano, Y., & Tsukita, S. (2002). Differential expression patterns of claudins, tight junction membrane proteins, in mouse nephron segments. *Journal of the American Society of Nephrology*, *13*, 875–886.

Koda, R., Zhao, L., Yaoita, E., Yoshida, Y., Tsukita, S., Tamura, A., Nameta, M., Zhang, Y., Fujinaka, H., Magdeldin, S., et al. (2011). Novel expression of claudin-5 in glomerular podocytes. *Cell and Tissue Research*, *343*, 637–648.

Konishi, T., & Salt, A. N. (1983). Electrochemical profile for potassium ions across the cochlear hair cell membranes of normal and noise-exposed guinea pigs. *Hearing Research*, *11*, 219–233.

Konrad, M., Schaller, A., Seelow, D., Pandey, A. V., Waldegger, S., Lesslauer, A., Vitzthum, H., Suzuki, Y., Luk, J. M., Becker, C., et al. (2006). Mutations in the tight-junction gene claudin 19 (CLDN19) are associated with renal magnesium wasting, renal failure, and severe ocular involvement. *American Journal of Human Genetics*, *79*, 949–957.

Krug, S. M., Gunzel, D., Conrad, M. P., Rosenthal, R., Fromm, A., Amasheh, S., Schulzke, J. D., & Fromm, M. (2012). Claudin-17 forms tight junction channels with distinct anion selectivity. *Cellular and Molecular Life Sciences*, *69*, 2765–2778.

Kubo, A., Nagao, K., Yokouchi, M., Sasaki, H., & Amagai, M. (2009). External antigen uptake by Langerhans cells with reorganization of epidermal tight junction barriers. *The Journal of Experimental Medicine*, *206*, 2937–2946.

LaFemina, M. J., Rokkam, D., Chandrasena, A., Pan, J., Bajaj, A., Johnson, M., & Frank, J. A. (2010). Keratinocyte growth factor enhances barrier function without altering claudin expression in primary alveolar epithelial cells. *American Journal of Physiology Lung Cellular and Molecular Physiology*, *299*, L724–734.

LaFemina, M. J., Sutherland, K. M., Bentley, T., Gonzales, L. W., Allen, L., Chapin, C. J., Rokkam, D., Sweerus, K. A., Dobbs, L. G., Ballard, P. L., et al. (2014). Claudin-18 deficiency results in alveolar barrier dysfunction and impaired alveologenesis in mice. *American Journal of Respiratory Cell and Molecular Biology*, *51*, 550–558.

Lameris, A. L., Huybers, S., Kaukinen, K., Makela, T. H., Bindels, R. J., Hoenderop, J. G., & Nevalainen, P. I. (2013). Expression profiling of claudins in the human gastrointestinal tract in health and during inflammatory bowel disease. *Scandinavian Journal of Gastroenterology*, *48*, 58–69.

Laukoetter, M. G., Nava, P., Lee, W. Y., Severson, E. A., Capaldo, C. T., Babbin, B. A., Williams, I. R., Koval, M., Peatman, E., Campbell, J. A., et al. (2007). JAM-A regulates permeability and inflammation in the intestine in vivo. *The Journal of Experimental Medicine*, *204*, 3067–3076.

Li, G., Flodby, P., Luo, J., Kage, H., Sipos, A., Gao, D., Ji, Y., Beard, L. L., Marconett, C. N., DeMaio, L., et al. (2014). Knockout mice reveal key roles for claudin 18 in alveolar barrier properties and fluid homeostasis. *American Journal of Respiratory Cell and Molecular Biology*, *51*, 210–222.

Li, W. Y., Huey, C. L., & Yu, A. S. (2004). Expression of claudin-7 and -8 along the mouse nephron. *American Journal of Physiology Renal Physiology*, *286*, F1063–1071.

Liebner, S., Kniesel, U., Kalbacher, H., & Wolburg, H. (2000). Correlation of tight junction morphology with the expression of tight junction proteins in blood-brain barrier endothelial cells. *European Journal of Cell Biology*, *79*, 707–717.

Marasco, W. A., Phan, S. H., Krutzsch, H., Showell, H. J., Feltner, D. E., Nairn, R., Becker, E. L., & Ward, P. A. (1984). Purification and identification of formyl-methionyl-leucyl-phenylalanine as the major peptide neutrophil chemotactic factor produced by *Escherichia coli*. *The Journal of Biological Chemistry*, *259*, 5430–5439.

Marcial, M. A., Carlson, S. L., & Madara, J. L. (1984). Partitioning of paracellular conductance along the ileal crypt-villus axis: a hypothesis based on structural analysis with detailed consideration of tight junction structure-function relationships. *The Journal of Membrane Biology*, *80*, 59–70.

Matsumoto, K., Imasato, M., Yamazaki, Y., Tanaka, H., Watanabe, M., Eguchi, H., Nagano, H., Hikita, H., Tatsumi, T., Takehara, T., et al. (2014). Claudin 2 deficiency reduces bile flow and increases susceptibility to cholesterol gallstone disease in mice. *Gastroenterology*, *147*, 1134–1145 e1110.

Matthay, M. A., Folkesson, H. G., & Clerici, C. (2002). Lung epithelial fluid transport and the resolution of pulmonary edema. *Physiological Reviews*, *82*, 569–600.

Mazaud-Guittot, S., Meugnier, E., Pesenti, S., Wu, X., Vidal, H., Gow, A., & Le Magueresse-Battistoni, B. (2010). Claudin 11 deficiency in mice results in loss of the Sertoli cell epithelial phenotype in the testis. *Biology of Reproduction*, *82*, 202–213.

Meng, J., Holdcraft, R. W., Shima, J. E., Griswold, M. D., & Braun, R. E. (2005). Androgens regulate the permeability of the blood-testis barrier. *Proceedings of the National Academy of Sciences of the United States of America*, *102*, 16696–16700.

Merad, M., Ginhoux, F., & Collin, M. (2008). Origin, homeostasis and function of Langerhans cells and other langerin-expressing dendritic cells. *Nature Reviews Immunology*, *8*, 935–947.

Miyamoto, T., Morita, K., Takemoto, D., Takeuchi, K., Kitano, Y., Miyakawa, T., Nakayama, K., Okamura, Y., Sasaki, H., Miyachi, Y., et al. (2005). Tight junctions in Schwann cells of peripheral myelinated axons: a lesson from claudin-19-deficient mice. *The Journal of Cell Biology*, *169*, 527–538.

Morita, K., Sasaki, H., Fujimoto, K., Furuse, M., & Tsukita, S. (1999a). Claudin-11/OSP-based tight junctions of myelin sheaths in brain and Sertoli cells in testis. *The Journal of Cell Biology*, *145*, 579–588.

Morita, K., Sasaki, H., Furuse, M., & Tsukita, S. (1999b). Endothelial claudin: claudin-5/TMVCF constitutes tight junction strands in endothelial cells. *The Journal of Cell Biology*, *147*, 185–194.

Morita, K., Tsukita, S., & Miyachi, Y. (2004). Tight junction-associated proteins (occludin, ZO-1, claudin-1, claudin-4) in squamous cell carcinoma and Bowen's disease. *The British Journal of Dermatology*, *151*, 328–334.

Morozko, E. L., Nishio, A., Ingham, N. J., Chandra, R., Fitzgerald, T., Martelletti, E., Borck, G., Wilson, E., Riordan, G. P., Wangemann, P., et al. (2015). ILDR1 null mice, a model of human deafness DFNB42, show structural aberrations of tricellular tight junctions and degeneration of auditory hair cells. *Human Molecular Genetics*, *24*, 609–624.

Morrow, C. M., Tyagi, G., Simon, L., Carnes, K., Murphy, K. M., Cooke, P. S., Hofmann, M. C., & Hess, R. A. (2009). Claudin 5 expression in mouse seminiferous epithelium is dependent upon the transcription factor ets variant 5 and contributes to blood-testis barrier function. *Biology of Reproduction*, *81*, 871–879.

Muto, S., Hata, M., Taniguchi, J., Tsuruoka, S., Moriwaki, K., Saitou, M., Furuse, K., Sasaki, H., Fujimura, A., Imai, M., et al. (2010). Claudin-2-deficient mice are defective in the leaky and cation-selective paracellular permeability properties of renal proximal tubules. *Proceedings of the National Academy of Sciences of the United States of America*, *107*, 8011–8016.

Nakano, Y., Kim, S. H., Kim, H. M., Sanneman, J. D., Zhang, Y., Smith, R. J., Marcus, D. C., Wangemann, P., Nessler, R. A., & Banfi, B. (2009). A claudin-9-based ion permeability barrier is essential for hearing. *PLoS Genetics*, *5*, e1000610.

Nayak, G., Lee, S. I., Yousaf, R., Edelmann, S. E., Trincot, C., Van Itallie, C. M., Sinha, G. P., Rafeeq, M., Jones, S. M., Belyantseva, I. A., et al. (2013). Tricellulin deficiency affects tight junction architecture and cochlear hair cells. *The Journal of Clinical Investigation*, *123*, 4036–4049.

Niimi, T., Nagashima, K., Ward, J. M., Minoo, P., Zimonjic, D. B., Popescu, N. C., & Kimura, S. (2001). claudin-18, a novel downstream target gene for the T/EBP/NKX2.1 homeodomain transcription factor, encodes lung- and stomach-specific isoforms through alternative splicing. *Molecular and Cellular Biology*, *21*, 7380–7390.

Nin, F., Hibino, H., Doi, K., Suzuki, T., Hisa, Y., & Kurachi, Y. (2008). The endocochlear potential depends on two K^+ diffusion potentials and an electrical barrier in the stria vascularis of the inner ear. *Proceedings of the National Academy of Sciences of the United States of America*, *105*, 1751–1756.

Nitta, T., Hata, M., Gotoh, S., Seo, Y., Sasaki, H., Hashimoto, N., Furuse, M., & Tsukita, S. (2003). Size-selective loosening of the blood-brain barrier in claudin-5-deficient mice. *The Journal of Cell Biology*, *161*, 653–660.

Noah, T. K., Donahue, B., & Shroyer, N. F. (2011). Intestinal development and differentiation. *Experimental Cell Research*, *317*, 2702–2710.

Obermeier, B., Daneman, R., & Ransohoff, R. M. (2013). Development, maintenance and disruption of the blood-brain barrier. *Nature Medicine*, *19*, 1584–1596.

Ohtsuki, S., Yamaguchi, H., Katsukura, Y., Asashima, T., & Terasaki, T. (2008). mRNA expression levels of tight junction protein genes in mouse brain capillary endothelial cells highly purified by magnetic cell sorting. *Journal of Neurochemistry*, *104*, 147–154.

O'Neil, R. G., & Sansom, S. C. (1984). Electrophysiological properties of cellular and paracellular conductive pathways of the rabbit cortical collecting duct. *The Journal of Membrane Biology*, *82*, 281–295.

Pei, L., Solis, G., Nguyen, M. T., Kamat, N., Magenheimer, L., Zhuo, M., Li, J., Curry, J., McDonough, A. A., Fields, T. A., et al. (2016). Paracellular epithelial sodium transport maximizes energy efficiency in the kidney. *The Journal of Clinical Investigation*, *126*, 2509–2518.

Peltonen, S., Riehokainen, J., Pummi, K., & Peltonen, J. (2007). Tight junction components occludin, ZO-1, and claudin-1, -4 and -5 in active and healing psoriasis. *The British Journal of Dermatology*, *156*, 466–472.

Peters, A. (1961). A radial component of central myelin sheaths. *The Journal of Biophysical and Biochemical Cytology*, *11*, 733–735.

Peters, A. (1964). Further observations on the structure of myelin sheaths in the central nervous system. *The Journal of Cell Biology*, *20*, 281–296.

Pope, J. L., Bhat, A. A., Sharma, A., Ahmad, R., Krishnan, M., Washington, M. K., Beauchamp, R. D., Singh, A. B., & Dhawan, P. (2014). Claudin-1 regulates intestinal epithelial homeostasis through the modulation of Notch-signalling. *Gut*, *63*, 622–634.

Preisig, P. A., & Berry, C. A. (1985). Evidence for transcellular osmotic water flow in rat proximal tubules. *The American Journal of Physiology*, *249*, F124–131.

Quaggin, S. E., & Kreidberg, J. A. (2008). Development of the renal glomerulus: good neighbors and good fences. *Development*, *135*, 609–620.

Rahner, C., Mitic, L. L., & Anderson, J. M. (2001). Heterogeneity in expression and subcellular localization of claudins 2, 3, 4, and 5 in the rat liver, pancreas, and gut. *Gastroenterology*, *120*, 411–422.

Rao, R. K., & Samak, G. (2013). Bile duct epithelial tight junctions and barrier function. *Tissue Barriers*, *1*, e25718.

Reale, E., Luciano, L., & Spitznas, M. (1975). Zonulae occludentes of the myelin lamellae in the nerve fibre layer of the retina and in the optic nerve of the rabbit: a demonstration by the freeze-fracture method. *Journal of Neurocytology*, *4*, 131–140.

Rector, F. C., Jr. (1983). Sodium, bicarbonate, and chloride absorption by the proximal tubule. *The American Journal of Physiology*, *244*, F461–471.

Riazuddin, S., Ahmed, Z. M., Fanning, A. S., Lagziel, A., Kitajiri, S., Ramzan, K., Khan, S. N., Chattaraj, P., Friedman, P. L., Anderson, J. M., et al. (2006). Tricellulin is a tight-junction protein necessary for hearing. *American Journal of Human Genetics*, *79*, 1040–1051.

Ryan, A. F., Wickham, M. G., & Bone, R. C. (1979). Element content of intracochlear fluids, outer hair cells, and stria vascularis as determined by energy-dispersive roentgen ray analysis. *Otolaryngology and Head and Neck Surgery*, *87*, 659–665.

Saitou, M., Furuse, M., Sasaki, H., Schulzke, J. D., Fromm, M., Takano, H., Noda, T., & Tsukita, S. (2000). Complex phenotype of mice lacking occludin, a component of tight junction strands. *Molecular Biology of the Cell*, *11*, 4131–4142.

Sansom, S. C., Weinman, E. J., & O'Neil, R. G. (1984). Microelectrode assessment of chloride-conductive properties of cortical collecting duct. *The American Journal of Physiology*, *247*, F291–302.

Schnermann, J., Chou, C. L., Ma, T., Traynor, T., Knepper, M. A., & Verkman, A. S. (1998). Defective proximal tubular fluid reabsorption in transgenic aquaporin-1 null mice. *Proceedings of the National Academy of Sciences of the United States of America*, *95*, 9660–9664.

Schulzke, J. D., Gitter, A. H., Mankertz, J., Spiegel, S., Seidler, U., Amasheh, S., Saitou, M., Tsukita, S., & Fromm, M. (2005). Epithelial transport and barrier function in occludin-deficient mice. *Biochimica et Biophysica Acta*, *1669*, 34–42.

Schwarz, K., Simons, M., Reiser, J., Saleem, M. A., Faul, C., Kriz, W., Shaw, A. S., Holzman, L. B., & Mundel, P. (2001). Podocin, a raft-associated component of the glomerular slit diaphragm, interacts with CD2AP and nephrin. *The Journal of Clinical Investigation*, *108*, 1621–1629.

Simon, D. B., Lu, Y., Choate, K. A., Velazquez, H., Al-Sabban, E., Praga, M., Casari, G., Bettinelli, A., Colussi, G., Rodriguez-Soriano, J., et al. (1999). Paracellin-1, a renal tight junction protein required for paracellular Mg^{2+} resorption. *Science*, *285*, 103–106.

Smith, H. W. (1959). The fate of sodium and water in the renal tubules. *Bulletin of the New York Academy of Medicine*, *35*, 293–316.

Smith, B. E., & Braun, R. E. (2012). Germ cell migration across Sertoli cell tight junctions. *Science*, *338*, 798–802.

Sohet, F., Lin, C., Munji, R. N., Lee, S. Y., Ruderisch, N., Soung, A., Arnold, T. D., Derugin, N., Vexler, Z. S., Yen, F. T., et al. (2015). LSR/angulin-1 is a tricellular tight junction protein involved in blood-brain barrier formation. *The Journal of Cell Biology*, *208*, 703–711.

Sterkers, O., Ferrary, E., & Amiel, C. (1988). Production of inner ear fluids. *Physiological Reviews*, *68*, 1083–1128.

Sugawara, T., Iwamoto, N., Akashi, M., Kojima, T., Hisatsune, J., Sugai, M., & Furuse, M. (2013). Tight junction dysfunction in the stratum granulosum leads to aberrant stratum corneum barrier function in claudin-1-deficient mice. *Journal of Dermatological Science*, *70*, 12–18.

Tamura, A., Hayashi, H., Imasato, M., Yamazaki, Y., Hagiwara, A., Wada, M., Noda, T., Watanabe, M., Suzuki, Y., & Tsukita, S. (2011). Loss of claudin-15, but not claudin-2, causes Na^+ deficiency and glucose malabsorption in mouse small intestine. *Gastroenterology*, *140*, 913–923.

Tamura, A., Kitano, Y., Hata, M., Katsuno, T., Moriwaki, K., Sasaki, H., Hayashi, H., Suzuki, Y., Noda, T., Furuse, M., et al. (2008). Megaintestine in claudin-15-deficient mice. *Gastroenterology*, *134*, 523–534.

Tanaka, H., Takechi, M., Kiyonari, H., Shioi, G., Tamura, A., & Tsukita, S. (2015). Intestinal deletion of claudin-7 enhances paracellular organic solute flux and initiates colonic inflammation in mice. *Gut, 64*, 1529–1538.

Tasaki, I., & Spyropoulos, C. S. (1959). Stria vascularis as source of endocochlear potential. *Journal of Neurophysiology, 22*, 149–155.

Tatum, R., Zhang, Y., Salleng, K., Lu, Z., Lin, J. J., Lu, Q., Jeansonne, B. G., Ding, L., & Chen, Y. H. (2010). Renal salt wasting and chronic dehydration in claudin-7-deficient mice. *American Journal of Physiology Renal Physiology, 298*, F24–34.

Telgenhoff, D., Ramsay, S., Hilz, S., Slusarewicz, P., & Shroot, B. (2008). Claudin 2 mRNA and protein are present in human keratinocytes and may be regulated by all-trans-retinoic acid. *Skin Pharmacology and Physiology, 21*, 211–217.

Thorleifsson, G., Holm, H., Edvardsson, V., Walters, G. B., Styrkarsdottir, U., Gudbjartsson, D. F., Sulem, P., Halldorsson, B. V., de Vegt, F., d'Ancona, F. C., et al. (2009). Sequence variants in the CLDN14 gene associate with kidney stones and bone mineral density. *Nature Genetics, 41*, 926–930.

Treyer, A., & Musch, A. (2013). Hepatocyte polarity. *Comprehensive Physiology, 3*, 243–287.

Troy, T. C., Arabzadeh, A., Lariviere, N. M., Enikanolaiye, A., & Turksen, K. (2009). Dermatitis and aging-related barrier dysfunction in transgenic mice overexpressing an epidermal-targeted claudin 6 tail deletion mutant. *PLoS One, 4*, e7814.

Troy, T. C., Rahbar, R., Arabzadeh, A., Cheung, R. M., & Turksen, K. (2005). Delayed epidermal permeability barrier formation and hair follicle aberrations in Inv-Cldn6 mice. *Mechanisms of Development, 122*, 805–819.

Tsai, P. Y., Zhang, B., He, W. Q., Zha, J. M., Odenwald, M. A., Singh, G., Tamura, A., Shen, L., Sailer, A., Yeruva, S., et al. (2017). IL-22 upregulates epithelial claudin-2 to drive diarrhea and enteric pathogen clearance. *Cell Host & Microbe, 21*, 671–681 e674.

Turksen, K., & Troy, T. C. (2002). Permeability barrier dysfunction in transgenic mice overexpressing claudin 6. *Development, 129*, 1775–1784.

Van Itallie, C. M., Rogan, S., Yu, A., Vidal, L. S., Holmes, J., & Anderson, J. M. (2006). Two splice variants of claudin-10 in the kidney create paracellular pores with different ion selectivities. *American Journal of Physiology Renal Physiology, 291*, F1288–1299.

Wada, M., Tamura, A., Takahashi, N., & Tsukita, S. (2013). Loss of claudins 2 and 15 from mice causes defects in paracellular Na^+ flow and nutrient transport in gut and leads to death from malnutrition. *Gastroenterology, 144*, 369–380.

Wang, F., Daugherty, B., Keise, L. L., Wei, Z., Foley, J. P., Savani, R. C., & Koval, M. (2003). Heterogeneity of claudin expression by alveolar epithelial cells. *American Journal of Respiratory Cell and Molecular Biology, 29*, 62–70.

Wangemann, P. (2006). Supporting sensory transduction: cochlear fluid homeostasis and the endocochlear potential. *The Journal of Physiology, 576*, 11–21.

Watson, R. E., Poddar, R., Walker, J. M., McGuill, I., Hoare, L. M., Griffiths, C. E., & O'Neill, C. A. (2007). Altered claudin expression is a feature of chronic plaque psoriasis. *The Journal of Pathology, 212*, 450–458.

Wilcox, E. R., Burton, Q. L., Naz, S., Riazuddin, S., Smith, T. N., Ploplis, B., Belyantseva, I., Ben-Yosef, T., Liburd, N. A., Morell, R. J., et al. (2001). Mutations in the gene encoding tight junction claudin-14 cause autosomal recessive deafness DFNB29. *Cell, 104*, 165–172.

Will, C., Breiderhoff, T., Thumfart, J., Stuiver, M., Kopplin, K., Sommer, K., Gunzel, D., Querfeld, U., Meij, I. C., Shan, Q., et al. (2010). Targeted deletion of murine Cldn16 identifies extra- and intrarenal compensatory mechanisms of Ca^{2+} and Mg^{2+} wasting. *American Journal of Physiology Renal Physiology, 298*, F1152–1161.

Wolburg, H., Wolburg-Buchholz, K., Kraus, J., Rascher-Eggstein, G., Liebner, S., Hamm, S., Duffner, F., Grote, E. H., Risau, W., & Engelhardt, B. (2003). Localization of claudin-3 in tight junctions of the blood-brain barrier is selectively lost during experimental autoimmune encephalomyelitis and human glioblastoma multiforme. *Acta Neuropathologica*, *105*, 586–592.

Wray, C., Mao, Y., Pan, J., Chandrasena, A., Piasta, F., & Frank, J. A. (2009). Claudin-4 augments alveolar epithelial barrier function and is induced in acute lung injury. *American Journal of Physiology Lung Cellular and Molecular Physiology*, *297*, L219–227.

Yokouchi, M., Atsugi, T., Logtestijn, M. V., Tanaka, R. J., Kajimura, M., Suematsu, M., Furuse, M., Amagai, M., & Kubo, A. (2016). Epidermal cell turnover across tight junctions based on Kelvin's tetrakaidecahedron cell shape. *eLife*, *5*, e19593.

Yoshida, K., Yokouchi, M., Nagao, K., Ishii, K., Amagai, M., & Kubo, A. (2013). Functional tight junction barrier localizes in the second layer of the stratum granulosum of human epidermis. *Journal of Dermatological Science*, *71*, 89–99.

Yu, A. S., Cheng, M. H., Angelow, S., Gunzel, D., Kanzawa, S. A., Schneeberger, E. E., Fromm, M., & Coalson, R. D. (2009). Molecular basis for cation selectivity in claudin-2-based paracellular pores: identification of an electrostatic interaction site. *The Journal of General Physiology*, *133*, 111–127.

Zhao, L., Yaoita, E., Nameta, M., Zhang, Y., Cuellar, L. M., Fujinaka, H., Xu, B., Yoshida, Y., Hatakeyama, K., & Yamamoto, T. (2008). Claudin-6 localized in tight junctions of rat podocytes. *American Journal of Physiology Regulatory, Integrative and Comparative Physiology*, *294*, R1856–1862.

Chapter 8

Paracellular Channel in Human Disease

8.1 GENETIC BASIS OF HUMAN DISEASE

Prior to the genetic era, many efforts have been made to search for the factors that can allow distinguishing diseased subjects from healthy subjects. These factors often show potent biological effects and their measurement and replacement have proven critical in the diagnosis and treatment of human diseases. However, discerning whether they are causally related to the pathogenic processes of disease, or merely secondary consequences of disease, has turned out to be extremely difficult. The genetic approaches have the capacity to elucidate the fundamental pathogenic mechanisms for virtually every disease. By comparing genetic variations between diseased subjects and healthy subjects, geneticists can reliably identify the causal genes important for various disease traits. In general, human traits attributable to single-gene variation, known as Mendelian traits, are inherited via three routes of transmission: autosomal dominant, autosomal recessive, or sex-linked. Genetic inheritance of traits can be related to the exact physical positions on human chromosomes, which establishes the biological basis for linkage analysis and positional cloning (Botstein, White, Skolnick, & Davis, 1980; Wensink, Finnegan, Donelson, & Hogness, 1974).

8.2 DISEASE CAUSED BY MUTATION IN CLAUDIN

8.2.1 Claudin-1

8.2.1.1 Neonatal Ichthyosis and Sclerosing Cholangitis (NISCH) Syndrome

The claudin-1 gene is linked to NISCH syndrome in man (OMIM #607626) (Table 8.1). Baala and coworkers first described a novel autosomal recessive syndrome characterized by ichthyosis, scalp hypotrichosis, scarring alopecia, sclerosing cholangitis, and leukocyte vacuolization in four affected individuals from two small inbred Moroccan kindreds (Baala et al., 2002). They mapped NISCH to a 16.2-cM interval on chromosome 3q27-q28 by homozygosity mapping. From the linked interval, Hadj-Rabia and coworkers identified a 2-bp deletion in exon 1 of the claudin-1 gene (c.200delTT), which resulted in a premature stop codon and the absence of claudin-1 protein from the skin and the liver (Hadj-Rabia et al., 2004). Two new mutations in claudin-1 have later been

The Paracellular Channel. http://dx.doi.org/10.1016/B978-0-12-814635-4.00008-5

TABLE 8.1 Hereditary Diseases Caused by Paracellular Channel Dysfunction

Gene	Disease	References
Claudin-1	Neonatal ichthyosis and sclerosing cholangitis (NISCH) syndrome	Hadj-Rabia et al. (2004)
Claudin-5	Schizophrenia (weak association)	Sun et al. (2004)
Claudin-10	Hypohidrosis, electrolyte imbalance, lacrimal gland dysfunction, ichthyosis, and xerostomia (HELIX) syndrome	Bongers et al. (2017), Hadj-Rabia et al. (2018), Klar et al. (2017)
Claudin-14	Deafness, autosomal recessive 29 (DFNB29)	Wilcox et al. (2001)
	Nephrolithiasis	Thorleifsson et al. (2009)
Claudin-16	Familial hypomagnesemia with hypercalciuria and nephrocalcinosis (FHHNC)	Simon et al. (1999)
Claudin-19	FHHNC	Konrad et al. (2006)
Occludin	Band-like calcification with simplified gyration and polymicrogyria (BLC-PMG)	O'Driscoll et al. (2010)
JAM-C	Hemorrhagic destruction of the brain, subependymal calcification, and cataracts (HDBSCC)	Mochida et al. (2010)
ZO-2	Familial hypercholanemia (FHCA)	Carlton et al. (2003)
	Progressive familial intrahepatic cholestasis 4 (PFIC4)	Sambrotta et al. (2014)
	Deafness, autosomal dominant 51 (DFNA51)	Walsh et al. (2010)
Tricellulin	Deafness, autosomal recessive 49 (DFNB49)	Riazuddin et al. (2006)
Angulin-2	Deafness, autosomal recessive 42 (DFNB42)	Borck et al. (2011)

discovered from NISCH patients: c.358delG and c.181C > T (p.Gln61X), both of which are loss-of-function mutations due to frame shift or premature stop codon (Feldmeyer et al., 2006; Kirchmeier et al., 2014).

8.2.1.2 Skin and Liver Barrier Defects

NISCH patients show dry skin, with large-scale ichthyosis (Fig. 8.1). The organization of the epidermis and the morphology of keratinocyte TJs appear normal in these patients (Hadj-Rabia et al., 2004). Claudin-1 knockout mice developed dry and wrinkled skin due to epidermal water loss (Furuse et al., 2002). The

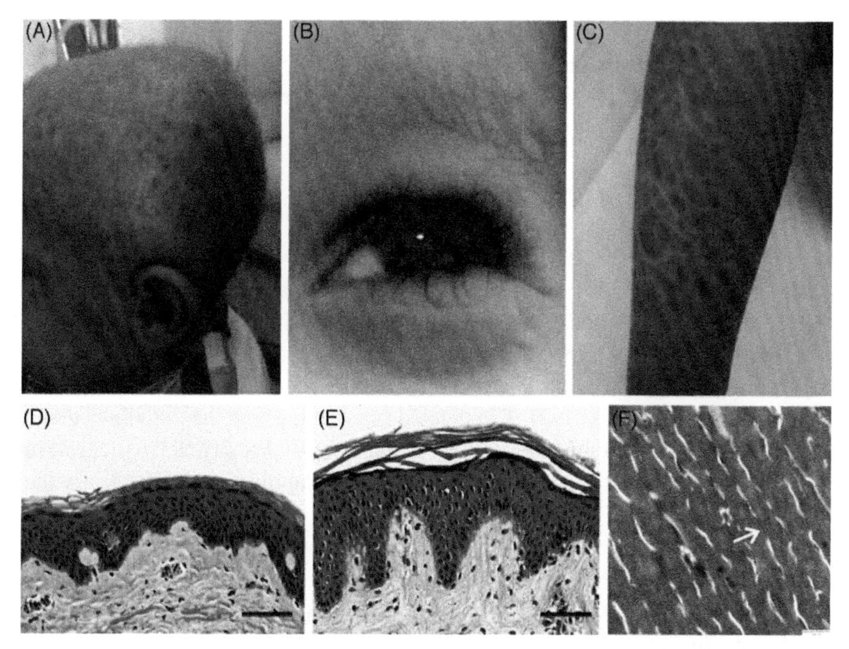

FIGURE 8.1 **Epidermal abnormality in NISCH.** (A) A 17-month old patient shows sparse scalp hair and lamellar ichthyosis with scaling on the head. (B) A 3-year old patient shows thin, intertwined, and curled eyelashes. (C) The forearm of the 3-year old patient shows large brownish adherent scales. (D) H&E staining of breast skin from a healthy control subject. Bar: 50 μm. (E) H&E staining of upper arm skin from the 3-year old patient. The patient's skin shows normal nonacanthotic epidermis covered by mildly orthohyperkeratotic stratum corneum. Keratinocytes exhibit normal morphology. The granular layer contains lower amounts of keratohyalin. Bar: 50 μm. (F) Electron microscopy reveals a large number of corneodesmosomes in the skin biopsy from the 3-year-old patient. Bar: 500 nm. *(Reproduced with permission from Kirchmeier, P., Sayar, E., Hotz, A., Hausser, I., Islek, A., Yilmaz, A., Artan, R., & Fischer, J. (2014). Novel mutation in the CLDN1 gene in a Turkish family with neonatal ichthyosis sclerosing cholangitis (NISCH) syndrome. The British Journal of Dermatology,* 170, *976–978.)*

paracellular permeability to small molecules (< 557 Da) was increased in the knockout mouse skin, which suggests that loss-of-function mutation in human claudin-1 gene may impair the epidermal TJ barrier to cause ichthyosis in NISCH patients. In the liver, TJs seal hepatocyte bile canaliculi to separate bile from plasma. Alterations in these TJs can cause chronic cholestatic liver diseases such as primary biliary cirrhosis and primary sclerosing cholangitis (Sakisaka et al., 2001). Claudin-1 is found in the TJ of the bile canaliculus in the mouse liver (Furuse, Fujita, Hiiragi, Fujimoto, & Tsukita, 1998). Because the claudin-1 knockout mouse died within 24 h after birth, any liver dysfunction in the knockout mouse would have been masked by its perinatal lethality (Furuse et al., 2002). In cultured rat hepatocytes, the expression level of claudin-1 was inversely correlated with the paracellular permeability to small molecules. Overexpression of claudin-1 decreased, while knockdown of claudin-1 increased, the paracellular permeability to molecules with sizes up to 4 kDa across the polarized hepatocytes, which suggests that increased paracellular permeability caused by loss-of-function mutation in claudin-1 may result in bile leakage, which in turn injures the liver and the bile duct (Grosse et al., 2012).

8.2.2 Claudin-5

8.2.2.1 Velo-Cardio-Facial Syndrome (VCFS)

VCFS (OMIM #192430) is a genetic disorder characterized by craniofacial anomalies and conotruncal heart malformations (Shprintzen et al., 1978). VCFS is caused by a 1.5- to 3.0-Mb hemizygous deletion of chromosome 22q11, suggesting that haploinsufficiency in this region is responsible for its etiology (Driscoll et al., 1992). The deleted interval harbors the claudin-5 gene, also known as transmembrane protein deleted in VCFS (TMVCF) (Sirotkin et al., 1997). Whether hemizygous deletion of claudin-5 contributes to the development of VCFS, however, remains unclear.

8.2.2.2 Schizophrenia

A single-nucleotide polymorphism (SNP) in the claudin-5 gene has been weakly associated with schizophrenia in a Chinese Han population and an Iranian population (Table 8.1) (Omidinia, Mashayekhi Mazar, Shahamati, Kianmehr, & Shahbaz Mohammadi, 2014; Sun et al., 2004; Wu et al., 2010; Ye et al., 2005). The SNP − rs10314 is a G to C base change in the 3′-UTR of the claudin-5 gene, which confers novel binding to two microRNAs: miR-125 and miR-3934. In human embryonic kidney 293 (HEK293) cells, the rs10314 variant gene expresses 50% less claudin-5 protein than the normal gene and the variant protein is mislocalized to the lysosome (Fig. 8.2) (Greene et al., 2017). Inducible knockdown of the claudin-5 protein in adult mouse brain caused schizophrenia-like phenotypes including impairments in learning and memory, anxiety-like behaviors, and sensorimotor gating defects (Greene et al., 2017). The claudin-5 protein is best known to regulate the paracellular permeability in the blood-brain barrier (BBB). The vascular permeability to small molecules (< 800 Da) was markedly increased in the claudin-5 knockout mouse brain (Nitta et al., 2003). A biomarker study has implicated that BBB hyperpermeability represents an important risk factor for schizophrenia in man (Falcone et al., 2010). Therefore, claudin-5 may modify the susceptibility to schizophrenia by regulating the BBB integrity and permeability.

8.2.3 Claudin-10

8.2.3.1 Hypohidrosis, Electrolyte Imbalance, Lacrimal Gland Dysfunction, Ichthyosis, and Xerostomia (HELIX) Syndrome

HELIX syndrome (OMIM #617671) is an autosomal recessive disorder characterized by hypohidrosis, severe heat intolerance, hypermagnesemia,

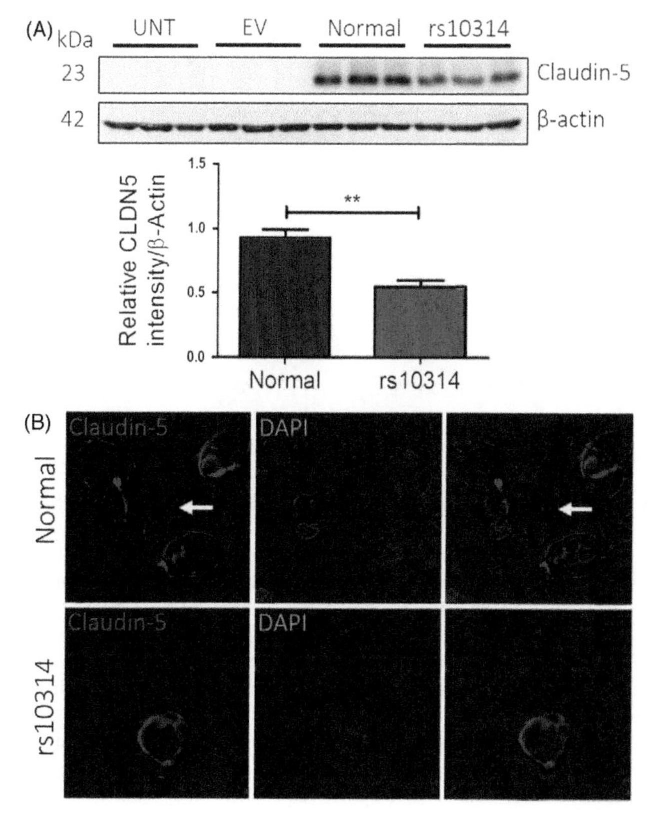

FIGURE 8.2 Protein expression and trafficking defects of claudin-5 variant rs10314. (A) Claudin-5 protein expression levels in HEK-293 cells expressing normal or rs10314 variant gene. Claudin-5 protein expression is significantly lower in cells expressing rs10314 variant than in cells expressing normal claudin-5. UNT, untransfected; EV, empty vector. *Double Asterisk*, $p < 0.01$. (B) Claudin-5 protein localization in HEK-293 cells expressing normal or rs10314 variant gene. Note that normal claudin-5 proteins are found at the cell junction (*arrow*), but variant claudin-5 proteins are trapped inside the intracellular vesicles, including the lysosome. *(Reproduced with permission from Greene, C., Kealy, J., Humphries, M.M., Gong, Y., Hou, J., Hudson, N. et al. (2017). Dose-dependent expression of claudin-5 is a modifying factor in schizophrenia. Molecular Psychiatry. Oct 10. doi:10.1038/mp.2017.156. [Epub ahead of print].)*

hypokalemia, polyuria and polydipsia, and secondary hyperaldosteronism and hyperparathyroidism, as well as ichthyosis, alacrimia (inability to produce tears), xerostomia (dry mouth), and severe enamel wear (Fig. 8.3) (Hadj-Rabia et al., 2018; Klar et al., 2017). Linkage analyses have identified three homozygous missense mutations (c.2T > C, p.M1T; c.144C > G, p.N48K; c.386C > T, p.S131L) in the claudin-10 gene from four consanguineous Pakistani families with HELIX syndrome (Table 8.1) (Hadj-Rabia et al., 2018; Klar et al., 2017). Four new claudin-10 mutations were discovered by exome sequencing from two unrelated patients with a hypokalemic-alkalotic-hypermagnesemic salt-losing nephropathy, which resembles the renal aspect of HELIX syndrome

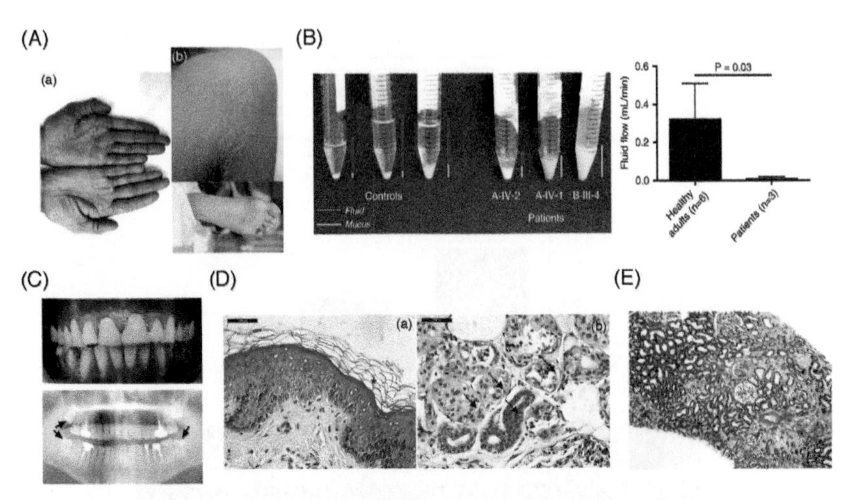

FIGURE 8.3 Clinical and histological phenotypes of HELIX syndrome. (A) Skin features of HELIX syndrome include: (a) palmar hyperlinearity; (b) fine white scaly ichthyosis predominately on the axillae; and (c) plantar keratoderma. (B) Saliva secretion in patients and healthy subjects. *Left panel*: saliva was collected for 15 min under resting condition in three patients and three healthy subjects (controls). The watery and mucous components of the saliva were separated by centrifugation for 10 min. In controls, saliva was mainly composed of fluid (*red bar*), whereas mucus was a minor component (*green bar*). In patients, the amount of secreted fluid was little in one (A-IV-2) or none in two subjects (A-IV-1 and B-III-4), whereas the amount of mucus was much greater than in controls. *Right panel*: the fluid flow rate was measured under resting condition in six healthy subjects and three patients. The fluid flow rate was reduced by 98% in patients compared to controls. (C) Dental features in HELIX syndrome. *Upper panel*: clinical view of patient showing poor dental condition with severe enamel wear restored by resin composites on upper incisors, associated with insufficient oral hygiene, generalized gingival inflammation and localized gingival recession on lower central incisors. *Lower panel*: panoramic radiograph revealing generalized enamel wear with prosthetic restorations and large carious lesions on premolars and molars. Note that the enamel is still present on the second upper molars and well-detectable on the nonerupted third molar germs (*arrows*). (D) Histology of patient skin biopsies includes: (a) slight epidermal hyperplasia with follicular ostial dilatation and basket-weave keratin (*Asterisk*); (b) abnormally high number of dilated eccrine sweat glands. Epithelial cells of eccrine sweat glands are not cohesive, showing an acantholytic appearance (*arrows*). (E) Histology of patient kidney biopsy reveals that the renal tissue contains both normal and fibrotic glomeruli. Renal tubular fibrosis is predominantly located at the superficial cortex. Tubules are slightly dilated. No sign of inflammation or vascular damage is visible. *(Reproduced with permission from Hadj-Rabia S., Brideau G., Al-Sarraj Y., Maroun R.C., Figueres M.L., Leclerc-Mercier S., Olinger E., Baron S., Chaussain C., Nochy D., Taha R.Z., Knebelmann B., Joshi V., Curmi P.A., Kambouris M., Vargas-Poussou R., Bodemer C., Devuyst O., Houillier P., El-Shanti H. (2018). Multiplex epithelium dysfunction due to CLDN10 mutation: the HELIX syndrome. Genet Med. 20(2), 190–201.)*

(Bongers et al., 2017). These variants are carried by each patient in compound heterozygosity: (c.446C > G, p.P149R/c.465–1G > A, p.E157_Y192del) and (c.446C > G, p.P149R/c.217G > A, p.D73N). Claudin-10 exists in two alternatively spliced forms: claudin-10a and -10b. Claudin-10a and -10b differ in exon 1, which encodes the first transmembrane domain and the first extracellular loop domain in the claudin-10 protein (Van Itallie et al., 2006). In contrast to the kidney specific expression of claudin-10a, claudin-10b is ubiquitously expressed in multiple organs including the brain, the gastrointestinal tract, the heart, the kidney, the lung, and the spleen (Van Itallie et al., 2006). The functions of claudin-10a and claudin-10b are vastly different. While claudin-10a confers anion permeability to the paracellular pathway, claudin-10b makes a paracellular cation channel in several epithelial cell models (Van Itallie et al., 2006). Among the claudin-10 mutations, p.M1T, p.N48K, and p.D73N are unique to claudin-10b, whereas p.S131L, p.P149R, and p.E157_Y192del affect both isoforms. The mutations—p.M1T and p.E157_Y192del are loss-of-function mutations because they either abolish the start codon or delete the entire fourth transmembrane domain in claudin-10. The N48 residue is part of the claudin consensus motif (W-G/NLW-C-C) in the first extracellular loop domain important for stabilizing the claudin structure. Its substitution reduces the paracellular permeability of claudin-10b by disrupting the claudin *trans*-interaction (Klar et al., 2017). The residues—D73, S131, and P149 are evolutionarily conserved amino acids in the claudin-10 protein (Bongers et al., 2017; Hadj-Rabia et al., 2018). Among them, the S131L mutation reduces the membrane abundance level of claudin-10b (Hadj-Rabia et al., 2018). The D73N and P149R mutant proteins are, nonetheless, expressed normally in the TJ (Bongers et al., 2017). How these two mutations affect claudin-10 function will need further investigation.

8.2.3.2 Lacrimal, Salivary, Eccrine Sweat Gland and Kidney Transport Failures

The lacrimal, salivary, and eccrine sweat glands all contain secretory epithelial cells. In the secretory epithelium, the Na-K-Cl cotransporter NKCC1 and the Na/K-ATPase are located on the basolateral membrane. NKCC1 allows electroneutral influx of K^+, Cl^-, and Na^+ into the cell. Then, Cl^- is secreted across the apical membrane into the lumen via the cystic fibrosis transmembrane conductance regulator (CFTR) whereas Na^+ is recycled via the Na/K-ATPase into the interstitium and secreted into the lumen through the TJ owing to the lumen-negative transepithelial voltage. Because claudin-10b makes a paracellular cation channel that can permeate Na^+ (Van Itallie et al., 2006), the phenotype of loss of saliva, tears, and sweat in HELIX syndrome might be a consequence of Na^+ transport failure across the TJ. In the kidney, claudin-10b is expressed by the thick ascending limb, where approximately 70% of Mg^{++} reabsorption takes place (Breiderhoff et al., 2012). Deletion of claudin-10b in the mouse kidney caused hypermagnesemia, hypomagnesiuria, and hypocalciuria due to reduced paracellular Na^+ permeability in the thick ascending limb (Breiderhoff et al., 2012). The knockout mouse was also polyuric and polydipsic, but without hypokalemia. The renal phenotypes of claudin-10b knockout mouse recapitulate many aspects of HELIX syndrome, suggesting that the human disease is in part caused by failure to reabsorb Na^+ via the TJ in the thick ascending limb of the kidney. Notably, several HELIX syndrome mutations also affect claudin-10a, which is expressed by the proximal tubule (Breiderhoff et al., 2012). Therefore, the spectrum of HELIX syndrome phenotypes will be much wider than what can be explained by the claudin-10b knockout mouse studies.

8.2.4 Claudin-14

8.2.4.1 Disease in the Inner Ear

8.2.4.1.1 Deafness, Autosomal Recessive 29 (DFNB29)

DFNB29 (OMIM #614035) is caused by homozygous mutation in the claudin-14 gene (Table 8.1). Wilcox and coworkers first identified two homozygous mutations in claudin-14: c.398delT and c.254T > A (p.V85D) from two large consanguineous Pakistani families with DFNB29 (Wilcox et al., 2001). Several new mutations in claudin-14 were later discovered from Pakistani, Spanish, and Greek patients of DFNB29, including c.167G > A (p.W56X), c.242G > A (p.R81H), c.259_260TC > AT (p.S87I), c.281C > T (p.A94V), c.301G > A (p.G101R), and c.694G > A (p.G232R) (Bashir et al., 2013; Lee et al., 2012; Wattenhofer et al., 2005). Among these mutations, c.398delT causes frame shift and c.167G > A (p.W56X) results in a premature stop codon. Therefore, both are loss-of-function mutations. The mutations—c.254T > A (p.V85D) and c.301G > A (p.G101R) are missense mutations, which cause intracellular trafficking defects (Wattenhofer et al., 2005). The remainder are evolutionarily conserved mutations but their functional roles have not been determined (Bashir et al., 2013; Lee et al., 2012).

8.2.4.1.2 Hair Cell Degeneration

The claudin-14 gene expression in the mouse inner ear is spatio-temporally regulated. At postnatal day 4, claudin-14 is found primarily in the inner and outer hair cells and to a lesser extent, in the supporting cells. By postnatal day 8, claudin-14 expression is observed throughout the organ of Corti (Wilcox et al., 2001). The claudin-14 knockout mice showed normal levels of endocochlear potential but were deaf due to hair cell degeneration (Fig. 7.26) (Ben-Yosef et al., 2003). Because claudin-14 impeded paracellular permeation of cations including K^+, the deafness in claudin-14 knockout mice and in DFNB29 patients might be a result of paracellular leakage of K^+ through hair cell TJs from the endolymph to the perilymph, which would depolarize hair cells and cause cell death (Konishi & Salt, 1983).

8.2.4.2 Disease in the Kidney

8.2.4.2.1 Nephrolithiasis

A genome-wide association study (GWAS) has identified four common, synonymous SNPs in the claudin-14 gene, which are associated by genome-wide criteria with nephrolithiasis in an Icelandic population and a Dutch population

(Table 8.1) (Thorleifsson et al., 2009). These SNPs are rs219778[T], with OR = 1.23 and $P = 1.7 \times 10^{-12}$; rs219779[C], with OR = 1.23 and $P = 1.7 \times 10^{-12}$; rs219780[C], with OR = 1.25 and $P = 4.0 \times 10^{-12}$; and rs219781[C], with OR = 1.23 and $P = 4.0 \times 10^{-12}$. Two of them are exonic, causing silent mutations in the claudin-14 protein: rs219779[C] (R81R) and rs219780[C] (T229T). The urinary Ca^{++} excretion levels are significantly higher in homozygous carriers of rs219780[C] compared with noncarriers. Approximately 62% of the general population are homozygous for rs219780[C] and estimated to inherit 1.64 times greater risk of developing kidney stone disease compared with noncarriers (Thorleifsson et al., 2009). Several new SNPs in the claudin-14 gene have later been associated with urinary Ca^{++} excretion levels or urinary excretion ratios of Mg^{++} over Ca^{++} by candidate gene association approach and genome-wide association approach (Corre et al., 2017; Toka, Genovese, Mount, Pollak, & Curhan, 2013). These association studies point to a regulatory role of claudin-14 in renal homeostasis of divalent cations.

8.2.4.2.2 Negative Regulator of Paracellular Cation Channel

The claudin-14 gene is predominantly expressed by the thick ascending limb in the kidney, where approximately 25% of Ca^{++} reabsorption and 70% of Mg^{++} reabsorption take place (Gong et al., 2012). Because claudin-14 impeded paracellular cation permeation, transgenic overexpression of claudin-14 in mouse thick ascending limbs caused increased urinary excretion of Ca^{++} and Mg^{++} (Gong & Hou, 2014). The Ca^{++} reabsorption in the thick ascending limb is normally inhibited by hypercalcemia, as part of the feedback control mechanism to excrete excessive Ca^{++}. Hypercalcemia has been shown to upregulate the renal claudin-14 gene expression. When experimental mice were fed a high Ca^{++} diet to induce hypercalcemia, the claudin-14 gene expression levels became dramatically increased in the thick ascending limbs (Fig. 8.4) (Gong et al., 2012). Gong and coworkers have further demonstrated that hypercalcemia regulates the claudin-14 gene expression by a microRNA-based mechanism. Two microRNA molecules – miR-9 and miR-374 bind to the 3′ untranslated region of the claudin-14 transcript, suppress its protein translation, and induce its mRNA decay under normocalcemic condition (Gong et al., 2012). Hypothetically, the association between claudin-14 and nephrolithiasis might be explained by the derangement of claudin-14 gene expression that somehow escapes the microRNA suppression to enter the hypercalcemic state.

8.2.5 Claudin-16

8.2.5.1 Familial Hypomagnesemia With Hypercalciuria and Nephrocalcinosis (FHHNC)

8.2.5.1.1 Genetic Linkage

Michelis and coworkers first described an autosomal recessive renal disorder featuring hypomagnesemia, hypercalciuria, and nephrocalcinosis, which

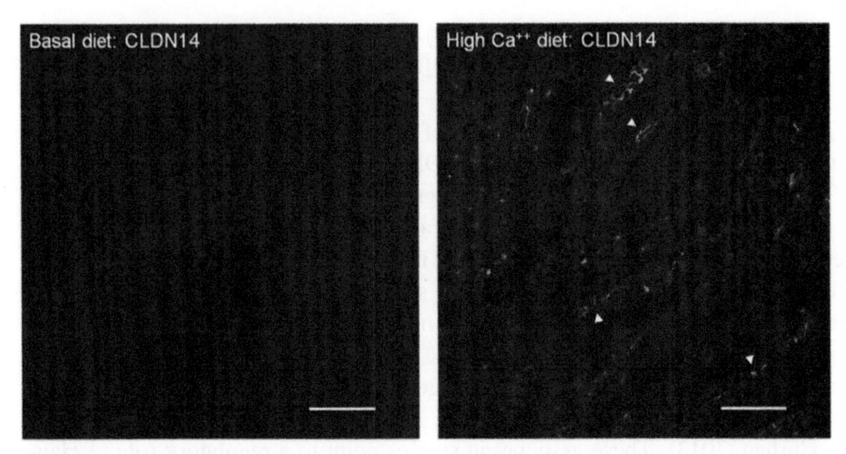

FIGURE 8.4 Claudin-14 gene regulation by hypercalcemia in the kidney. Immunostaining performed on mouse kidney sagittal sections shows CLDN14 protein expression in the TJs (*arrowheads*) of thick ascending limb tubules from mice fed with high Ca^{++} diet but not basal diet. Bar: 50 μm. *(Reproduced with permission from Gong, Y., Renigunta, V., Himmerkus, N., Zhang, J., Renigunta, A., Bleich, M., & Hou, J. (2012). Claudin-14 regulates renal Ca^{++} transport in response to CaSR signalling via a novel microRNA pathway. The EMBO Journal,* 31, *1999–2012.)*

was later termed FHHNC (OMIM #248250) (Michelis, Drash, Linarelli, De Rubertis, & Davis, 1972). Since then, numerous affected kindreds have been reported (Manz, Scharer, Janka, & Lombeck, 1978; Praga et al., 1995; Rodriguez-Soriano & Vallo, 1994). Diseased persons develop profound renal Mg^{++} wasting, which results in severe hypomagnesemia that is not corrected by oral or intravenous Mg^{++} supplementation. They also develop renal Ca^{++} wasting and renal parenchymal calcification. In early childhood, FHHNC patients may show recurrent urinary tract infection, polyuria and polydipsia, nephrolithiasis, and failure to thrive. Other clinical features include elevated serum parathyroid hormone levels before the onset of chronic renal failure, incomplete distal tubular acidosis, hypocitraturia, and hyperuricemia. On average, one-third of FHHNC patients reach end-stage renal failure during the first two decades of life. By linkage analysis, Lifton and coworkers mapped FHHNC to chromosome 3q27 and identified the claudin-16 gene (formerly known as paracellin-1) (Table 8.1) (Simon et al., 1999).

8.2.5.1.2 Mutation With Biological Significance

Numerous mutations have been found to cosegregate with the disease (Al-Haggar et al., 2009; Godron et al., 2012; Kang et al., 2005; Konrad et al., 2008; Kutluturk et al., 2006; Muller et al., 2006a; Muller et al., 2003; Muller, Kausalya, Meij, & Hunziker, 2006b; Sanjad, Hariri, Habbal, & Lifton, 2007; Simon et al., 1999; Weber et al., 2000; Weber et al., 2001b). These mutations include premature termination, splice site alteration, frame shift, and amino acid substitution. Most of the missense mutations affect the transmembrane domains and the extracellular loops (Fig. 8.5). Notably, patients from Germany and Eastern European

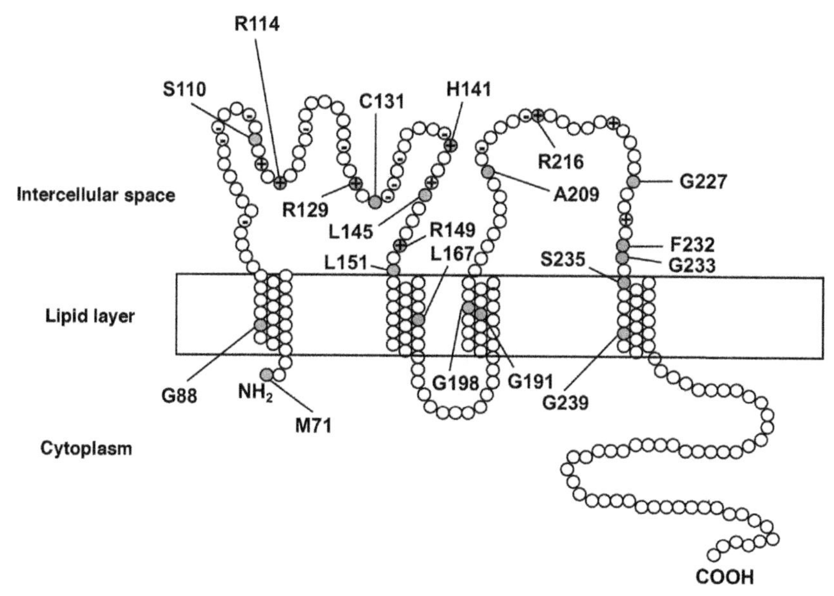

FIGURE 8.5 Schematic diagram showing FHHNC mutations in claudin-16. Amino acid residues affected by missense mutations are shown in *gray*. The charged amino acid residues in the first and second extracellular loop domains are labeled with charge signs. *(Reproduced with permission from Konrad, M., Hou, J., Weber, S., Dotsch, J., Kari, J.A., Seeman, T. et al. (2008). CLDN16 genotype predicts renal decline in familial hypomagnesemia with hypercalciuria and nephrocalcinosis. Journal of the American Society of Nephrology: JASN, 19, 171–181.)*

countries carry a common founder mutation c.453G > T (p.L151F), which occurs in approximately 50% of the mutant alleles (Weber et al., 2001b). There are two particularly interesting mutations. Sequence comparison reveals that the human claudin-16 gene encodes a 305 amino acid protein that possesses two in-frame start codons (ATG: encoding methionine M1 and M71, respectively) and that the second ATG corresponds to the start codon in the mouse claudin-16 gene (Weber et al., 2001a). Hou and coworkers have shown that the human claudin-16 protein migrates as two separate bands (33 kDa and 27 kDa) with different electrophoretic mobility (Fig. 4.10A). The truncated form (Δ70: amino acids 71-305), matching the 27 kDa band, is correctly localized in the TJ, but the full-length protein, matching the 33 kDa band, is mistargeted to the lysosome for degradation (Fig. 4.10B) (Hou, Paul, & Goodenough, 2005). As a result, the missense mutation c.212T > C (p.M71T) abolishes the second start codon in claudin-16 and causes the trafficking defect (Konrad et al., 2008). The missense mutation c.908C > G (p.T303R) affects the carboxyl-terminal PDZ-binding domain in claudin-16 important for its interaction with ZO-1, the TJ plaque protein (Muller et al., 2003). Interaction with ZO-1 has been known for anchoring the claudin protein to the TJ (Chapter 2, Section 2.5.1.1) (Umeda et al., 2006). The T303R mutant protein is consequently removed from the TJ for lysosomal degradation (Muller et al., 2003). In contrast to other claudin-16 mutations, patients with the T303R mutation show childhood hypercalciuria and nephrocalcinosis but no significant disturbance in renal Mg^{++} metabolism (Muller et al., 2003). More intriguingly, the hypercalciuria is self-limiting and often disappears after puberty. Perhaps, FHHNC is a heterogeneous disease. Some mutations in claudin-16 may only partially abrogate its function or even confer new function to the paracellular channel it makes.

8.2.5.1.3 Genotype-Phenotype Correlation

FHHNC is frequently associated with progressive renal failure, which suggests an additional role for claudin-16 in the maintenance of renal function. Hou and coworkers analyzed the paracellular transport function of 29 reported human claudin-16 mutations and noticed that whereas most mutations resulted in a complete loss of function, some mutations retained a substantial residual function (Fig. 8.6A) (Hou et al., 2005). Konrad, Hou and coworkers investigated 71

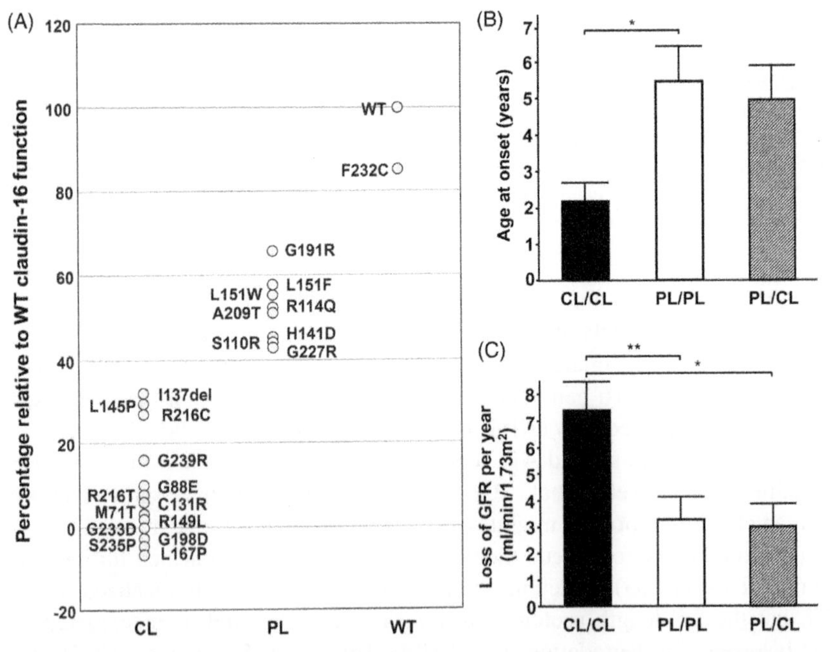

FIGURE 8.6 Correlative study of claudin-16 function with renal function in FHHNC. (A) Electrophysiological recordings of individual claudin-16 missense mutations reveal differences in paracellular permeabilities. Values are shown as percentage relative to wildtype (WT) claudin-16 function. (B) Age at first manifestation of renal failure. (C) Progression of renal failure expressed as loss of GFR (ml/min per 1.73 m^2) per year. Data are shown as means ± standard error of mean. *Asterisk, $P < 0.05$; Double Asterisk, $P < 0.01$. (Reproduced with permission from Konrad, M., Hou, J., Weber, S., Dotsch, J., Kari, J.A., Seeman, T. et al. (2008). CLDN16 genotype predicts renal decline in familial hypomagnesemia with hypercalciuria and nephrocalcinosis. Journal of the American Society of Nephrology: JASN, 19, 171–181.)*

FHHNC patients carrying these 29 claudin-16 mutations and revealed that 23 patients who carried mutations resulting in complete losses of function of both alleles were significantly younger at the onset of renal failure than 46 patients who carried at least one mutant allele retaining partial function (Fig. 8.6B) (Konrad et al., 2008). In addition, those carrying complete loss-of-function mutations showed a more rapid decline in glomerular filtration rate (GFR) than those who carried mutations retaining residual gene function (Fig. 8.6C) (Konrad et al., 2008). How paracellular channel dysfunction leads to renal failure, however, is largely elusive.

8.2.5.1.4 Pharmacological Rescue of Trafficking Defect

A substantial number of FHHNC mutations in claudin-16 cause intracellular trafficking defects (Hou et al., 2005; Kausalya et al., 2006). The mutant proteins fail to reach the cell surface but remain sequestered in various intracellular membranes such as the endoplasmic reticulum, the Golgi apparatus, the endosome, and the lysosome. Pharmacological chaperones, i.e. chemical compounds that facilitate the folding of membrane proteins, can rescue the trafficking defects of some misfolded membrane proteins (Ulloa-Aguirre, Janovick, Brothers, & Conn, 2004). Kausalya and coworkers have shown that two pharmacological chaperones, thapsigargin and 4-phenylbutyrate, effectively rescued the cell surface localization of several misfolded claudin-16 proteins harboring FHHNC mutations, in particular those affecting the transmembrane domains (Kausalya et al., 2006).

8.2.5.2 Paracellular Na^+ Channel for Mg^{++}

The claudin-16 gene is predominantly expressed in the thick ascending limb of the kidney, which handles approximately 25% of Ca^{++} reabsorption and 70% of Mg^{++} reabsorption (Simon et al., 1999). Hou and coworkers have generated a claudin-16 deficient mouse model by using RNA interference technology to knock down the claudin-16 gene expression in the kidney (Hou et al., 2007). The knockdown mouse phenocopied the main features of human FHHNC, including hypomagnesemia, hypercalciuria, and nephrocalcinosis, which confirms that the human disease is caused by the loss of claudin-16 function. Whether claudin-16 makes a selective divalent cation channel is highly controversial. Hou and coworkers showed that claudin-16 permeated monovalent cations, including Na^+, as well as divalent cations, including Mg^{++}, but the monovalent cation permeability was much higher than the divalent cation permeability (Hou et al., 2005). Others have reported Ca^{++} or Mg^{++} specific permeability for claudin-16 (Ikari et al., 2004; Kausalya et al., 2006). Hou and coworkers have revealed that when claudin-16 was removed from the mouse thick ascending limb, it was the cation-to-anion selectivity (P_{Na}/P_{Cl}), but not the divalent-to-monovalent cation selectivity (P_{Mg}/P_{Na}), that showed significant decreases (Hou et al., 2007). Perhaps, mutations in claudin-16 affect the driving force rather than the absolute permeability of divalent cations. The driving force depends upon the lumen-positive transepithelial voltage (V_{te}), which is generated by two additive mechanisms: (1) transcellular reabsorption of NaCl through the apical Na-K-2Cl cotransporter, coupled with K^+ secretion through the apical K channel, gives rise to a spontaneous lumen-positive potential (~ +8 mV) (Burg & Green, 1973); (2) dilution of the luminal fluid through constant NaCl reabsorption creates a diffusion potential through the cation-selective TJ, which adds approximately +30 mV to the transepithelial potential (Fig. 7.14C) (Greger, 1985). When thick ascending limbs were perfused with symmetrical NaCl solutions, there was no difference in V_{te} between claudin-16 knockdown and wildtype mice, which indicated that the first mechanism was normal in the absence of claudin-16 (Table 7.4) (Hou et al., 2007). When a NaCl gradient was applied from the basolateral space to the lumen, the V_{te} of thick ascending limb was +18 mV in wildtype mouse, but only +6.6 mV in claudin-16 knockdown mouse, which indicated that ablation of claudin-16 dissipated the lumen-positive V_{te} via the second mechanism (Table 7.4) (Hou et al., 2007). Paradoxically, Will and coworkers have noticed that in a knockout mouse model for claudin-16, the paracellular permeability to divalent cations such as Ca^{++} and Mg^{++} was reduced, but the monovalent cation permeability remained unchanged (Will et al., 2010). Very likely, the debate over claudin-16's propensity for divalent cation permeability will continue into the future. Finally, it is worth emphasizing that neither claudin-16 deficient mouse model developed renal failure, which is in sharp contrast to human cases. The pathophysiology of renal failure, which is far less common in other electrolyte imbalance disorders, remains largely unclear. Claudin-16 has been speculated to play roles in cell proliferation and differentiation (Lee, Huang, & Ward, 2006). A naturally occurring bovine knockout model for claudin-16 showed early-onset renal failure associated with renal tubular dysplasia (Hirano et al., 2000; Ohba et al., 2000).

8.2.6 Claudin-19

8.2.6.1 Familial Hypomagnesemia With Hypercalciuria and Nephrocalcinosis (FHHNC) and Ocular Involvement

Meier, Rodriguez-Soriano and coworkers reported a small cohort of patients from one Swiss and eight Spanish families showing the renal phenotypes indistinguishable from FHHNC (Meier et al., 1979; Rodriguez-Soriano & Vallo, 1994). The affected individuals in these families also have severe visual impairment, characterized by macular colobomata, significant myopia, and horizontal nystagmus. Konrad and coworkers genotyped these patients but did not find any mutation in claudin-16. By linkage analysis, they mapped the disease, now termed FHHNC with ocular involvement (OMIM #248190), to a new locus on chromosome 1p34.2 and identified three mutations in another claudin gene, claudin-19 (Table 8.1) (Konrad et al., 2006). Numerous new mutations have later been reported to cosegregate with the disease (Al-Shibli et al., 2013; Claverie-Martin et al., 2013; Ekinci, Karabas, & Konrad, 2012; Hou et al., 2008; Konrad et al., 2006; Naeem, Hussain, & Akhtar, 2011; Sharma et al., 2016; Yuan et al., 2015). These mutations include frame shift, nonsense, and missense mutations. Among the missense mutations, c.59G > A (p.G20D) and c.169C > G (p.Q57E) cause intracellular trafficking defects; c.269T > C (p.L90P) and c.367G > C (p.G123R) reduce the binding affinity of claudin-19 to claudin-16 protein (Hou et al., 2008; Konrad et al., 2006).

8.2.6.2 Double Claudin Deletion in Tight Junction

The claudin-19 gene is predominantly expressed in the thick ascending limb of the kidney (Konrad et al., 2006). Transgenic knockdown mouse model for claudin-19 developed the FHHNC manifestations of reduced plasma Mg^{++} levels and excessive renal losses of Ca^{++} and Mg^{++} (Hou et al., 2009). The phenotypic similarities of claudin-19 knockdown mouse to claudin-16 knockdown mouse can be explained by the interaction of the two claudin proteins. Intracellular and intercellular claudin interactions are essential for their polymerization into the TJ (Chapter 2, Section 2.4). Hou and coworkers have found strong intracellular interaction between claudin-16 and claudin-19, and more importantly, they have demonstrated that FHHNC-causing mutations in either claudin disrupted the claudin-16 and claudin-19 interaction (Hou et al., 2008). In mouse thick ascending limbs, removing one claudin from the cells caused the other to be delocalized from the TJ (Fig. 7.16) (Hou et al., 2009). Conceptually, FHHNC might be a consequence of claudin-16 and claudin-19 dissociation that abolishes the transport function of both claudins on the TJ level. In the eye, the claudin-19 protein is found specifically in the retina, with a strong expression in the retinal pigment epithelium (RPE) (Konrad et al., 2006). In human RPE cultures, knockdown of claudin-19 significantly decreased the transepithelial resistance, indicating a loss of paracellular barrier function (Peng, Rao, Adelman, & Rizzolo, 2011). How paracellular barrier defect contributes to the ocular abnormality is not known at present.

8.3 DISEASE CAUSED BY MUTATION IN OTHER TIGHT JUNCTION GENE

8.3.1 Occludin

Band-like calcification with simplified gyration and polymicrogyria (BLC-PMG) is a rare autosomal recessive neurological disease demonstrating clinical and neuroradiological features that are often interpreted as sequelae of congenital infection with Toxoplasma gondii, rubella, cytomegalovirus, or herpes simplex (so called TORCH syndrome), therefore termed as pseudo-TORCH syndrome 1 (OMIM #251290) (Fig. 8.7). By homozygosity mapping and copy number analyses, O'Driscoll and coworkers have identified intragenic deletions and mutations in the occludin gene from six families with BLC-PMG (Table 8.1)

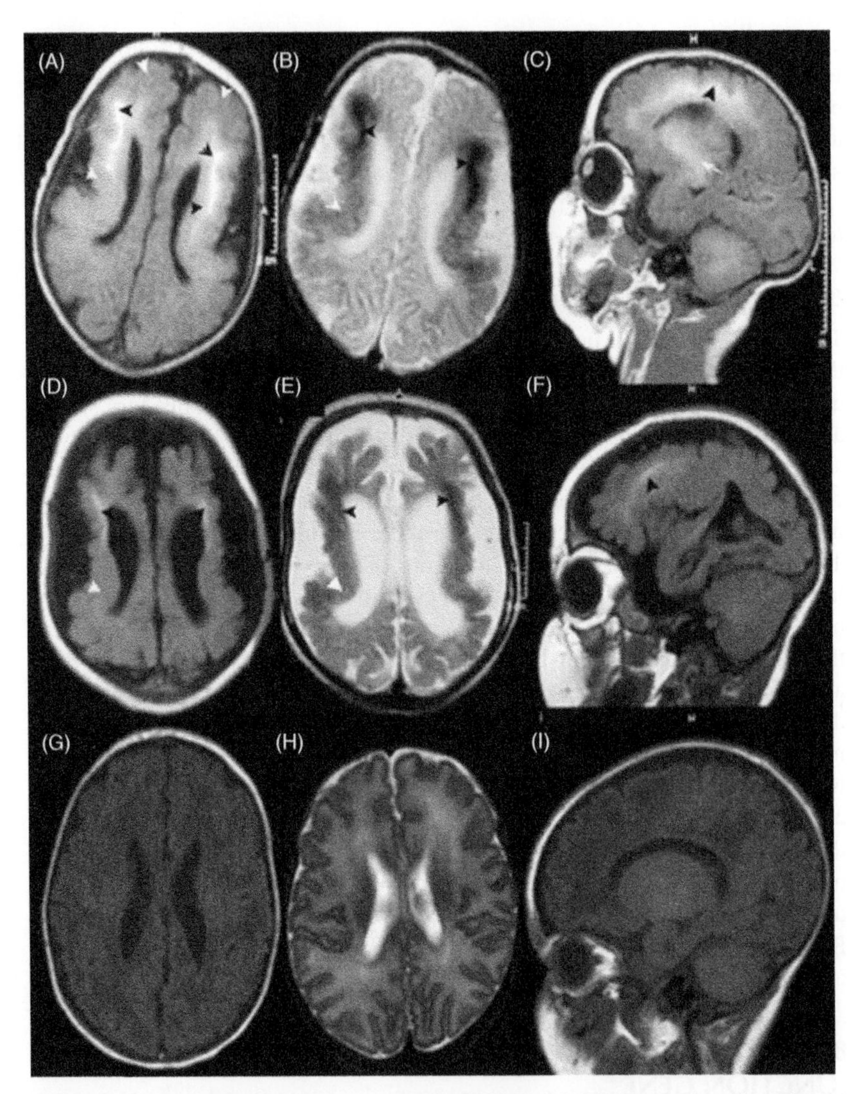

FIGURE 8.7 Neuroradiological features of BLC-PMG. Serial MRI images of a patient taken at ages of 2 weeks (A–C) and 9 months (D–F) show progressive cerebral atrophy. T1 (A, D) and T2 (B, E) weighted axial images also show severe reduction in cerebral volume, deep cortical (*black arrowheads*) and basal ganglion (*white arrow*) calcification, and bilateral fronto-parietal PMG (*white arrowheads*). Control MRI images (G–I) are taken from a 9-month-old healthy subject. *(Reproduced with permission from O'Driscoll, M.C., Daly, S.B., Urquhart, J.E., Black, G.C., Pilz, D.T., Brockmann, K. et al. (2010). Recessive mutations in the gene encoding the tight junction protein occludin cause band-like calcification with simplified gyration and polymicrogyria. American Journal of Human Genetics, 87, 354–364.)*

(O'Driscoll et al., 2010). Most of these genetic alterations cause exon skipping or deletion, premature termination, and frame shift, which are compatible with the loss of occludin gene function. Occludin belongs to the TJ associated Marvel domain containing protein (TAMP) family (Chapter 2, Section 2.6). Its mRNA and protein are expressed throughout the brain, primarily by the neurons in the cerebral cortex and the cerebellum, and by the endothelial cells making the blood-brain barrier (BBB) (Hirase et al., 1997; O'Driscoll et al., 2010). The occludin knockout mouse developed progressive cerebellar and basal ganglion calcification, frequently along the walls of capillaries and postcapillary venules (Saitou et al., 2000). Postmortem analyses of a patient with BLC-PMG revealed prominent calcification surrounding small blood vessels in deep cerebral and cerebellar cortices (Fig. 8.8), which resembled the pathology in the mouse model of occludin deficiency (O'Driscoll et al., 2010). Occludin has been speculated to regulate the paracellular permeability across the BBB (Wosik et al., 2007). The neuropathological studies of human patients and transgenic mice suggest that in utero cerebrovascular insults due to increased BBB permeability may underlie the pathogenesis of BLC-PMG.

8.3.2 JAM-C

Mochida and coworkers reported a large consanguineous family from the United Arab Emirates in which eight individuals inherited an autosomal recessive neurological disorder characterized by hemorrhagic destruction of the brain, subependymal calcification, and cataracts (HDBSCC, OMIM #613730) (Fig. 8.9) (Mochida et al., 2010). Linkage analysis mapped the disease to a 1.9-cM interval on chromosome 11q25, where a homozygous mutation in the JAM-C gene was identified (Table 8.1) (Mochida et al., 2010). This mutation, occurring at the first base position of intron 5 (c.747 + 1G > T), abolishes the canonical splice donor site and transcribes an alternatively spliced variant that results in frameshift and premature termination. JAM-C belongs to the junctional adhesion molecule (JAM) family exclusively localized in the TJs of epithelial and endothelial cells (Chapter 2, Section 2.7). JAM-C is highly expressed in the brain, particularly by the vascular endothelial cells in the parenchyma, the meninges, and the choroid plexus (Arrate, Rodriguez, Tran, Brock, & Cunningham, 2001; Wyss et al., 2012). Additional JAM-C expression can be found in the epithelial cells of the choroid plexus and the ependymal cells lining the ventricles. The JAM-C knockout mouse developed severe hydrocephalus accompanied by hippocampal dislocation, cortical malformations due to reactive gliosis, and hemorrhages in different brain compartments (Wyss et al., 2012). The phenotypic similarities between human HDBSCC and mouse JAM-C deficiency are notable. Both models show brain hemorrhages and enlarged lateral ventricles. JAM-C has been shown to regulate the paracellular permeability in the endothelial cells (Orlova, Economopoulou, Lupu, Santoso, & Chavakis, 2006). Surprisingly, the

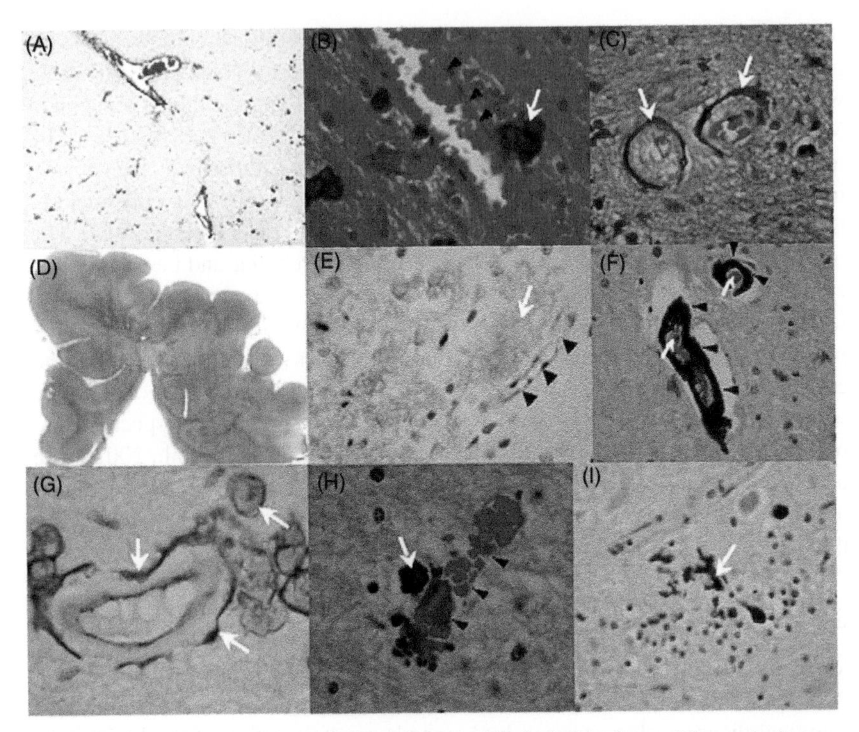

FIGURE 8.8 Histology of BLC-PMG. Postmortem histological analyses of the cerebral cortex of a patient reveal linear bands of calcification surrounding blood vessels at low magnification (A), focal calcification adjacent to (*arrow*), but not within the vessel wall (*arrowheads*) at high magnification (B), and rim of calcification (*arrows*) surrounding two blood vessels in cross-section (C). (D) Low-magnification view showing fused gyri indicative of PMG. (E) CD31 stained section showing endothelial staining (*arrowheads*) adjacent to a large region of calcification (*arrow*). Histology of the cerebellum is shown in F–I. (F) Calcification (*arrowheads*) surrounding two vessels of different sizes (*arrows*). (G) CD31-stained intact endothelium (*brown*) within a ring of calcification (*arrows*). (H) Calcification in a cell (*arrow*) in close apposition to a normal blood vessel (*arrowheads*). (I) "Staghorn" appearance suggesting calcification in a Purkinje cell (*arrow*). *(Reproduced with permission from O'Driscoll, M.C., Daly, S.B., Urquhart, J.E., Black, G.C., Pilz, D.T., Brockmann, K. et al. (2010). Recessive mutations in the gene encoding the tight junction protein occludin cause band-like calcification with simplified gyration and polymicrogyria. American Journal of Human Genetics, 87, 354–364.)*

endothelial re-expression of JAM-C failed to rescue the phenotype of hydrocephalus in the JAM-C knockout mouse, pointing to a nonendothelial role of JAM-C in the pathogenesis of HDBSCC (Wyss et al., 2012). Wyss and coworkers discovered that the drainage of cerebrospinal fluid (CSF) from the lateral ventricles to the third ventricle was markedly reduced in the JAM-C knockout mouse brain, which might be due to blockade at the level of the foramina of Monro or gross reduction in the CSF drainage triggered by subarachnoid hemorrhages (Wyss et al., 2012).

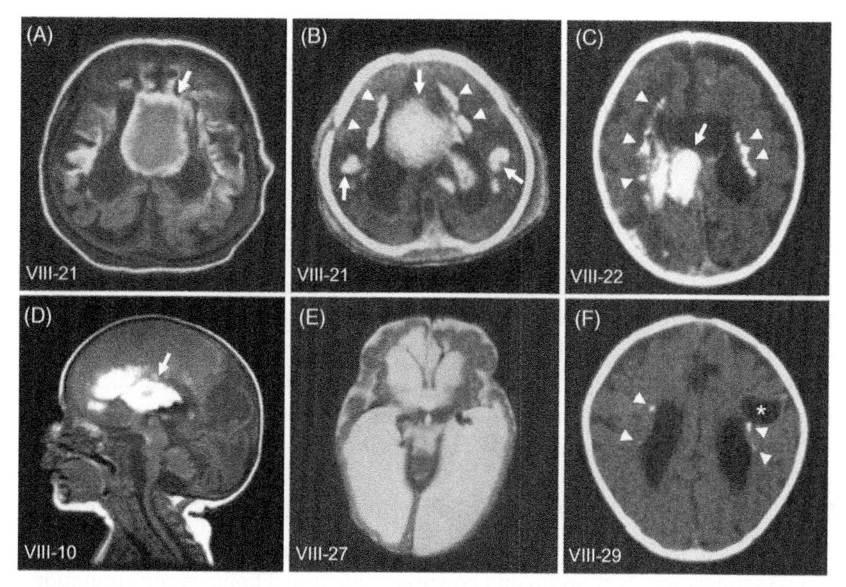

FIGURE 8.9 **Neuroradiological features of HDBSCC.** T1-weighted axial brain MRI (A) and brain CT (B) of patient VIII-21, brain CT of patient VIII-22 (C), and T1-weighted sagittal brain MRI of patient VIII-10 (D) show multifocal intraparenchymal hemorrhages, predominantly in the cerebral white matter and the basal ganglia (*arrows*). T2-weighted axial brain MRI of patient VIII-27 (E) shows massive cystic destruction of the cerebral white matter and the basal ganglia, resulting in enlarged lateral ventricles. Brain CT of patient VIII-29 (F) shows a porencephalic cyst centered in the left frontal subcortical white matter (*asterisk*), enlarged lateral ventricles, and reduced white matter volume, indicative of destructive pathology. Subependymal calcification is evident on CT (B, C and F, *arrowheads*). *(Reproduced with permission from Mochida, G.H., Ganesh, V.S., Felie, J.M., Gleason, D., Hill, R.S., Clapham, K.R. et al. (2010). A homozygous mutation in the tight-junction protein JAM3 causes hemorrhagic destruction of the brain, subependymal calcification, and congenital cataracts. American Journal of Human Genetics, 87, 882–889.)*

8.3.3 ZO-2

8.3.3.1 Familial Hypercholanemia (FHCA)

FHCA (OMIM #607748) is an autosomal recessive liver disorder characterized by elevated serum bile acid concentration, itching, and fat malabsorption. Carlton and coworkers studied this disease in 17 affected individuals from 12 families of Amish descent and identified a chromosomal region (9q12-q13) shared identically by descent and in linkage disequilibrium with FHCA (Carlton et al., 2003). This genomic region contains a candidate gene ZO-2, also known as TJ protein 2 (TJP2) (Table 8.1). DNA sequencing revealed that 11 out of the 17 affected individuals carried a homozygous mutation in ZO-2: c.143T > C (p.V48A) (Carlton et al., 2003). ZO-2 is a PDZ domain containing protein that anchors the TJ onto the cytoskeleton (Chapter 2, Section 2.5.1.1). ZO-2 is expressed by the hepatocyte in the liver and localized in the TJ of the bile canaliculus (Jesaitis & Goodenough, 1994). The V48A mutation, occurring in the first PDZ domain, reduces the domain stability and the ligand binding affinity of ZO-2 protein, suggesting a loss-of-function role in FHCA (Carlton et al., 2003). The ZO-2 knockout mouse died because of growth arrest during early gastrulation phase (Xu et al., 2008). The embryonic lethal phenotype of the knockout mouse contrasts with the human disease, which indicates that human V48A mutation is not a null mutation. Moreover, the penetrance of mutant allele is incomplete. Three unaffected siblings from the same families are homozygous carriers of the mutation (Carlton et al., 2003). In the liver, TJs maintain a 200 – 10,000-fold bile acid concentration gradient from bile to plasma. Biopsy from an FHCA patient revealed morphological TJ abnormalities in the liver, including TJ expansion and elongation (Carlton et al., 2003). These changes in TJ ultrastructure may result in paracellular leakage of bile acids to plasma, thereby causing hypercholanemia.

8.3.3.2 Progressive Familial Intrahepatic Cholestasis 4 (PFIC4)

PFIC refers to a large class of autosomal recessive, polygenic liver diseases. By exome sequencing, Sambrotta and coworkers identified homozygous mutations in the ZO-2 gene from 12 unrelated but consanguineous patients with PFIC4 (Table 8.1) (OMIM #615878) (Sambrotta et al., 2014). The mutations include exon deletion, splice site alteration, frame shift, and premature termination, all abolishing the protein translation (Sambrotta et al., 2014). ZO-2 has been demonstrated to bind to claudins and regulate their TJ localization (Itoh et al., 1999). Liver biopsies from PFIC4 patients revealed that claudin-1 failed to localize to cholangiocyte cell borders and to biliary canalicular TJs, despite normal protein expression levels (Fig. 8.10) (Sambrotta et al., 2014). Abnormal claudin-1 localization in the absence of ZO-2 may explain the phenotypic similarity of PFIC4 to NISCH in the liver, the latter being caused by claudin-1 mutations (Section 8.2.1). On the other hand, while PFIC4 and FHCA (Section 8.3.3.1) are both due to recessive mutations in ZO-2, FHCA manifests itching and fat malabsorption but rarely progresses to cholestasis as PFIC4. Increased paracellular permeability in the bile canaliculus has been hypothesized to underlie both diseases. Nevertheless, additional mechanisms may exist to provide binding partners to ZO-2 via non-PDZ domains, which are functional in FHCA but absent in PFIC4.

8.3.3.3 Deafness, Autosomal Dominant 51 (DFNA51)

Walsh and coworkers reported an Israeli kindred with autosomal dominant, adult-onset, progressive deafness DFNA51 (OMIM #613558) and identified an inverted genomic duplication of 270 kb containing the entire locus of the ZO-2 gene (Table 8.1) (Walsh et al., 2010). In the organ of Corti, ZO-2 is localized predominantly in the TJs connecting the hair cells and the supporting cells (Walsh et al., 2010). Studies of DFNA51 patient lymphoblasts revealed that the

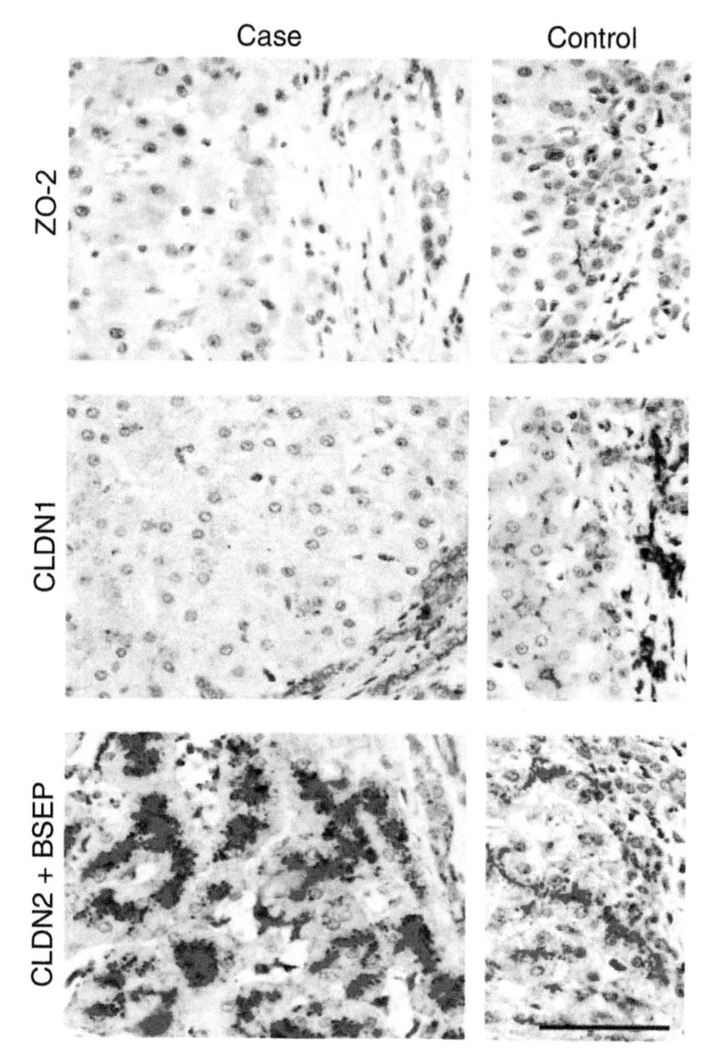

FIGURE 8.10 **Claudin-1 protein delocalization in PFIC4.** Immunohistochemical staining of ZO-2, CLDN1, and CLDN2 + BSEP was performed on patient (case) and healthy subject (control) liver biopsies. ZO-2 protein is absent from cholangiocyte and bile canaliculus in patient biopsy. CLDN1 protein is markedly reduced from cholangiocyte cell border and biliary canalicular TJ in patient biopsy. CLDN2 proteins cluster inside the cytoplasm of hepatocytes adjoining bile canaliculi (highlighted by BSEP, *red*) in both patient and healthy subject biopsies. Bar: 100 μm. *(Reproduced with permission from Sambrotta, M., Strautnieks, S., Papouli, E., Rushton, P., Clark, B.E., Parry, D.A. et al. (2014). Mutations in TJP2 cause progressive cholestatic liver disease. Nature Genetics, 46, 326–328.)*

mRNA and protein levels of ZO-2 were increased by approximately twofold, consistent with overexpression of the gene resulting from genomic duplication (Walsh et al., 2010). ZO-2 has been shown to decrease the phosphorylation of GSK-3β, a serine/threonine kinase involved in the signaling pathway of apoptosis (Tapia et al., 2009). Hair cell apoptosis is implicated in age-related hearing loss in mice (Someya et al., 2009). Hypothetically, ZO-2 overexpression may trigger hair cell apoptosis by genetic reprogramming in DFNA51 patients (Walsh et al., 2010).

8.3.4 Tricellular tight junction gene

8.3.4.1 Tricellulin

Autosomal recessive deafness 49 (DFNB49, OMIM #610153) was initially discovered from two large consanguineous Pakistani families and mapped to an 11-cM interval on chromosome 5q12.3-q14.1 by linkage analysis (Ramzan et al., 2005). Riazuddin and coworkers identified six additional DFNB49 families and established genetic linkage to the tricellulin gene, also known as marvelD2 (Table 8.1) (Riazuddin et al., 2006). Four homozygous mutations in the tricellulin gene cosegregate with the disease. These mutations, including splice site alteration, frame shift and premature termination, remove all or most of the carboxyl-terminal domain in the tricellulin protein (Riazuddin et al., 2006). Tricellulin belongs to the TJ-associated MARVEL domain containing protein (TAMP) family (Chapter 2, Section 2.6). It is localized in the tTJs of the epithelial cells in the stria vascularis and of the mechanosensory hair cells in the organ of Corti (Riazuddin et al., 2006). The carboxyl-terminal domain in tricellulin facilitates its interaction with ZO-1, the TJ plaque protein, and this interaction presumably anchors tricellulin onto the tTJ. A nonsense mutation, c.1498C > T (p.R500X) from DFNB49 patients truncates the carboxyl-terminal domain in tricellulin and reduces its binding affinity to ZO-1 (Fig. 8.11). The transgenic knockin mouse carrying an orthologous mutation to p.R500X in DFNB49 was generated to model the human disease (Nayak et al., 2013). The knockin mouse developed congenital deafness due to hair cell degeneration. Tricellulin plays an important role in regulating the permeability of small ions and solutes through the tTJ (Chapter 6, Section 6.2.3.1). Removal of tricellulin from the tTJ due to the p.R500X mutation disrupted the tTJ ultrastructure in the mouse hair cells (Fig. 8.12), which is postulated to cause paracellular K^+ leakage from the endolymph to the perilymph (Nayak et al., 2013). Increases in perilymphatic K^+ concentration are known to be toxic to the hair cells (Konishi & Salt, 1983).

8.3.4.2 Angulin-2

Autosomal recessive deafness 42 (DFNB42, OMIM #609646) was first reported in a consanguineous Pakistani family and mapped to a 21.6-cM region on chromosome 3q13.31-q22.3 by linkage analysis (Aslam et al., 2005). From

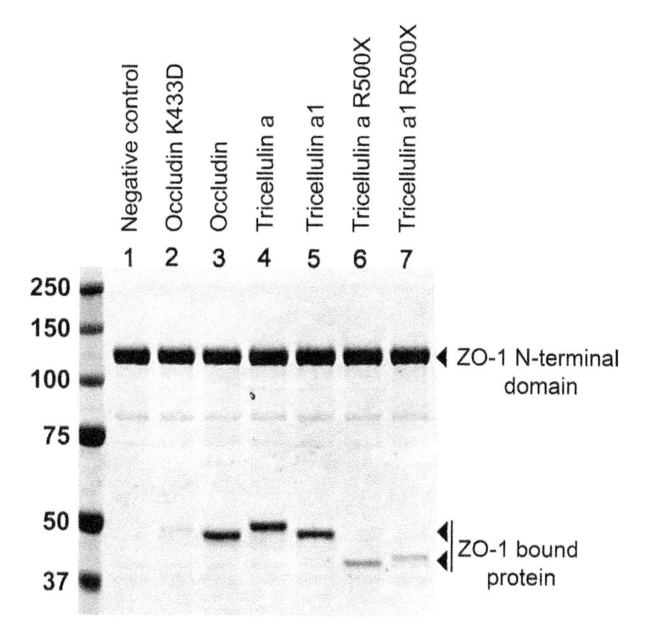

FIGURE 8.11 **Disruption of tricellulin interaction with ZO-1 in DFNB49.** In vitro binding assays were performed on purified recombinant proteins. Proteins were resolved by SDS-PAGE and stained with Coomassie Brilliant Blue. Lane 1, negative loading control having only the N-terminal half of ZO-1 immobilized on agarose beads (ZO-1N; amino acids 1–888). Lanes 2 and 3, GST-fusion proteins encoding the carboxyl-terminal domain of human occludin (amino acids 371–522) and a mutant occludin (K433D) that cannot bind to ZO-1N, used as positive and negative binding controls, respectively. Lanes 4–7, GST-fusion proteins encoding tricellulin isoform a or a1 showing strong binding affinity to ZO-1N. The p.R500X mutation in tricellulin decreases its binding to ZO-1N. The molecular weight markers are shown in kDa on the left side. *(Reproduced with permission from Riazuddin, S., Ahmed, Z.M., Fanning, A.S., Lagziel, A., Kitajiri, S., Ramzan, K. et al. (2006). Tricellulin is a tight-junction protein necessary for hearing. American Journal of Human Genetics, 79, 1040–1051.)*

the linked interval, Borck and coworkers identified 10 homozygous mutations in the angulin-2 gene, also known as immunoglobulin [Ig]-like domain-containing receptor [ILDR] 1, which cosegregated with DFNB42 (Table 8.1) (Borck et al., 2011). These mutations include missense, nonsense, frameshift, and splice-site mutations. Seven of the mutations result in protein truncation, consistent with the loss of protein function. Angulin-2 is a member of the angulin family of tTJ associated proteins, which also include angulin-1 (also known as lipolysis-stimulated lipoprotein receptor [LSR] or ILDR3) and angulin-3 (also known as ILDR2, LISCH-like, or C1orf32) (Chapter 6, Section 6.2.2.2). In the organ of Corti, the angulin-2 mRNA and protein levels are low in the hair cells but high in the supporting cells (Borck et al., 2011; Morozko et al., 2015). The angulin-2 knockout mouse developed early-onset deafness due to rapid degeneration of the hair cells (Morozko et al., 2015).

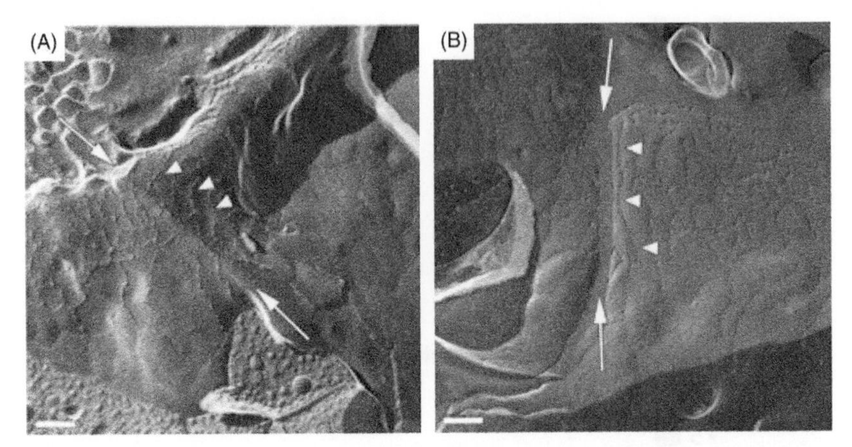

FIGURE 8.12 Abnormal tTJ ultrastructure in DFNB49 mouse model. Freeze-fracture replica electron microscopy reveals the ultrastructure of bicellular and tricellular TJs in the inner hair cells of Tricellulin$^{R497X/+}$ (A) and Tricellulin$^{R497X/R497X}$ (B) mice. (A) In the Tricellulin$^{R497X/+}$ mice, the tTJ strand appears as a "fishbone" (*arrows*), where the elements of bTJ intersect the tTJ (*arrowheads*). (B) In the Tricellulin$^{R497X/R497X}$ mice, the central sealing elements of tTJ are irregular and appear to be formed of a chain of disconnected particles (*arrows*). The bTJ strands do not intersect the tTJ; instead, they make a 90 degree turn and run parallel to the tTJ (*arrowheads*). Bar: 200 nm. *(Reproduced with permission from Nayak, G., Lee, S.I., Yousaf, R., Edelmann, S.E., Trincot, C., Van Itallie, C.M. et al. (2013). Tricellulin deficiency affects tight junction architecture and cochlear hair cells. The Journal of Clinical Investigation, 123, 4036–4049.)*

Angulins are known to recruit tricellulin to the tTJ via protein interactions (Masuda et al., 2011). In the angulin-2 knockout mouse cochlea, tricellulin was delocalized from the tTJ after the first postnatal week, which preceded the development of hearing loss (Morozko et al., 2015). The phenotypic similarity of angulin-2 to tricellulin deficiency in the inner ear might be a result of protein dissociation within the tTJ, which disrupts the paracellular permeation barrier as a potential cause of deafness.

REFERENCES

Al-Haggar, M., Bakr, A., Tajima, T., Fujieda, K., Hammad, A., Soliman, O., Darwish, A., Al-Said, A., Yahia, S., & Abdel-Hady, D. (2009). Familial hypomagnesemia with hypercalciuria and nephrocalcinosis: unusual clinical associations and novel claudin16 mutation in an Egyptian family. *Clinical and Experimental Nephrology*, *13*, 288–294.

Al-Shibli, A., Konrad, M., Altay, W., Al Masri, O., Al-Gazali, L., & Al Attrach, I. (2013). Familial hypomagnesemia with hypercalciuria and nephrocalcinosis (FHHNC): report of three cases with a novel mutation in CLDN19 gene. *Saudi Journal of Kidney Diseases and Transplantation: An Official Publication of the Saudi Center for Organ Transplantation, Saudi Arabia*, *24*, 338–344.

Arrate, M. P., Rodriguez, J. M., Tran, T. M., Brock, T. A., & Cunningham, S. A. (2001). Cloning of human junctional adhesion molecule 3 (JAM3) and its identification as the JAM2 counter-receptor. *The Journal of Biological Chemistry*, *276*, 45826–45832.

Aslam, M., Wajid, M., Chahrour, M. H., Ansar, M., Haque, S., Pham, T. L., Santos, R. P., Yan, K., Ahmad, W., & Leal, S. M. (2005). A novel autosomal recessive nonsyndromic hearing impairment locus (DFNB42) maps to chromosome 3q13 31-q22. 3. *American Journal of Medical Genetics Part A 133a*, 18–22.

Baala, L., Hadj-Rabia, S., Hamel-Teillac, D., Hadchouel, M., Prost, C., Leal, S. M., et al. (2002). Homozygosity mapping of a locus for a novel syndromic ichthyosis to chromosome 3q27-q28. *The Journal of Investigative Dermatology*, *119*, 70–76.

Bashir, Z. E., Latief, N., Belyantseva, I. A., Iqbal, F., Riazuddin, S. A., Khan, S. N., Friedman, T. B., Riazuddin, S., & Riazuddin, S. (2013). Phenotypic variability of CLDN14 mutations causing DFNB29 hearing loss in the Pakistani population. *Journal of Human Genetics*, *58*, 102–108.

Ben-Yosef, T., Belyantseva, I. A., Saunders, T. L., Hughes, E. D., Kawamoto, K., Van Itallie, C. M., et al. (2003). Claudin 14 knockout mice, a model for autosomal recessive deafness DFNB29, are deaf due to cochlear hair cell degeneration. *Human Molecular Genetics*, *12*, 2049–2061.

Bongers, E., Shelton, L. M., Milatz, S., Verkaart, S., Bech, A. P., Schoots, J., et al. (2017). A novel hypokalemic-alkalotic salt-losing tubulopathy in patients with CLDN10 Mutations. *Journal of the American Society of Nephrology: JASN*, *28*, 3118–3128.

Borck, G., Ur Rehman, A., Lee, K., Pogoda, H. M., Kakar, N., von Ameln, S., et al. (2011). Loss-of-function mutations of ILDR1 cause autosomal-recessive hearing impairment DFNB42. *American Journal of Human Genetics*, *88*, 127–137.

Botstein, D., White, R. L., Skolnick, M., & Davis, R. W. (1980). Construction of a genetic linkage map in man using restriction fragment length polymorphisms. *American Journal of Human Genetics*, *32*, 314–331.

Breiderhoff, T., Himmerkus, N., Stuiver, M., Mutig, K., Will, C., Meij, I. C., Bachmann, S., Bleich, M., Willnow, T. E., & Muller, D. (2012). Deletion of claudin-10 (Cldn10) in the thick ascending limb impairs paracellular sodium permeability and leads to hypermagnesemia and nephrocalcinosis. *Proceedings of the National Academy of Sciences of the United States of America*, *109*, 14241–14246.

Burg, M. B., & Green, N. (1973). Function of the thick ascending limb of Henle's loop. *The American Journal of Physiology*, *224*, 659–668.

Carlton, V. E., Harris, B. Z., Puffenberger, E. G., Batta, A. K., Knisely, A. S., Robinson, D. L., et al. (2003). Complex inheritance of familial hypercholanemia with associated mutations in TJP2 and BAAT. *Nature Genetics*, *34*, 91–96.

Claverie-Martin, F., Garcia-Nieto, V., Loris, C., Ariceta, G., Nadal, I., Espinosa, L., et al. (2013). Claudin-19 mutations and clinical phenotype in Spanish patients with familial hypomagnesemia with hypercalciuria and nephrocalcinosis. *PloS One*, *8*, e53151.

Corre, T., Olinger, E., Harris, S. E., Traglia, M., Ulivi, S., Lenarduzzi, S., et al. (2017). Common variants in CLDN14 are associated with differential excretion of magnesium over calcium in urine. *Pflugers Archiv: European Journal of Physiology*, *469*, 91–103.

Driscoll, D. A., Spinner, N. B., Budarf, M. L., McDonald-McGinn, D. M., Zackai, E. H., Goldberg, R. B., et al. (1992). Deletions and microdeletions of 22q112 in velo-cardio-facial syndrome. *American Journal of Medical Genetics*, *44*, 261–268.

Ekinci, Z., Karabas, L., & Konrad, M. (2012). Hypomagnesemia-hypercalciuria-nephrocalcinosis and ocular findings: a new claudin-19 mutation. *The Turkish Journal of Pediatrics*, *54*, 168–170.

Falcone, T., Fazio, V., Lee, C., Simon, B., Franco, K., Marchi, N., & Janigro, D. (2010). Serum S100B: A potential biomarker for suicidality in adolescents? *PloS One*, *5*, e11089.

Feldmeyer, L., Huber, M., Fellmann, F., Beckmann, J. S., Frenk, E., & Hohl, D. (2006). Confirmation of the origin of NISCH syndrome. *Human Mutation*, *27*, 408–410.

Furuse, M., Fujita, K., Hiiragi, T., Fujimoto, K., & Tsukita, S. (1998). Claudin-1 and -2: novel integral membrane proteins localizing at tight junctions with no sequence similarity to occludin. *The Journal of Cell Biology*, *141*, 1539–1550.

Furuse, M., Hata, M., Furuse, K., Yoshida, Y., Haratake, A., Sugitani, Y., Noda, T., Kubo, A., & Tsukita, S. (2002). Claudin-based tight junctions are crucial for the mammalian epidermal barrier: a lesson from claudin-1-deficient mice. *The Journal of Cell Biology*, *156*, 1099–1111.

Godron, A., Harambat, J., Boccio, V., Mensire, A., May, A., Rigothier, C., et al. (2012). Familial hypomagnesemia with hypercalciuria and nephrocalcinosis: phenotype-genotype correlation and outcome in 32 patients with CLDN16 or CLDN19 mutations. *Clinical Journal of the American Society of Nephrology: CJASN*, *7*, 801–809.

Gong, Y., & Hou, J. (2014). Claudin-14 underlies Ca + +-sensing receptor-mediated Ca + + metabolism via NFAT-microRNA-based mechanisms. *Journal of the American Society of Nephrology: JASN*, *25*, 745–760.

Gong, Y., Renigunta, V., Himmerkus, N., Zhang, J., Renigunta, A., Bleich, M., & Hou, J. (2012). Claudin-14 regulates renal Ca(+)(+) transport in response to CaSR signalling via a novel microRNA pathway. *The EMBO Journal*, *31*, 1999–2012.

Greene, C., Kealy, J., Humphries, M. M., Gong, Y., Hou, J., Hudson, N., et al. (2017). Dose-dependent expression of claudin-5 is a modifying factor in schizophrenia. *Molecular psychiatry*. Oct 10. doi:10.1038/mp.2017.156. [Epub ahead of print].

Greger, R. (1985). Ion transport mechanisms in thick ascending limb of Henle's loop of mammalian nephron. *Physiological Reviews*, *65*, 760–797.

Grosse, B., Cassio, D., Yousef, N., Bernardo, C., Jacquemin, E., & Gonzales, E. (2012). Claudin-1 involved in neonatal ichthyosis sclerosing cholangitis syndrome regulates hepatic paracellular permeability. *Hepatology*, *55*, 1249–1259 Baltimore, Md.

Hadj-Rabia, S., Baala, L., Vabres, P., Hamel-Teillac, D., Jacquemin, E., Fabre, M., et al. (2004). Claudin-1 gene mutations in neonatal sclerosing cholangitis associated with ichthyosis: a tight junction disease. *Gastroenterology*, *127*, 1386–1390.

Hadj-Rabia, S., Brideau, G., Al-Sarraj, Y., Maroun, R. C., Figueres, M. L., Leclerc-Mercier, S., Olinger, E., Baron, S., Chaussain, C., Nochy, D., Taha, R. Z., Knebelmann, B., Joshi, V., Curmi, P. A., Kambouris, M., Vargas-Poussou, R., Bodemer, C., Devuyst, O., Houillier, P., & El-Shanti, H. (2018). Multiplex epithelium dysfunction due to CLDN10 mutation: the HELIX syndrome. *Genet Med*, *20*(2), 190–201.

Hirano, T., Kobayashi, N., Itoh, T., Takasuga, A., Nakamaru, T., Hirotsune, S., & Sugimoto, Y. (2000). Null mutation of PCLN-1/Claudin-16 results in bovine chronic interstitial nephritis. *Genome Research*, *10*, 659–663.

Hirase, T., Staddon, J. M., Saitou, M., Ando-Akatsuka, Y., Itoh, M., Furuse, M., Fujimoto, K., Tsukita, S., & Rubin, L. L. (1997). Occludin as a possible determinant of tight junction permeability in endothelial cells. *Journal of Cell Science*, *110*(Pt 14), 1603–1613.

Hou, J., Paul, D. L., & Goodenough, D. A. (2005). Paracellin-1 and the modulation of ion selectivity of tight junctions. *Journal of Cell Science*, *118*, 5109–5118.

Hou, J., Shan, Q., Wang, T., Gomes, A. S., Yan, Q., Paul, D. L., Bleich, M., & Goodenough, D. A. (2007). Transgenic RNAi depletion of claudin-16 and the renal handling of magnesium. *The Journal of Biological Chemistry*, *282*, 17114–17122.

Hou, J., Renigunta, A., Konrad, M., Gomes, A. S., Schneeberger, E. E., Paul, D. L., Waldegger, S., & Goodenough, D. A. (2008). Claudin-16 and claudin-19 interact and form a cation-selective tight junction complex. *The Journal of Clinical Investigation*, *118*, 619–628.

Hou, J., Renigunta, A., Gomes, A. S., Hou, M., Paul, D. L., Waldegger, S., & Goodenough, D. A. (2009). Claudin-16 and claudin-19 interaction is required for their assembly into tight junctions and for renal reabsorption of magnesium. *Proceedings of the National Academy of Sciences of the United States of America, 106*, 15350–15355.

Ikari, A., Hirai, N., Shiroma, M., Harada, H., Sakai, H., Hayashi, H., Suzuki, Y., Degawa, M., & Takagi, K. (2004). Association of paracellin-1 with ZO-1 augments the reabsorption of divalent cations in renal epithelial cells. *The Journal of Biological Chemistry, 279*, 54826–54832.

Itoh, M., Furuse, M., Morita, K., Kubota, K., Saitou, M., & Tsukita, S. (1999). Direct binding of three tight junction-associated MAGUKs, ZO-1, ZO-2, and ZO-3, with the COOH termini of claudins. *The Journal of Cell Biology, 147*, 1351–1363.

Jesaitis, L. A., & Goodenough, D. A. (1994). Molecular characterization and tissue distribution of ZO-2, a tight junction protein homologous to ZO-1 and the Drosophila discs-large tumor suppressor protein. *The Journal of Cell Biology, 124*, 949–961.

Kang, J. H., Choi, H. J., Cho, H. Y., Lee, J. H., Ha, I. S., Cheong, H. I., & Choi, Y. (2005). Familial hypomagnesemia with hypercalciuria and nephrocalcinosis associated with CLDN16 mutations. *Pediatric Nephrology, 20*, 1490–1493 Berlin, Germany.

Kausalya, P. J., Amasheh, S., Gunzel, D., Wurps, H., Muller, D., Fromm, M., & Hunziker, W. (2006). Disease-associated mutations affect intracellular traffic and paracellular Mg2+ transport function of Claudin-16. *The Journal of Clinical Investigation, 116*, 878–891.

Kirchmeier, P., Sayar, E., Hotz, A., Hausser, I., Islek, A., Yilmaz, A., Artan, R., & Fischer, J. (2014). Novel mutation in the CLDN1 gene in a Turkish family with neonatal ichthyosis sclerosing cholangitis (NISCH) syndrome. *The British Journal of Dermatology, 170*, 976–978.

Klar, J., Piontek, J., Milatz, S., Tariq, M., Jameel, M., Breiderhoff, T., et al. (2017). Altered paracellular cation permeability due to a rare CLDN10B variant causes anhidrosis and kidney damage. *PLoS Genetics, 13*, e1006897.

Konishi, T., & Salt, A. N. (1983). Electrochemical profile for potassium ions across the cochlear hair cell membranes of normal and noise-exposed guinea pigs. *Hearing Research, 11*, 219–233.

Konrad, M., Schaller, A., Seelow, D., Pandey, A. V., Waldegger, S., Lesslauer, A., et al. (2006). Mutations in the tight-junction gene claudin 19 (CLDN19) are associated with renal magnesium wasting, renal failure, and severe ocular involvement. *American Journal of Human Genetics, 79*, 949–957.

Konrad, M., Hou, J., Weber, S., Dotsch, J., Kari, J. A., Seeman, T., et al. (2008). CLDN16 genotype predicts renal decline in familial hypomagnesemia with hypercalciuria and nephrocalcinosis. *Journal of the American Society of Nephrology: JASN, 19*, 171–181.

Kutluturk, F., Temel, B., Uslu, B., Aral, F., Azezli, A., Orhan, Y., Konrad, M., & Ozbey, N. (2006). An unusual patient with hypercalciuria, recurrent nephrolithiasis, hypomagnesemia and G227R mutation of Paracellin-1: An unusual patient with hypercalciuria and hypomagnesemia unresponsive to thiazide diuretics. *Hormone Research, 66*, 175–181.

Lee, D. B., Huang, E., & Ward, H. J. (2006). Tight junction biology and kidney dysfunction. *American Journal of Physiology Renal physiology, 290*, F20–34.

Lee, K., Ansar, M., Andrade, P. B., Khan, B., Santos-Cortez, R. L., Ahmad, W., & Leal, S. M. (2012). Novel CLDN14 mutations in Pakistani families with autosomal recessive non-syndromic hearing loss. *American Journal of Medical Genetics Part A 158a*, 315–321.

Manz, F., Scharer, K., Janka, P., & Lombeck, J. (1978). Renal magnesium wasting, incomplete tubular acidosis, hypercalciuria and nephrocalcinosis in siblings. *European Journal of Pediatrics, 128*, 67–79.

Masuda, S., Oda, Y., Sasaki, H., Ikenouchi, J., Higashi, T., Akashi, M., Nishi, E., & Furuse, M. (2011). LSR defines cell corners for tricellular tight junction formation in epithelial cells. *Journal of Cell Science, 124*, 548–555.

Meier, W., Blumberg, A., Imahorn, W., De Luca, F., Wildberger, H., & Oetliker, O. (1979). Idiopathic hypercalciuria with bilateral macular colobomata: a new variant of oculo-renal syndrome. *Helvetica Paediatrica Acta, 34*, 257–269.

Michelis, M. F., Drash, A. L., Linarelli, L. G., De Rubertis, F. R., & Davis, B. B. (1972). Decreased bicarbonate threshold and renal magnesium wasting in a sibship with distal renal tubular acidosis: (Evaluation of the pathophysiological role of parathyroid hormone). *Metabolism: Clinical and Experimental, 21*, 905–920.

Mochida, G. H., Ganesh, V. S., Felie, J. M., Gleason, D., Hill, R. S., Clapham, K. R., et al. (2010). A homozygous mutation in the tight-junction protein JAM3 causes hemorrhagic destruction of the brain, subependymal calcification, and congenital cataracts. *American Journal of Human Genetics, 87*, 882–889.

Morozko, E. L., Nishio, A., Ingham, N. J., Chandra, R., Fitzgerald, T., Martelletti, E., et al. (2015). ILDR1 null mice, a model of human deafness DFNB42, show structural aberrations of tricellular tight junctions and degeneration of auditory hair cells. *Human molecular genetics, 24*, 609–624.

Muller, D., Kausalya, P. J., Claverie-Martin, F., Meij, I. C., Eggert, P., Garcia-Nieto, V., & Hunziker, W. (2003). A novel claudin 16 mutation associated with childhood hypercalciuria abolishes binding to ZO-1 and results in lysosomal mistargeting. *American Journal of Human Genetics, 73*, 1293–1301.

Muller, D., Kausalya, P. J., Bockenhauer, D., Thumfart, J., Meij, I. C., Dillon, M. J., van't Hoff, W., & Hunziker, W. (2006a). Unusual clinical presentation and possible rescue of a novel claudin-16 mutation. *The Journal of Clinical Endocrinology and Metabolism, 91*, 3076–3079.

Muller, D., Kausalya, P. J., Meij, I. C., & Hunziker, W. (2006b). Familial hypomagnesemia with hypercalciuria and nephrocalcinosis: Blocking endocytosis restores surface expression of a novel Claudin-16 mutant that lacks the entire C-terminal cytosolic tail. *Human Molecular Genetics, 15*, 1049–1058.

Naeem, M., Hussain, S., & Akhtar, N. (2011). Mutation in the tight-junction gene claudin 19 (CLDN19) and familial hypomagnesemia, hypercalciuria, nephrocalcinosis (FHHNC) and severe ocular disease. *American Journal of Nephrology, 34*, 241–248.

Nayak, G., Lee, S. I., Yousaf, R., Edelmann, S. E., Trincot, C., Van Itallie, C. M., et al. (2013). Tricellulin deficiency affects tight junction architecture and cochlear hair cells. *The Journal of Clinical Investigation, 123*, 4036–4049.

Nitta, T., Hata, M., Gotoh, S., Seo, Y., Sasaki, H., Hashimoto, N., Furuse, M., & Tsukita, S. (2003). Size-selective loosening of the blood-brain barrier in claudin-5-deficient mice. *The Journal of Cell Biology, 161*, 653–660.

O'Driscoll, M. C., Daly, S. B., Urquhart, J. E., Black, G. C., Pilz, D. T., Brockmann, K., et al. (2010). Recessive mutations in the gene encoding the tight junction protein occludin cause band-like calcification with simplified gyration and polymicrogyria. *American Journal of Human Genetics, 87*, 354–364.

Ohba, Y., Kitagawa, H., Kitoh, K., Sasaki, Y., Takami, M., Shinkai, Y., & Kunieda, T. (2000). A deletion of the paracellin-1 gene is responsible for renal tubular dysplasia in cattle. *Genomics, 68*, 229–236.

Omidinia, E., Mashayekhi Mazar, F., Shahamati, P., Kianmehr, A., & Shahbaz Mohammadi, H. (2014). Polymorphism of the CLDN5 gene and Schizophrenia in an Iranian Population. *Iranian Journal of Public Health, 43*, 79–83.

Orlova, V. V., Economopoulou, M., Lupu, F., Santoso, S., & Chavakis, T. (2006). Junctional adhesion molecule-C regulates vascular endothelial permeability by modulating VE-cadherin-mediated cell-cell contacts. *The Journal of Experimental Medicine, 203*, 2703–2714.

Peng, S., Rao, V. S., Adelman, R. A., & Rizzolo, L. J. (2011). Claudin-19 and the barrier properties of the human retinal pigment epithelium. *Investigative Ophthalmology & Visual Science, 52*, 1392–1403.

Praga, M., Vara, J., Gonzalez-Parra, E., Andres, A., Alamo, C., Araque, A., Ortiz, A., & Rodicio, J. L. (1995). Familial hypomagnesemia with hypercalciuria and nephrocalcinosis. *Kidney International, 47*, 1419–1425.

Ramzan, K., Shaikh, R. S., Ahmad, J., Khan, S. N., Riazuddin, S., Ahmed, Z. M., Friedman, T. B., Wilcox, E. R., & Riazuddin, S. (2005). A new locus for nonsyndromic deafness DFNB49 maps to chromosome 5q12.3-q14. 1. *Human Genetics, 116*, 17–22.

Riazuddin, S., Ahmed, Z. M., Fanning, A. S., Lagziel, A., Kitajiri, S., Ramzan, K., et al. (2006). Tricellulin is a tight-junction protein necessary for hearing. *American Journal of Human Genetics, 79*, 1040–1051.

Rodriguez-Soriano, J., & Vallo, A. (1994). Pathophysiology of the renal acidification defect present in the syndrome of familial hypomagnesaemia-hypercalciuria. *Pediatric Nephrology, 8*, 431–435 Berlin, Germany.

Saitou, M., Furuse, M., Sasaki, H., Schulzke, J. D., Fromm, M., Takano, H., Noda, T., & Tsukita, S. (2000). Complex phenotype of mice lacking occludin, a component of tight junction strands. *Molecular Biology of the Cell, 11*, 4131–4142.

Sakisaka, S., Kawaguchi, T., Taniguchi, E., Hanada, S., Sasatomi, K., Koga, H., et al. (2001). Alterations in tight junctions differ between primary biliary cirrhosis and primary sclerosing cholangitis. *Hepatology, 33*, 1460–1468 Baltimore, Md.

Sambrotta, M., Strautnieks, S., Papouli, E., Rushton, P., Clark, B. E., Parry, D. A., et al. (2014). Mutations in TJP2 cause progressive cholestatic liver disease. *Nat Genet, 46*, 326–328.

Sanjad, S. A., Hariri, A., Habbal, Z. M., & Lifton, R. P. (2007). A novel PCLN-1 gene mutation in familial hypomagnesemia with hypercalciuria and atypical phenotype. *Pediatric Nephrology, 22*, 503–508 Berlin, Germany.

Sharma, S., Place, E., Lord, K., Leroy, B. P., Falk, M. J., & Pradhan, M. (2016). Claudin 19-based familial hypomagnesemia with hypercalciuria and nephrocalcinosis in a sibling pair. *Clinical Nephrology, 85*, 346–352.

Shprintzen, R. J., Goldberg, R. B., Lewin, M. L., Sidoti, E. J., Berkman, M. D., Argamaso, R. V., & Young, D. (1978). A new syndrome involving cleft palate, cardiac anomalies, typical facies, and learning disabilities: velo-cardio-facial syndrome. *The Cleft Palate Journal, 15*, 56–62.

Simon, D. B., Lu, Y., Choate, K. A., Velazquez, H., Al-Sabban, E., Praga, M., et al. (1999). Paracellin-1, a renal tight junction protein required for paracellular Mg2+ resorption. *Science (New York, NY), 285*, 103–106.

Sirotkin, H., Morrow, B., Saint-Jore, B., Puech, A., Das Gupta, R., Patanjali, S. R., Skoultchi, A., Weissman, S. M., & Kucherlapati, R. (1997). Identification, characterization, and precise mapping of a human gene encoding a novel membrane-spanning protein from the 22q11 region deleted in velo-cardio-facial syndrome. *Genomics, 42*, 245–251.

Someya, S., Xu, J., Kondo, K., Ding, D., Salvi, R. J., Yamasoba, T., et al. (2009). Age-related hearing loss in C57BL/6J mice is mediated by Bak-dependent mitochondrial apoptosis. *Proceedings of the National Academy of Sciences of the United States of America, 106*, 19432–19437.

Sun, Z. Y., Wei, J., Xie, L., Shen, Y., Liu, S. Z., Ju, G. Z., et al. (2004). The CLDN5 locus may be involved in the vulnerability to schizophrenia. *European Psychiatry: The Journal of the Association of European Psychiatrists, 19*, 354–357.

Tapia, R., Huerta, M., Islas, S., Avila-Flores, A., Lopez-Bayghen, E., Weiske, J., Huber, O., & Gonzalez-Mariscal, L. (2009). Zona occludens-2 inhibits cyclin D1 expression and cell proliferation and exhibits changes in localization along the cell cycle. *Molecular Biology of the Cell, 20*, 1102–1117.

Thorleifsson, G., Holm, H., Edvardsson, V., Walters, G. B., Styrkarsdottir, U., Gudbjartsson, D. F., et al. (2009). Sequence variants in the CLDN14 gene associate with kidney stones and bone mineral density. *Nature Genetics, 41*, 926–930.

Toka, H. R., Genovese, G., Mount, D. B., Pollak, M. R., & Curhan, G. C. (2013). Frequency of rare allelic variation in candidate genes among individuals with low and high urinary calcium excretion. *PloS One, 8*, e71885.

Ulloa-Aguirre, A., Janovick, J. A., Brothers, S. P., & Conn, P. M. (2004). Pharmacologic rescue of conformationally-defective proteins: implications for the treatment of human disease. *Traffic, 5*, 821–837 Copenhagen, Denmark.

Umeda, K., Ikenouchi, J., Katahira-Tayama, S., Furuse, K., Sasaki, H., Nakayama, M., Matsui, T., Tsukita, S., Furuse, M., & Tsukita, S. (2006). ZO-1 and ZO-2 independently determine where claudins are polymerized in tight-junction strand formation. *Cell, 126*, 741–754.

Van Itallie, C. M., Rogan, S., Yu, A., Vidal, L. S., Holmes, J., & Anderson, J. M. (2006). Two splice variants of claudin-10 in the kidney create paracellular pores with different ion selectivities. *American Journal of Physiology Renal Physiology, 291*, F1288–1299.

Walsh, T., Pierce, S. B., Lenz, D. R., Brownstein, Z., Dagan-Rosenfeld, O., Shahin, H., et al. (2010). Genomic duplication and overexpression of TJP2/ZO-2 leads to altered expression of apoptosis genes in progressive nonsyndromic hearing loss DFNA51. *American Journal of Human Genetics, 87*, 101–109.

Wattenhofer, M., Reymond, A., Falciola, V., Charollais, A., Caille, D., Borel, C., et al. (2005). Different mechanisms preclude mutant CLDN14 proteins from forming tight junctions in vitro. *Human Mutation, 25*, 543–549.

Weber, S., Hoffmann, K., Jeck, N., Saar, K., Boeswald, M., Kuwertz-Broeking, E., et al. (2000). Familial hypomagnesaemia with hypercalciuria and nephrocalcinosis maps to chromosome 3q27 and is associated with mutations in the PCLN-1 gene. *European Journal of Human Genetics: EJHG, 8*, 414–422.

Weber, S., Schlingmann, K. P., Peters, M., Nejsum, L. N., Nielsen, S., Engel, H., et al. (2001a). Primary gene structure and expression studies of rodent paracellin-1. *Journal of the American Society of Nephrology: JASN, 12*, 2664–2672.

Weber, S., Schneider, L., Peters, M., Misselwitz, J., Ronnefarth, G., Boswald, M., et al. (2001b). Novel paracellin-1 mutations in 25 families with familial hypomagnesemia with hypercalciuria and nephrocalcinosis. *Journal of the American Society of Nephrology: JASN, 12*, 1872–1881.

Wensink, P. C., Finnegan, D. J., Donelson, J. E., & Hogness, D. S. (1974). A system for mapping DNA sequences in the chromosomes of Drosophila melanogaster. *Cell, 3*, 315–325.

Wilcox, E. R., Burton, Q. L., Naz, S., Riazuddin, S., Smith, T. N., Ploplis, B., et al. (2001). Mutations in the gene encoding tight junction claudin-14 cause autosomal recessive deafness DFNB29. *Cell, 104*, 165–172.

Will, C., Breiderhoff, T., Thumfart, J., Stuiver, M., Kopplin, K., Sommer, K., et al. (2010). Targeted deletion of murine Cldn16 identifies extra- and intrarenal compensatory mechanisms of Ca2+ and Mg2+ wasting. *American Journal of Physiology Renal Physiology, 298*, F1152–1161.

Wosik, K., Cayrol, R., Dodelet-Devillers, A., Berthelet, F., Bernard, M., Moumdjian, R., Bouthillier, A., Reudelhuber, T. L., & Prat, A. (2007). Angiotensin II controls occludin function and is required for blood brain barrier maintenance: relevance to multiple sclerosis. *The Journal of Neuroscience : The Official Journal of the Society for Neuroscience, 27*, 9032–9042.

Wu, N., Zhang, X., Jin, S., Liu, S., Ju, G., Wang, Z., Liu, L., Ye, L., & Wei, J. (2010). A weak association of the CLDN5 locus with schizophrenia in Chinese case-control samples. *Psychiatry Research, 178*, 223.

Wyss, L., Schafer, J., Liebner, S., Mittelbronn, M., Deutsch, U., Enzmann, G., et al. (2012). Junctional adhesion molecule (JAM)-C deficient C57BL/6 mice develop a severe hydrocephalus. *PloS One*, *7*, e45619.

Xu, J., Kausalya, P. J., Phua, D. C., Ali, S. M., Hossain, Z., & Hunziker, W. (2008). Early embryonic lethality of mice lacking ZO-2, but Not ZO-3, reveals critical and nonredundant roles for individual zonula occludens proteins in mammalian development. *Molecular and Cellular Biology*, *28*, 1669–1678.

Ye, L., Sun, Z., Xie, L., Liu, S., Ju, G., Shi, J., et al. (2005). Further study of a genetic association between the CLDN5 locus and schizophrenia. *Schizophrenia Research*, *75*, 139–141.

Yuan, T., Pang, Q., Xing, X., Wang, X., Li, Y., Li, J., et al. (2015). First report of a novel missense CLDN19 mutations causing familial hypomagnesemia with hypercalciuria and nephrocalcinosis in a Chinese family. *Calcified Tissue International*, *96*, 265–273.

Chapter 9

Paracellular Channel as Drug Target

9.1 STRUCTURAL BASIS OF MOLECULAR ADHESION

9.1.1 Adherens Junction

In the epithelium or the endothelium, the adherens junction (AJ) is part of the tripartite junctional complex comprising the tight junction (TJ), the AJ, and the desmosome, which are aligned in the order of the apical to basal end of the lateral membrane (Farquhar & Palade, 1963). Classical cadherins mediate cell membrane adhesion at the AJ and the desmosome (Meng & Takeichi, 2009). The cadherin protein features an amino-terminal extracellular domain or ectodomain that is followed by a transmembrane domain and a carboxyl-terminal intracellular domain. The cadherin ectodomain plays an adhesive role whereas the intracellular domain anchors cadherin to the underlying cytoskeleton. The cadherin ectodomain contains five repeating secondary structures, termed extracellular cadherin (EC) domains and numbered from EC1 to EC5. Each EC domain is made of approximately 110 amino acids and resembles the immunoglobulin domain found in antibodies (Shapiro & Weis, 2009). The crystal structures of cadherin EC domains have been resolved (Fig. 9.1A) (Boggon et al., 2002; Nagar, Overduin, Ikura, & Rini, 1996; Shapiro et al., 1995). Each EC domain is composed of seven β-strands, numbered A − G and arranged as two antiparallel β-sheets. The connections between successive EC domains are rigidified by Ca^{++} ions to adopt a strong curvature (Fig. 9.1A). Single-molecule fluorescence microscopy reveals that cadherins bind in *trans* via dimerization of their EC1 domains (Zhang, Sivasankar, Nelson, & Chu, 2009). The primary model of EC1 dimerization is based upon the domain swapping theory, which states that the amino-terminal β-strand (the A strand) in EC1 domains swaps so that the A strand of one protomer replaces the A strand of the other protomer (Fig. 9.1A) (Chen, Posy, Ben-Shaul, Shapiro, & Honig, 2005). In this way, symmetric dimers are formed without altering the overall conformation of either EC1 domain. The molecular interfaces between the "main" and the "swapped" domains are virtually identical to those in the monomer, except that they are *inter*molecular in the dimer, but *intra*molecular in the monomer.

The Paracellular Channel. http://dx.doi.org/10.1016/B978-0-12-814635-4.00009-7

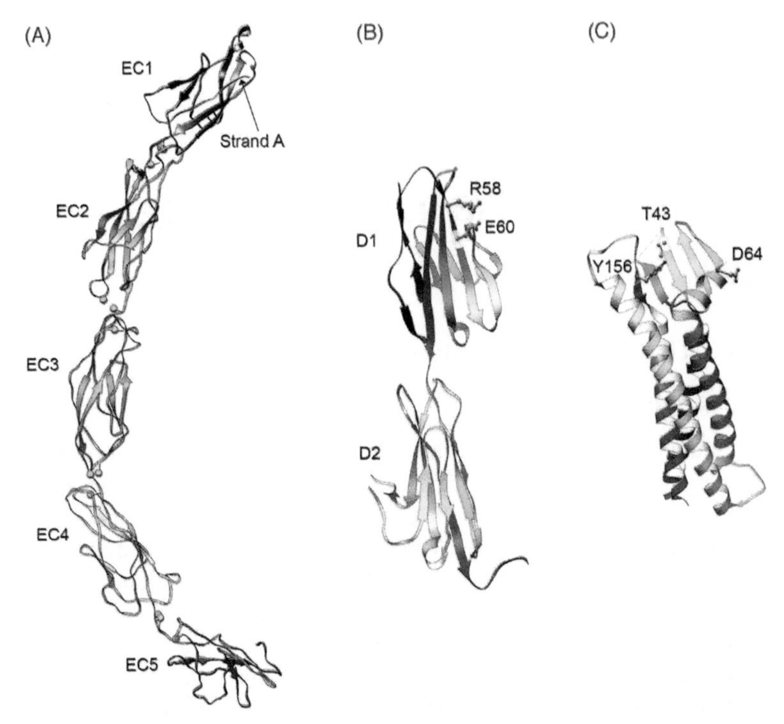

FIGURE 9.1 Molecular structures of cell junction proteins. (A) Crystal structure of *Xenopus* C-cadherin ectodomain in ribbon representation. The *green spheres* depict the bound Ca^{++} ions. The amino-terminal A strand in EC1 domain swaps with the A strand in another cadherin molecule during the dimerizing process. The depicted structure is based upon the X-ray analysis by Boggon et al. (2002). (B) Crystal structure of mouse JAM-A ectodomain in ribbon representation. The sites important for dimerization are labeled. The depicted X-ray structure is from Kostrewa et al. (2001). (C) Crystal structure of mouse claudin-15 in ribbon representation. The sites important for *trans*-interaction are labeled. T43 in claudin-15 is homologous to N44 in claudin-3; D64 in claudin-15 is homologous to D65 in claudin-2; Y156 in claudin-15 is homologous to Y158 in claudin-5. The depicted X-ray structure is from Suzuki et al. (2014). The color changes gradually from the N terminus (*blue*) to the C terminus (*red*) in all three panels.

9.1.2 Tight Junction

9.1.2.1 Junctional Adhesion Molecule (JAM)

JAMs belong to a family of single-pass transmembrane proteins found in the TJs of the epithelial cells and the endothelial cells (Chapter 2, Section 2.7). JAM contributes to the adhesive function of the TJ. JAM consists in an amino-terminal ectodomain, a transmembrane domain, and a carboxyl-terminal cytoplasmic domain with a PDZ binding motif (Martin-Padura et al., 1998). The ectodomain comprises two concatenated immunoglobulin-like domains, termed D1 and D2. The crystal structures of D1 and D2 domains in JAM-A have been resolved (Fig. 9.1B) (Kostrewa et al., 2001; Prota et al., 2003). The D1 and D2 domains each contain seven β-strands that fold into two antiparallel β-sheets. In solution, JAM-A assembles noncovalently into dimers (Bazzoni et al., 2000). The dimeric interfaces include two salt bridges connecting the D1 domain in two protomers, which are formed by Arg58 from one protomer and Glu60 from the other protomer (Fig. 9.1B). Notably, JAM dimers are oriented in *cis* on the surface of one cell according to this interaction model. Additional mechanisms may mediate *trans* JAM interaction between two cells.

9.1.2.2 Claudin

The crystal structures of several claudins have now been resolved (Chapter 2, Section 2.3). The most prominent feature is the β-sheet structure in the ectodomain, which comprises five β-strands, four contributed by the first extracellular loop (ECL1) and one by the second extracellular loop (ECL2) (Fig. 9.1C). *Trans* claudin interaction depends upon the β-sheet structure, thereby involving both extracellular loops. A few loci in ECL1 and ECL2 have been demonstrated to play important roles in *trans* claudin interaction. For example, mutation of Asn44 to threonine in ECL1 of claudin-3 conferred novel *trans*-binding affinity (Daugherty, Ward, Smith, Ritzenthaler, & Koval, 2007). Mutation of Asp65 to cysteine in ECL1 of claudin-2 resulted in the formation of *trans*-dimer under nonreducing condition (Angelow & Yu, 2009). Mutation of Tyr158 to alanine in ECL2 of claudin-5 reduced the junctional protein abundance of claudin-5 without affecting its overall protein abundance in the plasma membrane (Piontek et al., 2008). Intriguingly, a serine protease – prostasin was found to act upon Arg158 in claudin-4, which is homologous to Tyr158 in claudin-5, to break *trans* claudin-4 interaction (Gong et al., 2014).

9.2 SMALL-MOLECULE APPROACH

9.2.1 Calcium Chelator

Cereijido and coworkers first demonstrated that calcium chelation with Ethylenediaminetetraacetic acid (EDTA) opened the paracellular pathway in epithelial cells with an experimental approach that later became a standard assay known as the Ca^{++}-switch assay (Cereijido, Robbins, Dolan, Rotunno, & Sabatini, 1978). Electron microscopy revealed that EDTA treatment disrupted the apical cell junctions including AJ and TJ, and widened the intercellular space (Cereijido et al., 1978). It is clear from the crystal structure of cadherin that cell adhesion in the AJ depends upon the presence of Ca^{++} (Fig. 9.1A) (Boggon et al., 2002). Ca^{++}, via interaction with the linker regions in cadherin's ectodomain, holds the loosely connected EC domains. Removal of Ca^{++} leads to the disorientation of EC domains (Pokutta, Herrenknecht, Kemler, & Engel, 1994). While cell adhesion in the TJ is not Ca^{++} dependent, a crosstalk mechanism has long been speculated to operate between AJ and TJ. Taddei and coworkers provided genetic evidence that VE-cadherin directly regulated the gene expression of claudin-5 and the paracellular permeability via the FoxO1–β-catenin transcriptional program in the endothelial cells (Taddei et al., 2008).

9.2.2 Sodium Caprate

Sodium caprate is a sodium salt of medium chain fatty acid (MCFA). It can increase the paracellular permeabilities to ions and small solutes across both epithelial and endothelial cells (Anderberg, Lindmark, & Artursson, 1993; Del Vecchio et al., 2012; Krug et al., 2013; Preston, Slinn, Vinokourov, & Stanimirovic, 2008). In Sweden, sodium caprate has been clinically approved as an absorption enhancer for rectal delivery of ampicillin (Lindmark et al., 1997). Mechanistically, sodium caprate reduces *trans* claudin-5 interaction and destabilizes claudin-5 localization in the TJ (Fig. 9.2) (Del Vecchio et al., 2012). Sodium caprate also induces the contraction of the perijunctional actomyosin ring via intracellular signaling pathways (Anderberg et al., 1993). The contractile force of the perijunctional actomyosin ring is an important modulator of the paracellular permeability (Chapter 2, Section 2.8.1). Interestingly, sodium caprate has been found to regulate the tricellular TJ permeability, which appears to be independent from its effect on bicellular TJ permeability (Krug et al., 2013). The tricellular TJ protein, tricellulin, was rapidly delocalized from the tricellular junction in response to sodium caprate treatment (Krug et al., 2013).

9.2.3 Chitosan

Chitosan is a linear polysaccharide and a polycation at acidic pH. Chitosan increases the paracellular permeabilities to ions and solutes with sizes up to 10 kDa in several intestinal epithelial cell models (Rosenthal et al., 2012; Smith, Wood, & Dornish, 2004; Yeh et al., 2011). Electron microscopy revealed that chitosan treatment disrupted the structural integrity of TJ, allowing lanthanum (a heavy cation tracer, 139 Da) to penetrate into the lateral space (Fig. 9.3) (Sonaje et al., 2012). Several TJ proteins including JAM-A, claudin-4, and ZO-1 were mislocalized by chitosan treatment via various signaling molecules such as integrin, protein kinase C, and Cl^-/HCO_3^- exchanger (Hsu et al., 2013; Hsu et al., 2012; Rosenthal et al., 2012; Smith, Dornish, & Wood, 2005; Sonaje et al., 2012).

9.2.4 Histamine

Histamine has been found to alter the paracellular permeability in the airway epithelium and the vascular endothelium (Flynn, Itani, Moninger, & Welsh, 2009; Gardner et al., 1996). The histamine effect is multifaceted. Histamine directly regulates the claudin proteins, thereby altering the ion selectivity and permeability of the paracellular pathway (Fig. 9.4) (Flynn et al., 2009). Histamine also disrupts the binding between cadherin and β-catenin, which destabilizes the AJ and affects the paracellular permeation indirectly (Guo, Breslin, Wu,

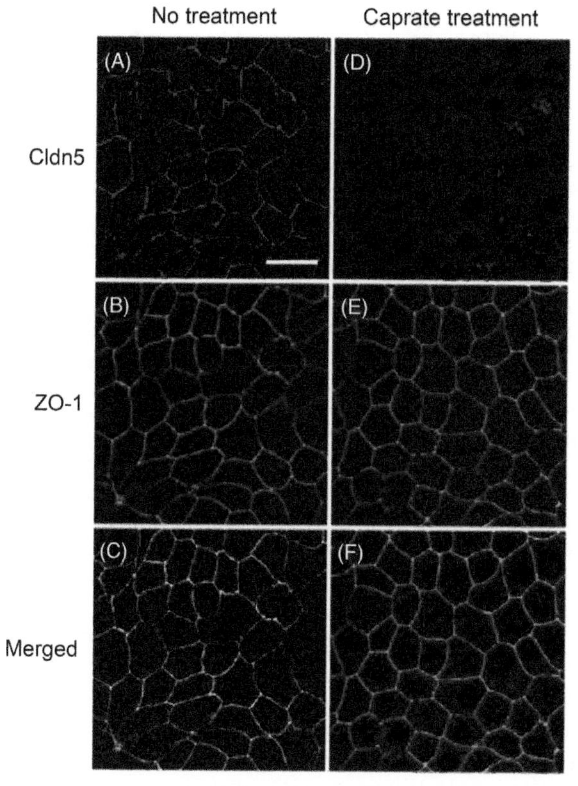

FIGURE 9.2 Sodium caprate effect on claudin-5. In MDCK-II cells transfected with claudin-5, treatment with 7.5 mM sodium caprate for 30 min significantly reduced the TJ abundance level of claudin-5 but not ZO-1. (A–C) Control cells receiving no treatment. (D–F) Sodium caprate treated cells. (A, D) Cells stained for claudin-5. (B, E) Cells stained for ZO-1. (C, F) Merged channel. Bar: 20 μm. *(Reproduced with permission from Del Vecchio, G., Tscheik, C., Tenz, K., Helms, H.C., Winkler, L., Blasig, R., & Blasig, I.E. (2012). Sodium caprate transiently opens claudin-5-containing barriers at tight junctions of epithelial and endothelial cells. Molecular Pharmaceutics, 9, 2523–2533).*

Gottardi, & Yuan, 2008). Finally, histamine induces myosin II regulatory light chain phosphorylation by activating myosin light chain kinase (MLCK) (Srinivas, Satpathy, Guo, & Anandan, 2006). MLCK is known to regulate the dynamic behavior of TJ proteins by contracting the perijunctional actomyosin ring (Chapter 2, Section 2.8.3).

9.3 CYTOKINE

9.3.1 Bradykinin

Bradykinin is a 9-amino acid peptide (RPPGFSPFR) and produced from proteolytic cleavage of its precursor, high-molecular-weight kininogen by the

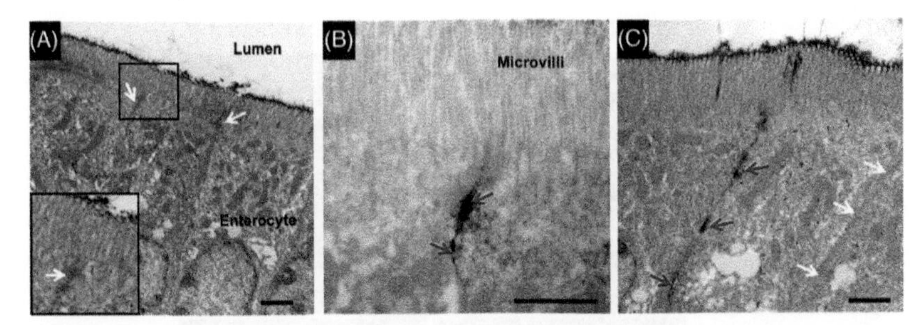

FIGURE 9.3 Chitosan effect on TJ barrier. Transmission electron micrographs showing lanthanum-stained intestinal segments before (A) and after (B, C) oral chitosan treatment in mice. (A) Before the treatment, lanthanum was mostly observed on the surface of the microvilli without entering the paracellular space demarcated by TJs (*white arrows*). (B, C) After the treatment, lanthanum was found in the paracellular space (*red arrows*), indicating the opening of TJs. However, the effect was not homogenous. There were TJs resistant to chitosan treatment (*white arrows*). Bar: 1 μm. *(Reproduced with permission from Sonaje, K., Chuang, E.Y., Lin, K.J., Yen, T.C., Su, F.Y., Tseng, M.T., & Sung, H.W. (2012). Opening of epithelial tight junctions and enhancement of paracellular permeation by chitosan: Microscopic, ultrastructural, and computed-tomographic observations. Molecular Pharmaceutics, 9, 1271–1279).*

serine protease kallikrein in the plasma. Bradykinin elicits many physiological responses including vasodilation and natriuresis (Marceau & Regoli, 2004). Bradykinin and its agonist, RPM-7 (also known as Cereport), were found to induce a transient increase in paracellular permeability across the brain vascular endothelium (Liu, Xue, Liu, & Wang, 2008; Sanovich et al., 1995; Zhou et al., 2014). Mechanistically, bradykinin treatment triggers intracellular Ca^{++} increases and activates the cAMP signaling pathway to downregulate the protein abundance and the TJ localization of claudin-5, occludin, and ZO-1 (Liu et al., 2008; Zhou et al., 2014).

9.3.2 Tumor Necrosis Factor α

TNFα is produced mainly by activated macrophages. It is synthesized as a 26-kDa single-pass transmembrane protein and secreted in a soluble form of 17 kDa via proteolytic cleavage of the extracellular domain by TNFα converting enzyme (TACE, also known as ADAM17) (Palladino, Bahjat, Theodorakis, & Moldawer, 2003). In the intestine, TNFα regulates the paracellular permeability via myosin II regulatory light chain phosphorylation and TJ structural remodeling (Clayburgh, Musch, Leitges, Fu, & Turner, 2006; Schmitz et al., 1999). Occludin was endocytosed within 90 min after TNFα treatment in the mouse intestine (Fig. 9.5) (Marchiando et al., 2010). TNFα also induced a dose-dependent increase in endothelial cell permeability (Anda et al., 1997; Royall et al., 1989). While the TNFα effect on the endothelium involved both transcellular and paracellular pathways, the paracellular alteration was mainly caused by reduced expression of the TJ proteins including claudin-5, occludin,

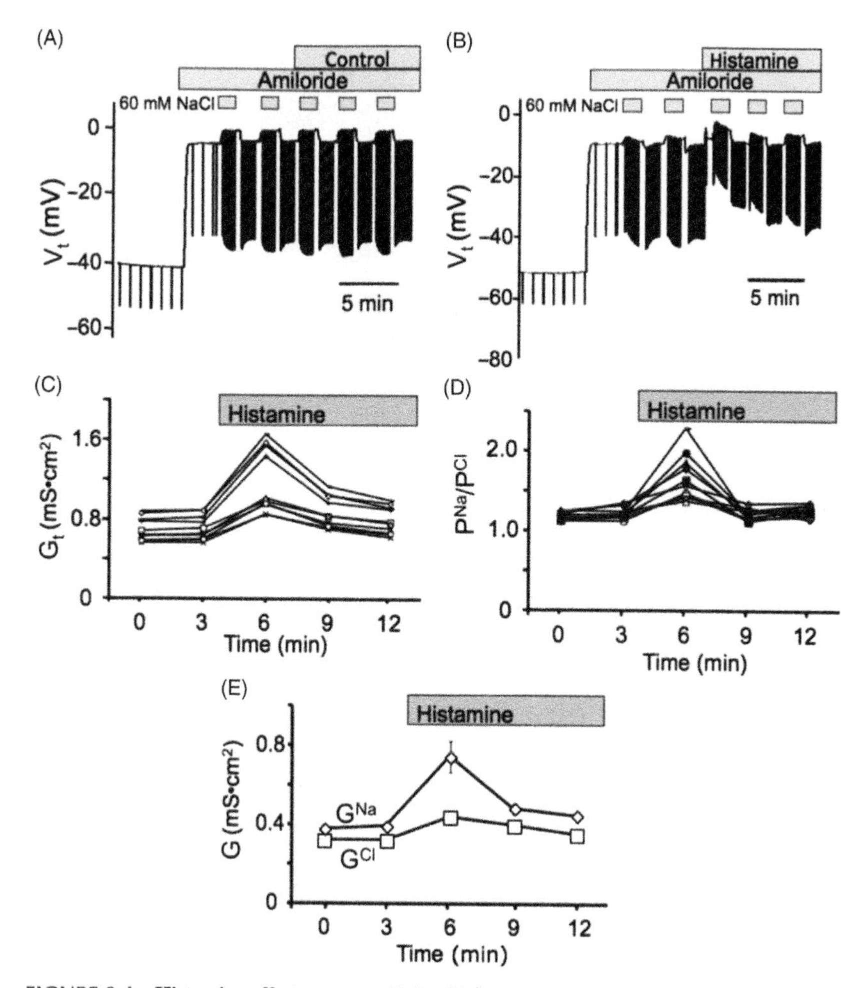

FIGURE 9.4 **Histamine effect on paracellular Na^+ permeability.** Recordings were performed on human airway epithelia. (A and B) Traces of transepithelial voltage (V_{te}) in response to repeated reductions of apical solution NaCl concentration from 135 to 60 mM. Epithelia were treated with basolateral vehicle control (A) or 100 μM histamine (B) during the time indicated by bars. (C–E) Experiments such as those shown in A and B were used to calculate transepithelial conductance (G_{te}) (C), paracellular ion selectivity (P_{Na}/P_{Cl}) (D), and paracellular Na^+ and Cl^- conductance (G_{Na} and G_{Cl}) (E). In C and D, each set of symbol and line is from an independent experiment. Because these epithelia lack CFTR and ENaC has been inhibited by amiloride, histamine-induced changes in G_{te} reflect predominantly changes in paracellular conductance (G_p). Histamine has little effect on V_{te}. Although there is substantial variability among epithelial cultures, histamine increases G_{te} in all cultures. *(Reproduced with permission from Flynn, A.N., Itani, O.A., Moninger, T.O., & Welsh, M.J. (2009). Acute regulation of tight junction ion selectivity in human airway epithelia. Proceedings of the National Academy of Sciences of the United States of America, 106, 3591–3596).*

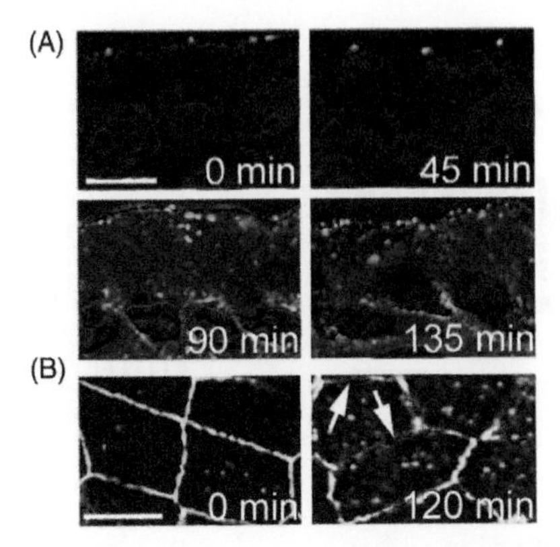

FIGURE 9.5 **Occludin endocytosis in response to TNFα treatment.** Jejunum was harvested from wildtype mice at the indicated time points after intraperitoneal injection of 5 μg TNFα and labeled for occludin (*green*), F-actin (*red*), and nuclei (*blue*). Sagittal view (A) and *en face* view (B) of jejunal sections reveal that occludin endocytosis begins 90 min after TNFα treatment. Note that TNFα treatment renders large regions in the TJ completely lacking occludin (*arrows*). Bar: 10 μm. *(Reproduced with permission from Marchiando, A.M., Shen, L., Graham, W.V., Weber, C.R., Schwarz, B.T., Austin, J.R., 2nd, Raleigh, D.R., Guan, Y., Watson, A.J., Montrose, M.H., et al. (2010). Caveolin-1-dependent occludin endocytosis is required for TNF-induced tight junction regulation in vivo. The Journal of Cell Biology, 189, 111–126).*

and ZO-1 as a result of TNFα-triggered NADPH oxidation event (Rochfort, Collins, Murphy, & Cummins, 2014; Rochfort & Cummins, 2015).

9.4 PROTEASE

9.4.1 Trypsin and Trypsin-Like Protease

Low levels of trypsin and the trypsin-like proteases such as channel activating protease 1 (CAP1, also known as prostasin) and CAP3 (also known as matriptase) are potent inducers of TJ formation in a variety of epithelia (Buzza et al., 2013; Buzza et al., 2010; Lynch et al., 1995). Trypsin, when added to the basolateral surface of Madin-Darby Canine Kidney II (MDCK-II) cells, elicited a rapid decrease in paracellular ionic permeability due to *de novo* formation of aberrant TJ strands (Lynch et al., 1995). The CAP1 knockout mouse died within 60 days after birth because of impaired epidermal barrier function (Leyvraz et al., 2005). The TJ protein−occludin was absent from the knockout mouse skin, which resulted in increased paracellular permeability to small molecules and uncontrolled epidermal water loss (Fig. 9.6). In the kidney epithelial cells, CAP1 selectively reduced the paracellular Cl^- permeability by disrupting the *trans*-interaction of claudin-4, a paracellular anion channel (Fig. 5.3) (Gong et al., 2014). The CAP3 hypomorphic mutant mouse developed a leaky gut barrier (Buzza et al., 2010). Loss of CAP3 promoted the incorporation of claudin-2, a paracellular cation channel, into the TJ via the PKCζ signaling cascade (Buzza et al., 2010).

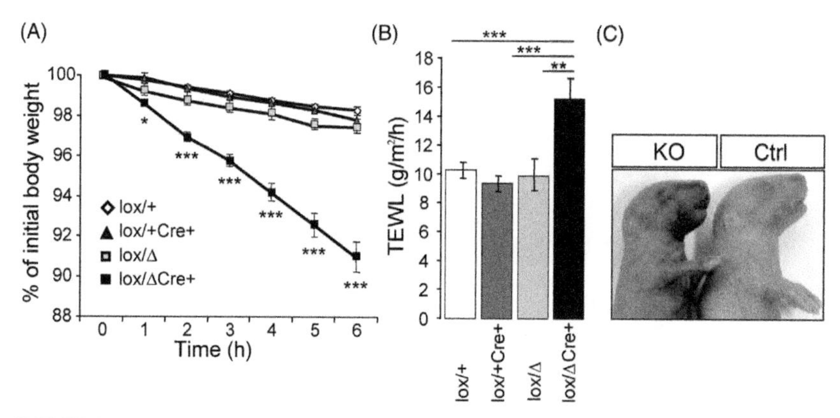

FIGURE 9.6 **Disruption of epidermal barrier function in CAP1 knockout mouse.** (A) Dehydration assay over time. Data are presented as percentages of initial body weight in knockout ($CAP1^{lox/\Delta}$/K14-Cre; *black squares*; $n = 5$) and control ($CAP1^{lox/\Delta}$ and $CAP1^{lox/+}$/ ± K14-Cre; *white diamonds*, *black triangle* and *gray squares*; $n = 5$-8) mice. *Asterisk*, $p < 0.05$; *Triple Asterisk*, $p < 0.001$. (B) Transepidermal water loss (TEWL) measured on ventral skin is decreased in knockout (*black bar*; $n = 6$) pups compared with control pups ($CAP1^{lox/\Delta}$; *light gray bar*; $n = 9$; $CAP1^{lox/+}$/ K14-Cre; *dark gray bar*; $n = 11$; $CAP1^{lox/+}$; white bar; $n = 15$). *Double Asterisk*, $p < 0.01$; *Triple Asterisk*, $p < 0.001$. (C) Barrier dependent dye exclusion assay in knockout (left; $n = 8$) and control (right; $n = 5$) pups. Representative photographs reveal dye (toluidine blue, 374 Da) penetration of ventral epidermal barrier in knockout, but not in control pups. *(Reproduced with permission from Leyvraz, C., Charles, R.P., Rubera, I., Guitard, M., Rotman, S., Breiden, B., Sandhoff, K., and Hummler, E. (2005). The epidermal barrier function is dependent on the serine protease CAP1/Prss8. The Journal of Cell Biology, 170, 487–496).*

9.4.2 Matrix Metalloprotease

The matrix metalloproteases (MMPs) are Zn^{++}- and Ca^{++}-dependent endopeptidases that function to cleave the extracellular matrix (Sternlicht & Werb, 2001). How MMPs regulate the paracellular pathway is best exemplified in the context of blood-brain barrier (BBB) breakdown during ischemic stroke. The BBB refers to the tightly controlled permeability of the brain vasculature (Chapter 7, Section 7.2.1). Shortly after an ischemic stroke attack, the vascular permeability becomes increased around the ischemic area, which causes bleeding into the parenchyma when cerebral blood flow is restored to the damaged vasculature (Jickling et al., 2014). The brain-derived MMP-2 and the leukocyte-derived MMP-9 are the major regulators of BBB permeability in response to ischemic stroke. While there is no direct evidence that these two MMPs act upon the TJ, the protein levels of MMP-2 and MMP-9 are inversely correlated to the protein levels of claudin-5, occludin, and ZO-1 in the mouse brains with ischemic stroke (Feng et al., 2011; Liu, Jin, Liu, & Liu, 2012; Yang, Estrada, Thompson, Liu, & Rosenberg, 2007). In the MMP-9 knockout mouse brain, there was a pronounced reduction in the degree of TJ protein degradation, the vascular permeability, and the infarct size, compared with the wildtype mouse brain injured by the same level of ischemic stroke (Fig. 9.7) (Asahi et al., 2001).

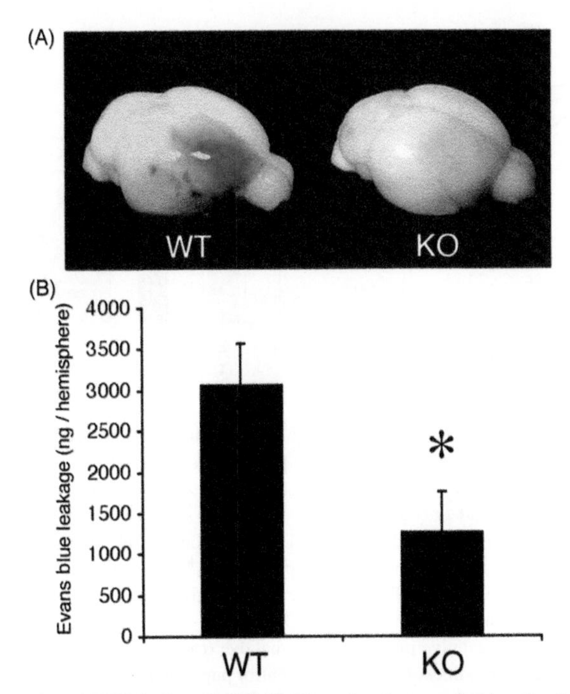

FIGURE 9.7 Reduced BBB leakage in MMP-9 knockout mouse brain after ischemic stroke. (A) Representative brain images showing Evans blue (960 Da) leakage in wildtype (WT) and MMP-9 knockout (KO) mice injured by the same level of ischemic stroke. (B) Quantitation of Evans blue reveals that BBB leakage is significantly lower in knockout mouse brain than in wildtype mouse brain. $N = 7$ per group; *Asterisk*, $p < 0.05$. *(Reproduced with permission from Asahi, M., Wang, X., Mori, T., Sumii, T., Jung, J.C., Moskowitz, M.A., Fini, M.E., and Lo, E.H. (2001). Effects of matrix metalloproteinase-9 gene knock-out on the proteolysis of blood-brain barrier and white matter components after cerebral ischemia. The Journal of Neuroscience, 21, 7724–7732).*

9.5 PEPTIDOMIMETIC

9.5.1 Cadherin Peptidomimetic

A peptidomimetic is a synthetic molecule designed to mimic a structural domain in a natural protein. It is now well-known that *trans* cadherin interaction is mediated by the EC1 domain (Nose, Tsuji, & Takeichi, 1990). There are two potentially important motifs in the EC1 domain of cadherin purported to take part in the interaction: (1) A-D-T located in the hinge region connecting β-strand C to D; (2) H-A-V located in β-strand F (Fig. 9.1A). The peptidomimetics containing the A-D-T or H-A-V motif can increase the paracellular permeability in several epithelial and endothelial cell models (Table 9.1) (Makagiansar, Avery, Hu, Audus, & Siahaan, 2001; Pal, Audus, & Siahaan, 1997; Sinaga et al., 2002). Nevertheless, neither motif is compatible with the domain swapping theory proposed to explain the cadherin *trans*-interaction based upon its crystal structure (Chen et al., 2005).

TABLE 9.1 TJ Modulating Peptidomimetics

Peptide Derived From	Motif Sequence	Peptide Sequence	Reference
Human E-cadherin	ADT	ADTPPV	Sinaga et al. (2002)
Human E-cadherin	HAV	SHAVSS	Makagiansar et al. (2001)
Mouse claudin-1	ECL1:53-80	SCVSQSTGQIQCKVFD-SLLNLNSTLQAT	Mrsny et al. (2008)
Rat claudin-1	ECL1:53-81	SSVSQSTGQIQSKVFD-SLLNLNSTLQATR[a]	Zwanziger et al. (2012)
Mouse claudin-3	ECL2: 145-149	DFYNP[b]	Baumgartner et al. (2011)
Mouse claudin-5	ECL2: 146-150	EFYDP[b]	Schlingmann et al. (2016)
Human occludin	ECL1: 90-103	C_{14}-DRGYGTSLLGGSVG[c]	Everett et al. (2006), Tavelin et al. (2003)
Human occludin	ECL1: 90-112	DRGYGTSLLGGSVGY-PYGGSGFG	Van Itallie & Anderson (1997)
Chick occludin	ECL2: 184-227	GVNPQAQMSS-GYYYSPLLAMCSQ AYGSTYLNQYIYHYCT-VDPQE[d]	Vietor, Bader, Paiha, and Huber (2001), Wong and Gumbiner (1997)
Rat occludin	ECL2: 209-230	GSQIYTICSQFYTPGGT-GLYVD	Chung et al. (2001)
Human occludin	ECL2: 210-228	SQIYALCNQFYTPAAT-GLYVD	Nusrat et al. (2005)
Human tricellulin	ECL2: 313-336	NDTNRGGLSYYPLFNT-PVNAVFSR[a]	Cording et al. (2017)

ECL1, 1st extracellular loop; *ECL2*, 2nd extracellular loop.
[a]*The two underlined serines substitute cysteines, which prevents the formation of intrachain disulfide bond.*
[b]*The peptide is synthesized in the D-amino acid form.*
[c]*A lipophilic amino acid moiety [H_2N—$CH(C_{12}H_{25})$—COOH] is conjugated to the N-terminus of the peptide to prevent its enzymatic degradation.*
[d]*The two underlined cysteines are modified by covalent linkage to acetamidomethyl group to prevent the formation of intrachain disulfide bond.*

9.5.2 Claudin Peptidomimetic

By examining the crystal lattice of claudin-15 molecules, Suzuki and coworkers have proposed a *cis*-assembly model for claudin polymerization (Chapter 2, Section 2.4.4) (Suzuki et al., 2014). In this model, a hydrophobic residue (Met68) from the ECL1 domain of one claudin snugly fits into a hydrophobic pocket formed by the residues (Phe146, Phe147, and Leu158) in the third

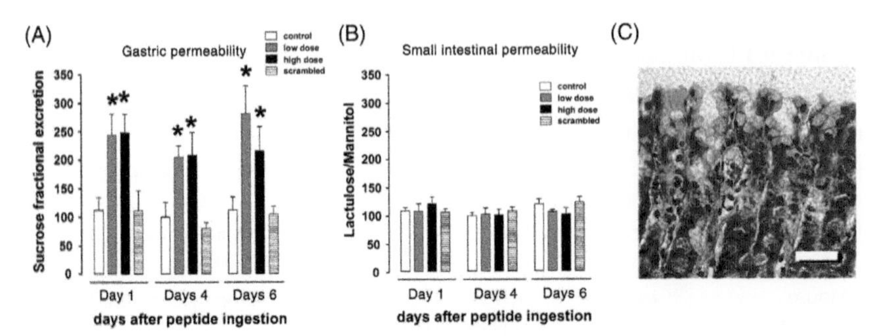

FIGURE 9.8 Claudin-1 peptidomimetic effect on gastric permeability in rat. (A and B) Rats were administered sucrose, lactulose, and mannitol alone (control, *white bars*) or along with claudin-1 peptidomimetic (CLDN$1_{53\text{-}80}$) at either 0.1 mg/kg body weight (low dose, *gray bars*), 1 mg/kg body weight (high dose, *black bars*), or with 1 mg/kg scrambled peptide (*stippled bars*) by oral gavage. Permeabilities of the stomach (A) and small intestine (B) were assessed by measuring urinary disaccharide content. CLDN$1_{53\text{-}80}$ induced a significant increase in gastric permeability but not in small intestinal permeability at both low and high doses. *Asterisk*, $p < 0.05$. (C) H&E-stained section of gastric tissue from rats treated with high dose CLDN$1_{53\text{-}80}$. Peptide administration caused no epithelial loss or altered mucosal architecture. Bar: 100 μm. *(Reproduced with permission from Mrsny, R.J., Brown, G.T., Gerner-Smidt, K., Buret, A.G., Meddings, J.B., Quan, C., Koval, M., & Nusrat, A. (2008). A key claudin extracellular loop domain is critical for epithelial barrier integrity. The American Journal of Pathology, 172, 905–915).*

transmembrane domain and the ECL2 domain of the adjacent claudin (Fig. 2.6). *Trans* claudin interaction involves additional loci in ECL1 and ECL2 domains (Chapter 9, Section 9.1.2.2). Compatible with the *cis*- and *trans*-interaction models of claudin, several peptidomimetics derived from the ECL1 or ECL2 domain in claudin show potent regulatory effects on the paracellular pathway in a variety of cells and tissues (Table 9.1). Mrsny and coworkers first developed a peptidomimetic (CLDN$1_{53\text{-}80}$) corresponding to amino acids 53-80 in the ECL1 domain of mouse claudin-1 protein (Mrsny et al., 2008). When applied to human intestinal epithelial cells, CLDN$1_{53\text{-}80}$ increased the paracellular permeabilities to ions and solutes with sizes of <3 kDa. When fed to experimental rats, CLDN$1_{53\text{-}80}$ enhanced the gastric absorption of sucrose (Fig. 9.8) (Mrsny et al., 2008). Unexpectedly, CLDN$1_{53\text{-}80}$ binds not only to claudin-1 but also to occludin. Prolonged incubation with CLDN$1_{53\text{-}80}$ caused TJ disintegration, manifested by actin rearrangement and delocalization of claudin-1, occludin, JAM-A, and ZO-1 (Mrsny et al., 2008). A modified version of claudin-1 peptidomimetic (CLDN$1_{53\text{-}81}$) has been developed to aid the paracellular permeation of antinociceptive agents through the perineurial barrier (Zwanziger et al., 2012). Again, the CLDN$1_{53\text{-}81}$ effect is promiscuous. It altered the localization of claudin-1, -2, -3, -4, -5, and occludin (Staat et al., 2015). If a peptidomimetic disrupts the claudin-1 interaction by competing for the ECL1 domain, then a peptidomimetic acting upon the ECL2 domain will exert a similar effect. There are two claudin peptidomimetics targeting the ECL2 domain: the claudin-3 peptidomimetic (CLDN$3_{145\text{-}149}$) corresponding to the motif – DFYNP (aa. 145-149) in mouse claudin-3 protein (Baumgartner, Beeman, Hodges, & Neville, 2011); and the claudin-5 peptidomimetic ($CLDN5_{146\text{-}150}$) corresponding to the motif – EFYDP (aa. 146-150) in mouse claudin-5 protein (Schlingmann et al., 2016). Both motifs are vital to the forming of the hydrophobic pocket needed to support claudin *cis*-assembly (Fig. 2.6). Apart from regulating the paracellular pathway, $CLDN3_{145\text{-}149}$ was found to induce apoptosis in mammary epithelial cells (Baumgartner et al., 2011). $CLDN5_{146\text{-}150}$ protected the alveolar barrier function by antagonizing endogenous claudin-5 proteins, whose expression levels were increased in response to chronic alcohol treatment. Overexpressed claudin-5 proteins rearranged the actin cytoskeleton to create the TJ "spikes," which are abnormal cell junction structures associated with increased paracellular permeabilities across the alveolar epithelium (Schlingmann et al., 2016).

9.5.3 Occludin Peptidomimetic

Occludin belongs to the TJ associated Marvel domain containing protein (TAMP) family (Chapter 2, Section 2.6). From the hydrophobicity plot, occludin is predicted to consist in four transmembrane domains and two extracellular loop (ECL1 and ECL2) domains. Various forms of occludin peptidomimetics have been developed, all deriving from amino acid sequences in the ECL1 or ECL2 domain (Table 9.1). The results, however, are somewhat conflicting. Van Itallie and Anderson showed that a peptidomimctic ($OCLN_{90\text{-}112}$) corresponding to amino acids 90-112 in the ECL1 domain of human occludin protein inhibited *trans* occludin interaction in fibroblasts ectopically expressing the occludin gene (Van Itallie & Anderson, 1997). A shorter and modified version of occludin peptidomimetic ($OCLN_{90\text{-}103}$) increased the paracellular permeabilities to ions and solutes with a wide range of molecular sizes, for example, mannitol (182 Da), 70-kDa or 2,000-kDa dextran in human intestinal and airway epithelial cells (Everett, Vanhook, Barozzi, Toth, & Johnson, 2006; Tavelin et al., 2003). The peptidomimetic derived from the ECL2 domain of human occludin is, however, without any significant effect in these abovementioned model systems. Paradoxically, Wong and Gumbiner found that the peptidomimetic corresponding to the sequence in the ECL2 but not the ECL1 domain of chick occludin protein altered the paracellular pathway in *Xenopus* kidney epithelial A6 cells (Wong & Gumbiner, 1997). This ECL2-based peptidomimetic, corresponding to amino acids 184-227 and termed $OCLN_{184\text{-}227}$, increased the paracellular permeabilities to a variety of solutes, including mannitol, insulin (5.8 kDa), 3-kDa dextran, and 40-kDa dextran. Mechanistically, $OCLN_{184\text{-}227}$ delocalized the occludin protein from the TJ, whereas the localization of ZO-1, ZO-2, or E-cadherin remained not affected (Wong & Gumbiner, 1997). Similar peptidomimetics derived from the ECL2 domain of rat or human occludin protein were later demonstrated to increase the paracellular permeability in rat Sertoli cells and human intestinal epithelial cells respectively (Chung, Mruk, Mo, Lee, & Cheng, 2001; Nusrat et al., 2005).

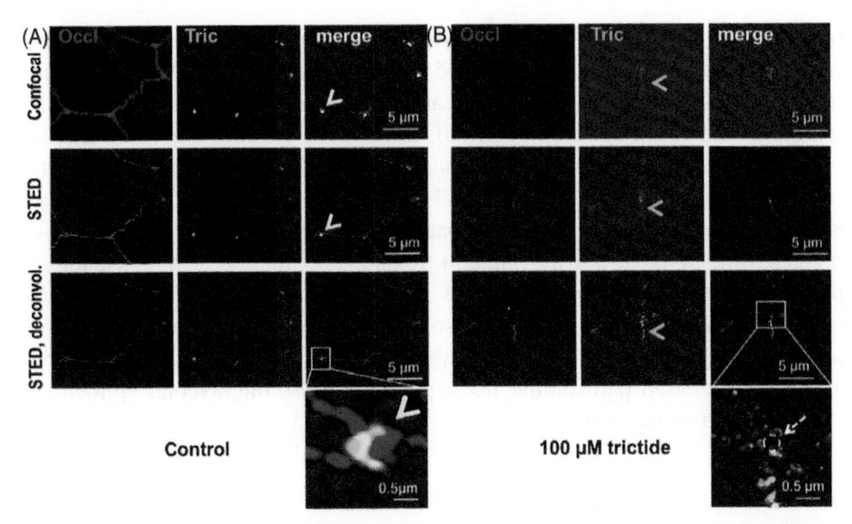

FIGURE 9.9 Tricellulin peptidomimetic effect on tTJ protein localization. (A) Control MDCK II cells stained for occludin (Occl) and tricellulin (Tric). Individual cells are imaged using confocal microscopy (top) or STED super-resolution microscopy (middle). The deconvoluted STED image shows further background reduction (bottom). The enlarged panel (*white box*, merged) illustrates partially overlapping fluorescent signals of Tric and Occl (*yellow arrowhead*). (B) MDCK II cells treated with tricellulin peptidomimetic—trictide (100 μM) for 16 h. Tricellulin spreads to the bicellular TJs (*green arrowheads*) to create a tricellulin-free area in the tTJ (*white dotted arrow* and *circle* in enlarged merged panel). STED, stimulated emission depletion microscopy. Bar: 5 μm in all panels except enlarged panels. Bar: 0.5 μm in enlarged panels. *(Reproduced with permission from Cording, J., Arslan, B., Staat, C., Dithmer, S., Krug, S.M., Kruger, A., Berndt, P., Gunther, R., Winkler, L., Blasig, I.E., et al. (2017). Trictide, a tricellulin-derived peptide to overcome cellular barriers. Annals of the New York Academy of Sciences, 1405(1), 89–101).*

9.5.4 Tricellulin Peptidomimetic

Tricellulin also belongs to the TAMP family and is a component of the tricellular tight junction (tTJ) (Chapter 6, Section 6.2.2.1). A peptidomimetic, termed trictide and corresponding to amino acids 313-336 in the ECL2 domain of human tricellulin protein, was found to increase the tTJ permeability by delocalizing tricellulin from the tTJ in the canine kidney epithelial MDCK-II cells (Table 9.1) (Fig. 9.9) (Cording et al., 2017).

9.6 INSIGHT FROM TOXICOLOGY

9.6.1 *Clostridium Botulinum* Hemagglutinin

Botulinum Neurotoxin (BoNT) is a 150-kDa protein produced by the bacterium *Clostridium botulinum* (Hill & Smith, 2013). Three hemagglutinins (HAs) (HA70, HA33, and HA17) assemble into a symmetric hetero-dodecameric complex of 470 kDa and function as critical virulence factors that facilitate the intestinal absorption of BoNT (Ito et al., 2011; Matsumura et al., 2008). In

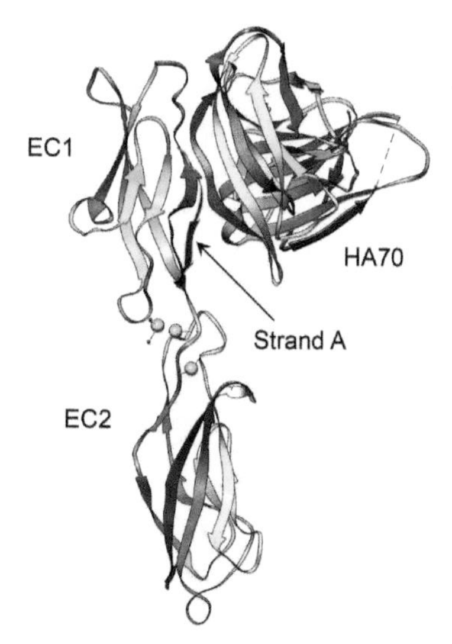

FIGURE 9.10 Molecular structure of HA–E-cadherin complex. Ribbon representation of the crystal structures of EC1-EC2 domains in mouse E-cadherin bound to HA70 in HA protein complex. Upon HA70 binding, the amino-terminal A strand in EC1 domain is pushed back to its monomeric conformation. *Green spheres* depict bound Ca^{++} ions. The color changes gradually from the N terminus (*blue*) to the C terminus (*red*) in each moiety. The presented structure is based upon the X-ray analysis by Lee et al. (2014).

intestinal epithelial cells, HAs directly bind to E-cadherin, disrupt cell adhesion, and dissociate the apical cell junction complex including the TJ (Sugawara et al., 2010). The crystal structure of HA bound E-cadherin has been resolved (Lee et al., 2014). Major conformational changes are observed in the amino-terminal region of the EC1 domain in E-cadherin upon HA binding (Fig. 9.10). EC1 is responsible for *trans* cadherin dimerization by swapping the amino-terminal β-strand (the A strand) (Chapter 9, Section 9.1.1). HA pushes the A strand in EC1 back to its monomeric conformation and blocks the EC1 *trans*-dimerization by occupying the molecular interface (Fig. 9.10). The interaction between HA and E-cadherin is abolished by Ca^{++} chelation, which suggests that the structural model requires Ca^{++}-coordinated orientation of the EC domains in E-cadherin (Sugawara et al., 2010).

9.6.2 *Clostridium Perfringens* Enterotoxin

CPE is a 35-kDa protein produced by the bacterium *Clostridium perfringens* and causes intestinal tissue necrosis (Freedman, Shrestha, & McClane, 2016). Claudin-3 and claudin-4 are the receptor proteins for CPE (Katahira, Inoue, Horiguchi, Matsuda, & Sugimoto, 1997a; Katahira et al., 1997b). The carboxyl-terminal domain of CPE (C-CPE) binds to claudins (Sonoda et al., 1999). When applied to the epithelial MDCK cells, C-CPE removed claudin-3 and

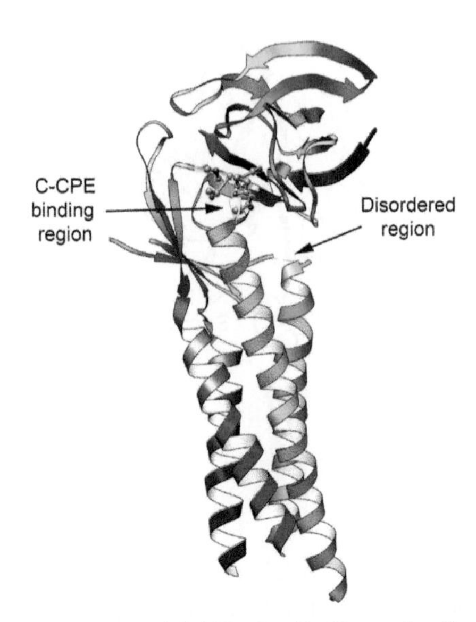

FIGURE 9.11 Molecular structure of C-CPE–claudin-19 complex. Crystal structures of mouse claudin-19 bound to C-CPE in ribbon representation. The side chains of residues (N150, P151, S152, T153, and P154) in the C-CPE binding region of claudin-19 are shown in ball and stick representation. The disordered region in claudin-19 corresponds to the extracellular helical structure in claudin-15. The color changes gradually from the N terminus (*blue*) to the C terminus (*red*) in each moiety. The depicted structure is based upon the X-ray analysis by Saitoh et al. (2015).

claudin-4 from the TJ, disintegrated the TJ strands, and increased the paracellular permeabilities to ions and solutes with sizes up to 10 kDa (Sonoda et al., 1999). The crystal structures of C-CPE bound claudin-4 and claudin-19 have been resolved (Saitoh et al., 2015; Shinoda et al., 2016). Both structures reveal conformational changes in the ECL1 and ECL2 domains of claudin when bound to C-CPE. The C-CPE bound claudin-19 structure offers critical insights into how C-CPE disassembles a TJ strand (Saitoh et al., 2015). According to the crystal structureof claudin-15, the extracellular helical structure connecting the β4 strand in the ECL1 domain to the 2nd transmembrane domain and the hydrophobic pocket formed by the residues (Phe146, Phe147, and Leu158) in the 3rd transmembrane domain and the ECL2 domain are important for claudin *cis*-assembly (Fig. 2.6). The region corresponding to the extracellular helical structure in claudin-15 is disordered in C-CPE bound claudin-19 (Fig. 9.11). Moreover, C-CPE directly binds to a region in the ECL2 domain of claudin-19, which is close to the hydrophobic pocket structure (Fig. 9.11). The binding of C-CPE may distort the hydrophobic pocket structure required for claudin *cis*-interaction. Therefore, C-CPE triggered TJ disassembly might be a result of the disruption of hydrophobic interactions that polymerize claudins in *cis* and the steric clashes that block claudin interactions in *trans*.

9.6.3 *Vibrio Cholerae* Zonula Occludens Toxin

Zot is a 45-kDa protein produced by the bacterium *Vibrio cholerae* (Fasano et al., 1991). In rabbit small intestines, Zot increased the paracellular permeabilities to ions and solutes including polyethylene glycol 4000 (molecular mass, 4 kDa) and horseradish peroxidase (molecular mass, 44 kDa) via a membrane receptor (Fasano et al., 1991; Fasano, Uzzau, Fiore, & Margaretten, 1997). The bound receptor activated the protein kinase C (PKC) to induce actin rearrangement in the perijunctional actomyosin ring (Fasano et al., 1995). By utilizing the cross-reactivity of an anti-Zot antibody to human intestinal tissues, Fasano and coworkers purified a eukaryotic protein of 47 kDa, which, they termed zonulin, was structurally and functionally similar to the bacterial Zot protein (Fasano et al., 2000; Wang, Uzzau, Goldblum, & Fasano, 2000). Zonulin turns out to be the precursor for haptoglobin-2 (pre-HP2), which gives rise to mature HP2, an abundant serum protein capable of scavenging free hemoglobin (Tripathi et al., 2009). Zonulin binds to the same receptor as Zot to modulate the paracellular pathway (Wang et al., 2000). While the receptor for zonulin or Zot has not been cloned, both proteins contain a peptide motif, FCAGMS, which is best known to bind to protease-activated receptor 2 (PAR2) and trigger its downstream signaling cascade (Tripathi et al., 2009).

9.6.4 *Helicobacter Pylori* Vacuolating Toxin

VacA is a 95-kDa protein produced by the bacterium *Helicobacter pylori* (Cover, 1996). While little is known about the effect of VacA on the gastric epithelium *in vivo*, VacA has been shown to modulate the paracellular pathway *in vitro* in a number of epithelial cell models. Independent of cell model, VacA significantly increased the paracellular permeabilities to ions and solutes with sizes of <350 Da (Papini et al., 1998).

9.6.5 *Dermatophagoides Pteronyssinus* Der p 1

Der p 1 is a cysteine protease allergen from the fecal pellet of the house dust mite *Dermatophagoides pteronyssinus* (Stewart, Thompson, & Simpson, 1989). In human airway epithelial cells, Der p 1 rapidly increased the paracellular permeability to mannitol by cleaving the occludin and claudin-1 proteins at their extracellular loop domains (Wan et al., 1999).

9.6.6 *Clostridium Perfringens* Iota Toxin

Clostridium perfringens iota toxin belongs to the family of binary actin-ADP ribosylating toxins that also include *Clostridium difficile* transferase, *Clostridium spiroforme* toxin, *Clostridium botulinum* C2 toxin, and *Bacillus cereus* vegetative insecticidal protein (Barth, Aktories, Popoff, & Stiles, 2004).

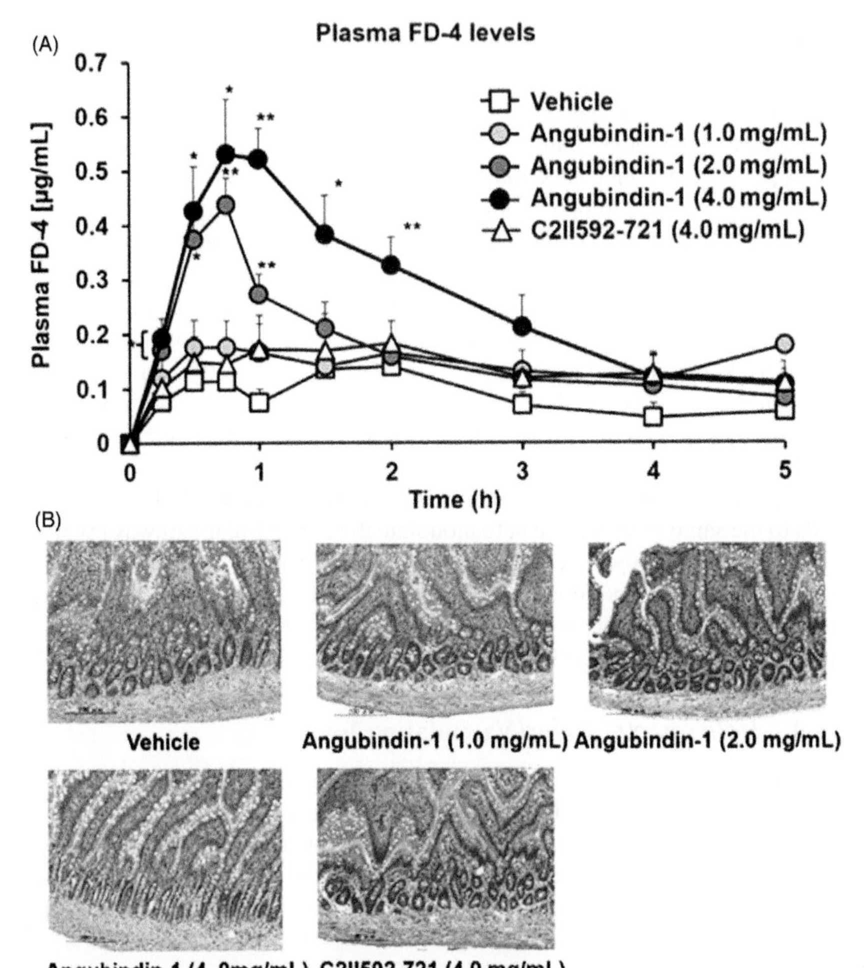

FIGURE 9.12 Angulin-1 peptidomimetic effect on mucosal absorption. (A) Rat jejunum was infused with 10 mg/ml 4-kDa fluorescent dextran (FD-4) in the presence of angulin-1 peptidomimetic (angubindin-1) or control peptidomimetic (C2II592-721) at the indicated concentrations. The plasma FD-4 levels were measured at the indicated time points. Angubindin-1 treatment significantly elevated the plasma FD-4 levels, suggesting increased intestinal permeability. *Asterisk*, $p < 0.05$, *Double Asterisk*, $p < 0.01$. B. Histology of angubindin-1- or C2II592-721-treated rat jejunum. After 5-h treatment with angubindin-1 or C2II592-721, the jejunum was fixed, sectioned, and stained with H&E. There was no pathological lesion after angubindin-1 treatment. *(Reproduced with permission from Krug, S.M., Hayaishi, T., Iguchi, D., Watari, A., Takahashi, A., Fromm, M., Nagahama, M., Takeda, H., Okada, Y., Sawasaki, T., et al. (2017). Angubindin-1, a novel paracellular absorption enhancer acting at the tricellular tight junction. Journal of Controlled Release: Official Journal of the Controlled Release Society, 260, 1–11).*

Both *Clostridium perfringens* iota toxin and *Clostridium difficile* transferase utilize angulin-1, also known as lipolysis-stimulated lipoprotein receptor (LSR), as the receptor protein for entry into host cells (Papatheodorou et al., 2011). Angulin-1 is an integral component of the tTJ (Chapter 6, Section 6.2.2.2). A peptidomimetic corresponding to amino acids 421-664 of the Ib component of iota-toxin, termed angubindin-1, was shown to increase the paracellular permeability across intestinal epithelial cells or tissues (Fig. 9.12) (Krug et al., 2017).

REFERENCES

Anda, T., Yamashita, H., Khalid, H., Tsutsumi, K., Fujita, H., Tokunaga, Y., & Shibata, S. (1997). Effect of tumor necrosis factor-alpha on the permeability of bovine brain microvessel endothelial cell monolayers. *Neurological Research, 19*, 369–376.

Anderberg, E. K., Lindmark, T., & Artursson, P. (1993). Sodium caprate elicits dilatations in human intestinal tight junctions and enhances drug absorption by the paracellular route. *Pharmaceutical Research, 10*, 857–864.

Angelow, S., & Yu, A. S. (2009). Structure-function studies of claudin extracellular domains by cysteine-scanning mutagenesis. *The Journal of Biological Chemistry, 284*, 29205–29217.

Asahi, M., Wang, X., Mori, T., Sumii, T., Jung, J. C., Moskowitz, M. A., Fini, M. E., & Lo, E. H. (2001). Effects of matrix metalloproteinase-9 gene knock-out on the proteolysis of blood-brain barrier and white matter components after cerebral ischemia. *The Journal of Neuroscience, 21*, 7724–7732.

Barth, H., Aktories, K., Popoff, M. R., & Stiles, B. G. (2004). Binary bacterial toxins: biochemistry, biology, and applications of common clostridium and bacillus proteins. *Microbiology and molecular biology reviews: MMBR, 68*, 373–402.

Baumgartner, H. K., Beeman, N., Hodges, R. S., & Neville, M. C. (2011). A D-peptide analog of the second extracellular loop of claudin-3 and -4 leads to mislocalized claudin and cellular apoptosis in mammary epithelial cells. *Chemical Biology & Drug Design, 77*, 124–136.

Bazzoni, G., Martinez-Estrada, O. M., Mueller, F., Nelboeck, P., Schmid, G., Bartfai, T., Dejana, E., & Brockhaus, M. (2000). Homophilic interaction of junctional adhesion molecule. *The Journal of Biological Chemistry, 275*, 30970–30976.

Boggon, T. J., Murray, J., Chappuis-Flament, S., Wong, E., Gumbiner, B. M., & Shapiro, L. (2002). C-cadherin ectodomain structure and implications for cell adhesion mechanisms. *Science, 296*, 1308–1313.

Buzza, M. S., Martin, E. W., Driesbaugh, K. H., Desilets, A., Leduc, R., & Antalis, T. M. (2013). Prostasin is required for matriptase activation in intestinal epithelial cells to regulate closure of the paracellular pathway. *The Journal of Biological Chemistry, 288*, 10328–10337.

Buzza, M. S., Netzel-Arnett, S., Shea-Donohue, T., Zhao, A., Lin, C. Y., List, K., Szabo, R., Fasano, A., Bugge, T. H., & Antalis, T. M. (2010). Membrane-anchored serine protease matriptase regulates epithelial barrier formation and permeability in the intestine. *Proceedings of the National Academy of Sciences of the United States of America, 107*, 4200–4205.

Cereijido, M., Robbins, E. S., Dolan, W. J., Rotunno, C. A., & Sabatini, D. D. (1978). Polarized monolayers formed by epithelial cells on a permeable and translucent support. *The Journal of Cell Biology, 77*, 853–880.

Chen, C. P., Posy, S., Ben-Shaul, A., Shapiro, L., & Honig, B. H. (2005). Specificity of cell-cell adhesion by classical cadherins: Critical role for low-affinity dimerization through beta-strand swapping. *Proceedings of the National Academy of Sciences of the United States of America, 102*, 8531–8536.

Chung, N. P., Mruk, D., Mo, M. Y., Lee, W. M., & Cheng, C. Y. (2001). A 22-amino acid synthetic peptide corresponding to the second extracellular loop of rat occludin perturbs the blood-testis barrier and disrupts spermatogenesis reversibly in vivo. *Biology of Reproduction, 65*, 1340–1351.

Clayburgh, D. R., Musch, M. W., Leitges, M., Fu, Y. X., & Turner, J. R. (2006). Coordinated epithelial NHE3 inhibition and barrier dysfunction are required for TNF-mediated diarrhea in vivo. *The Journal of Clinical Investigation, 116*, 2682–2694.

Cording, J., Arslan, B., Staat, C., Dithmer, S., Krug, S. M., Kruger, A., Berndt, P., Gunther, R., Winkler, L., Blasig, I. E., et al. (2017). Trictide, a tricellulin-derived peptide to overcome cellular barriers. *Annals of the New York Academy of Sciences, 1405*(1), 89–101.

Cover, T. L. (1996). The vacuolating cytotoxin of helicobacter pylori. *Molecular Microbiology, 20*, 241–246.

Daugherty, B. L., Ward, C., Smith, T., Ritzenthaler, J. D., & Koval, M. (2007). Regulation of heterotypic claudin compatibility. *The Journal of Biological Chemistry, 282*, 30005–30013.

Del Vecchio, G., Tscheik, C., Tenz, K., Helms, H. C., Winkler, L., Blasig, R., & Blasig, I. E. (2012). Sodium caprate transiently opens claudin-5-containing barriers at tight junctions of epithelial and endothelial cells. *Molecular Pharmaceutics, 9*, 2523–2533.

Everett, R. S., Vanhook, M. K., Barozzi, N., Toth, I., & Johnson, L. G. (2006). Specific modulation of airway epithelial tight junctions by apical application of an occludin peptide. *Molecular Pharmacology, 69*, 492–500.

Farquhar, M. G., & Palade, G. E. (1963). Junctional complexes in various epithelia. *The Journal of Cell Biology, 17*, 375–412.

Fasano, A., Baudry, B., Pumplin, D. W., Wasserman, S. S., Tall, B. D., Ketley, J. M., & Kaper, J. B. (1991). Vibrio cholerae produces a second enterotoxin, which affects intestinal tight junctions. *Proceedings of the National Academy of Sciences of the United States of America, 88*, 5242–5246.

Fasano, A., Fiorentini, C., Donelli, G., Uzzau, S., Kaper, J. B., Margaretten, K., Ding, X., Guandalini, S., Comstock, L., & Goldblum, S. E. (1995). Zonula occludens toxin modulates tight junctions through protein kinase C-dependent actin reorganization, in vitro. *The Journal of Clinical Investigation, 96*, 710–720.

Fasano, A., Not, T., Wang, W., Uzzau, S., Berti, I., Tommasini, A., & Goldblum, S. E. (2000). Zonulin, a newly discovered modulator of intestinal permeability, and its expression in coeliac disease. *Lancet, 355*, 1518–1519 (London, England).

Fasano, A., Uzzau, S., Fiore, C., & Margaretten, K. (1997). The enterotoxic effect of zonula occludens toxin on rabbit small intestine involves the paracellular pathway. *Gastroenterology, 112*, 839–846.

Feng, S., Cen, J., Huang, Y., Shen, H., Yao, L., Wang, Y., & Chen, Z. (2011). Matrix metalloproteinase-2 and -9 secreted by leukemic cells increase the permeability of blood-brain barrier by disrupting tight junction proteins. *PloS One, 6*, e20599.

Flynn, A. N., Itani, O. A., Moninger, T. O., & Welsh, M. J. (2009). Acute regulation of tight junction ion selectivity in human airway epithelia. *Proceedings of the National Academy of Sciences of the United States of America, 106*, 3591–3596.

Freedman, J. C., Shrestha, A., & McClane, B. A. (2016). Clostridium perfringens enterotoxin: Action, genetics, and translational applications. *Toxins*, 8.

Gardner, T. W., Lesher, T., Khin, S., Vu, C., Barber, A. J., & Brennan, W. A., Jr. (1996). Histamine reduces ZO-1 tight-junction protein expression in cultured retinal microvascular endothelial cells. *The Biochemical Journal, 320*(3), 717–721.

Gong, Y., Yu, M., Yang, J., Gonzales, E., Perez, R., Hou, M., Tripathi, P., Hering-Smith, K. S., Hamm, L. L., & Hou, J. (2014). The Cap1-claudin-4 regulatory pathway is important for renal chloride reabsorption and blood pressure regulation. *Proceedings of the National Academy of Sciences of the United States of America, 111*, E3766–3774.

Guo, M., Breslin, J. W., Wu, M. H., Gottardi, C. J., & Yuan, S. Y. (2008). VE-cadherin and beta-catenin binding dynamics during histamine-induced endothelial hyperpermeability. *American Journal of Physiology Cell Physiology, 294*, C977–984.

Hill, K. K., & Smith, T. J. (2013). Genetic diversity within clostridium botulinum serotypes, botulinum neurotoxin gene clusters and toxin subtypes. *Current Topics in Microbiology and Immunology, 364*, 1–20.

Hsu, L. W., Ho, Y. C., Chuang, E. Y., Chen, C. T., Juang, J. H., Su, F. Y., Hwang, S. M., & Sung, H. W. (2013). Effects of pH on molecular mechanisms of chitosan-integrin interactions and resulting tight-junction disruptions. *Biomaterials, 34*, 784–793.

Hsu, L. W., Lee, P. L., Chen, C. T., Mi, F. L., Juang, J. H., Hwang, S. M., Ho, Y. C., & Sung, H. W. (2012). Elucidating the signaling mechanism of an epithelial tight-junction opening induced by chitosan. *Biomaterials, 33*, 6254–6263.

Ito, H., Sagane, Y., Miyata, K., Inui, K., Matsuo, T., Horiuchi, R., Ikeda, T., Suzuki, T., Hasegawa, K., Kouguchi, H., et al. (2011). HA-33 facilitates transport of the serotype D botulinum toxin across a rat intestinal epithelial cell monolayer. *FEMS Immunology and Medical Microbiology, 61*, 323–331.

Jickling, G. C., Liu, D., Stamova, B., Ander, B. P., Zhan, X., Lu, A., & Sharp, F. R. (2014). Hemorrhagic transformation after ischemic stroke in animals and humans. *Journal of Cerebral Blood Flow and Metabolism: Official Journal of the International Society of Cerebral Blood Flow and Metabolism, 34*, 185–199.

Katahira, J., Inoue, N., Horiguchi, Y., Matsuda, M., & Sugimoto, N. (1997a). Molecular cloning and functional characterization of the receptor for clostridium perfringens enterotoxin. *The Journal of Cell Biology, 136*, 1239–1247.

Katahira, J., Sugiyama, H., Inoue, N., Horiguchi, Y., Matsuda, M., & Sugimoto, N. (1997b). Clostridium perfringens enterotoxin utilizes two structurally related membrane proteins as functional receptors in vivo. *The Journal of Biological Chemistry, 272*, 26652–26658.

Kostrewa, D., Brockhaus, M., D'Arcy, A., Dale, G. E., Nelboeck, P., Schmid, G., Mueller, F., Bazzoni, G., Dejana, E., Bartfai, T., et al. (2001). X-ray structure of junctional adhesion molecule: structural basis for homophilic adhesion via a novel dimerization motif. *The EMBO Journal, 20*, 4391–4398.

Krug, S. M., Amasheh, M., Dittmann, I., Christoffel, I., Fromm, M., & Amasheh, S. (2013). Sodium caprate as an enhancer of macromolecule permeation across tricellular tight junctions of intestinal cells. *Biomaterials, 34*, 275–282.

Krug, S. M., Hayaishi, T., Iguchi, D., Watari, A., Takahashi, A., Fromm, M., Nagahama, M., Takeda, H., Okada, Y., Sawasaki, T., et al. (2017). Angubindin-1, a novel paracellular absorption enhancer acting at the tricellular tight junction. *Journal of Controlled Release: Official Journal of the Controlled Release Society, 260*, 1–11.

Lee, K., Zhong, X., Gu, S., Kruel, A. M., Dorner, M. B., Perry, K., Rummel, A., Dong, M., & Jin, R. (2014). Molecular basis for disruption of E-cadherin adhesion by botulinum neurotoxin A complex. *Science, 344*, 1405–1410.

Leyvraz, C., Charles, R. P., Rubera, I., Guitard, M., Rotman, S., Breiden, B., Sandhoff, K., & Hummler, E. (2005). The epidermal barrier function is dependent on the serine protease CAP1/Prss8. *The Journal of Cell Biology, 170*, 487–496.

Lindmark, T., Soderholm, J. D., Olaison, G., Alvan, G., Ocklind, G., & Artursson, P. (1997). Mechanism of absorption enhancement in humans after rectal administration of ampicillin in suppositories containing sodium caprate. *Pharmaceutical Research, 14*, 930–935.

Liu, J., Jin, X., Liu, K. J., & Liu, W. (2012). Matrix metalloproteinase-2-mediated occludin degradation and caveolin-1-mediated claudin-5 redistribution contribute to blood-brain barrier damage in early ischemic stroke stage. *The Journal of Neuroscience: The Official Journal of the Society for Neuroscience, 32*, 3044–3057.

Liu, L. B., Xue, Y. X., Liu, Y. H., & Wang, Y. B. (2008). Bradykinin increases blood-tumor barrier permeability by down-regulating the expression levels of ZO-1, occludin, and claudin-5 and rearranging actin cytoskeleton. *Journal of Neuroscience Research, 86*, 1153–1168.

Lynch, R. D., Tkachuk-Ross, L. J., McCormack, J. M., McCarthy, K. M., Rogers, R. A., & Schneeberger, E. E. (1995). Basolateral but not apical application of protease results in a rapid rise of transepithelial electrical resistance and formation of aberrant tight junction strands in MDCK cells. *European Journal of Cell Biology, 66*, 257–267.

Makagiansar, I. T., Avery, M., Hu, Y., Audus, K. L., & Siahaan, T. J. (2001). Improving the selectivity of HAV-peptides in modulating E-cadherin-E-cadherin interactions in the intercellular junction of MDCK cell monolayers. *Pharmaceutical Research, 18*, 446–453.

Marceau, F., & Regoli, D. (2004). Bradykinin receptor ligands: therapeutic perspectives. *Nature Reviews Drug discovery, 3*, 845–852.

Marchiando, A. M., Shen, L., Graham, W. V., Weber, C. R., Schwarz, B. T., Austin, J. R., Raleigh, D. R., Guan, Y., Watson, A. J., Montrose, M. H., et al. (2010). Caveolin-1-dependent occludin endocytosis is required for TNF-induced tight junction regulation in vivo. *The Journal of Cell Biology, 189*, 111–126.

Martin-Padura, I., Lostaglio, S., Schneemann, M., Williams, L., Romano, M., Fruscella, P., Panzeri, C., Stoppacciaro, A., Ruco, L., Villa, A., et al. (1998). Junctional adhesion molecule, a novel member of the immunoglobulin superfamily that distributes at intercellular junctions and modulates monocyte transmigration. *The Journal of Cell Biology, 142*, 117–127.

Matsumura, T., Jin, Y., Kabumoto, Y., Takegahara, Y., Oguma, K., Lencer, W. I., & Fujinaga, Y. (2008). The HA proteins of botulinum toxin disrupt intestinal epithelial intercellular junctions to increase toxin absorption. *Cellular Microbiology, 10*, 355–364.

Meng, W., & Takeichi, M. (2009). Adherens junction: Molecular architecture and regulation. *Cold Spring Harbor Perspectives in Biology, 1*, a002899.

Mrsny, R. J., Brown, G. T., Gerner-Smidt, K., Buret, A. G., Meddings, J. B., Quan, C., Koval, M., & Nusrat, A. (2008). A key claudin extracellular loop domain is critical for epithelial barrier integrity. *The American Journal of Pathology, 172*, 905–915.

Nagar, B., Overduin, M., Ikura, M., & Rini, J. M. (1996). Structural basis of calcium-induced E-cadherin rigidification and dimerization. *Nature, 380*, 360–364.

Nose, A., Tsuji, K., & Takeichi, M. (1990). Localization of Specificity Determining Sites in Cadherin Cell Adhesion Molecules. *Cell, 61*, 147–155.

Nusrat, A., Brown, G. T., Tom, J., Drake, A., Bui, T. T., Quan, C., & Mrsny, R. J. (2005). Multiple protein interactions involving proposed extracellular loop domains of the tight junction protein occludin. *Molecular Biology of the Cell, 16*, 1725–1734.

Pal, D., Audus, K. L., & Siahaan, T. J. (1997). Modulation of cellular adhesion in bovine brain microvessel endothelial cells by a decapeptide. *Brain Research, 747*, 103–113.

Palladino, M. A., Bahjat, F. R., Theodorakis, E. A., & Moldawer, L. L. (2003). Anti-TNF-alpha therapies: The next generation. *Nature Reviews Drug Discovery, 2*, 736–746.

Papatheodorou, P., Carette, J. E., Bell, G. W., Schwan, C., Guttenberg, G., Brummelkamp, T. R., & Aktories, K. (2011). Lipolysis-stimulated lipoprotein receptor (LSR) is the host receptor for the binary toxin Clostridium difficile transferase (CDT). *Proceedings of the National Academy of Sciences of the United States of America, 108*, 16422–16427.

Papini, E., Satin, B., Norais, N., de Bernard, M., Telford, J. L., Rappuoli, R., & Montecucco, C. (1998). Selective increase of the permeability of polarized epithelial cell monolayers by helicobacter pylori vacuolating toxin. *The Journal of Clinical Investigation, 102*, 813–820.

Piontek, J., Winkler, L., Wolburg, H., Muller, S. L., Zuleger, N., Piehl, C., Wiesner, B., Krause, G., & Blasig, I. E. (2008). Formation of tight junction: Determinants of homophilic interaction between classic claudins. *The FASEB Journal, 22*, 146–158.

Pokutta, S., Herrenknecht, K., Kemler, R., & Engel, J. (1994). Conformational changes of the recombinant extracellular domain of E-cadherin upon calcium binding. *European Journal of Biochemistry, 223*, 1019–1026.

Preston, E., Slinn, J., Vinokourov, I., & Stanimirovic, D. (2008). Graded reversible opening of the rat blood-brain barrier by intracarotid infusion of sodium caprate. *Journal of Neuroscience Methods, 168*, 443–449.

Prota, A. E., Campbell, J. A., Schelling, P., Forrest, J. C., Watson, M. J., Peters, T. R., Aurrand-Lions, M., Imhof, B. A., Dermody, T. S., & Stehle, T. (2003). Crystal structure of human junctional adhesion molecule 1: Implications for reovirus binding. *Proceedings of the National Academy of Sciences of the United States of America, 100*, 5366–5371.

Rochfort, K. D., Collins, L. E., Murphy, R. P., & Cummins, P. M. (2014). Downregulation of blood-brain barrier phenotype by proinflammatory cytokines involves NADPH oxidase-dependent ROS generation: Consequences for interendothelial adherens and tight junctions. *PloS One, 9*, e101815.

Rochfort, K. D., & Cummins, P. M. (2015). Cytokine-mediated dysregulation of zonula occludens-1 properties in human brain microvascular endothelium. *Microvascular Research, 100*, 48–53.

Rosenthal, R., Gunzel, D., Finger, C., Krug, S. M., Richter, J. F., Schulzke, J. D., Fromm, M., & Amasheh, S. (2012). The effect of chitosan on transcellular and paracellular mechanisms in the intestinal epithelial barrier. *Biomaterials, 33*, 2791–2800.

Royall, J. A., Berkow, R. L., Beckman, J. S., Cunningham, M. K., Matalon, S., & Freeman, B. A. (1989). Tumor necrosis factor and interleukin 1 alpha increase vascular endothelial permeability. *The American Journal of Physiology, 257*, L399–410.

Saitoh, Y., Suzuki, H., Tani, K., Nishikawa, K., Irie, K., Ogura, Y., Tamura, A., Tsukita, S., & Fujiyoshi, Y. (2015). Tight junctions: Structural insight into tight junction disassembly by clostridium perfringens enterotoxin. *Science, 347*, 775–778 (New York, NY).

Sanovich, E., Bartus, R. T., Friden, P. M., Dean, R. L., Le, H. Q., & Brightman, M. W. (1995). Pathway across blood-brain barrier opened by the bradykinin agonist, RMP-7. *Brain Research, 705*, 125–135.

Schlingmann, B., Overgaard, C. E., Molina, S. A., Lynn, K. S., Mitchell, L. A., Dorsainvil White, S., Mattheyses, A. L., Guidot, D. M., Capaldo, C. T., & Koval, M. (2016). Regulation of claudin/zonula occludens-1 complexes by hetero-claudin interactions. *Nature Communications, 7*, 12276.

Schmitz, H., Fromm, M., Bentzel, C. J., Scholz, P., Detjen, K., Mankertz, J., Bode, H., Epple, H. J., Riecken, E. O., & Schulzke, J. D. (1999). Tumor necrosis factor-alpha (TNFalpha) regulates the epithelial barrier in the human intestinal cell line HT-29/B6. *Journal of Cell Science, 112*(1), 137–146.

Shapiro, L., Fannon, A. M., Kwong, P. D., Thompson, A., Lehmann, M. S., Grubel, G., Legrand, J. F., Als-Nielsen, J., Colman, D. R., & Hendrickson, W. A. (1995). Structural basis of cell-cell adhesion by cadherins. *Nature, 374*, 327–337.

Shapiro, L., & Weis, W. I. (2009). Structure and biochemistry of cadherins and catenins. *Cold Spring Harbor Perspectives in Biology*, *1*, a003053.

Shinoda, T., Shinya, N., Ito, K., Ohsawa, N., Terada, T., Hirata, K., Kawano, Y., Yamamoto, M., Kimura-Someya, T., Yokoyama, S., et al. (2016). Structural basis for disruption of claudin assembly in tight junctions by an enterotoxin. *Scientific Reports*, *6*, 33632.

Sinaga, E., Jois, S. D., Avery, M., Makagiansar, I. T., Tambunan, U. S., Audus, K. L., & Siahaan, T. J. (2002). Increasing paracellular porosity by e-cadherin peptides: Discovery of bulge and groove regions in the EC1-domain of E-cadherin. *Pharmaceutical Research*, *19*, 1170–1179.

Smith, J., Wood, E., & Dornish, M. (2004). Effect of chitosan on epithelial cell tight junctions. *Pharmaceutical Research*, *21*, 43–49.

Smith, J. M., Dornish, M., & Wood, E. J. (2005). Involvement of protein kinase C in chitosan glutamate-mediated tight junction disruption. *Biomaterials*, *26*, 3269–3276.

Sonaje, K., Chuang, E. Y., Lin, K. J., Yen, T. C., Su, F. Y., Tseng, M. T., & Sung, H. W. (2012). Opening of epithelial tight junctions and enhancement of paracellular permeation by chitosan: Microscopic, ultrastructural, and computed-tomographic observations. *Molecular Pharmaceutics*, *9*, 1271–1279.

Sonoda, N., Furuse, M., Sasaki, H., Yonemura, S., Katahira, J., Horiguchi, Y., & Tsukita, S. (1999). Clostridium perfringens enterotoxin fragment removes specific claudins from tight junction strands: Evidence for direct involvement of claudins in tight junction barrier. *The Journal of Cell Biology*, *147*, 195–204.

Srinivas, S. P., Satpathy, M., Guo, Y., & Anandan, V. (2006). Histamine-induced phosphorylation of the regulatory light chain of myosin II disrupts the barrier integrity of corneal endothelial cells. *Investigative Ophthalmology & Visual Science*, *47*, 4011–4018.

Staat, C., Coisne, C., Dabrowski, S., Stamatovic, S. M., Andjelkovic, A. V., Wolburg, H., Engelhardt, B., & Blasig, I. E. (2015). Mode of action of claudin peptidomimetics in the transient opening of cellular tight junction barriers. *Biomaterials*, *54*, 9–20.

Sternlicht, M. D., & Werb, Z. (2001). How matrix metalloproteinases regulate cell behavior. *Annual Review of Cell and Developmental Biology*, *17*, 463–516.

Stewart, G. A., Thompson, P. J., & Simpson, R. J. (1989). Protease antigens from house dust mite. *Lancet*, *2*, 154–155 London, England.

Sugawara, Y., Matsumura, T., Takegahara, Y., Jin, Y., Tsukasaki, Y., Takeichi, M., & Fujinaga, Y. (2010). Botulinum hemagglutinin disrupts the intercellular epithelial barrier by directly binding e-cadherin. *The Journal of Cell Biology*, *189*, 691–700.

Suzuki, H., Nishizawa, T., Tani, K., Yamazaki, Y., Tamura, A., Ishitani, R., Dohmae, N., Tsukita, S., Nureki, O., & Fujiyoshi, Y. (2014). Crystal structure of a claudin provides insight into the architecture of tight junctions. *Science*, *344*, 304–307 (New York, NY).

Taddei, A., Giampietro, C., Conti, A., Orsenigo, F., Breviario, F., Pirazzoli, V., Potente, M., Daly, C., Dimmeler, S., & Dejana, E. (2008). Endothelial adherens junctions control tight junctions by VE-cadherin-mediated upregulation of claudin-5. *Nature Cell Biology*, *10*, 923–934.

Tavelin, S., Hashimoto, K., Malkinson, J., Lazorova, L., Toth, I., & Artursson, P. (2003). A new principle for tight junction modulation based on occludin peptides. *Molecular Pharmacology*, *64*, 1530–1540.

Tripathi, A., Lammers, K. M., Goldblum, S., Shea-Donohue, T., Netzel-Arnett, S., Buzza, M. S., Antalis, T. M., Vogel, S. N., Zhao, A., Yang, S., et al. (2009). Identification of human zonulin, a physiological modulator of tight junctions, as prehaptoglobin-2. *Proceedings of the National Academy of Sciences of the United States of America*, *106*, 16799–16804.

Van Itallie, C. M., & Anderson, J. M. (1997). Occludin confers adhesiveness when expressed in fibroblasts. *Journal of Cell Science*, *110*(9), 1113–1121.

Vietor, I., Bader, T., Paiha, K., & Huber, L. A. (2001). Perturbation of the tight junction permeability barrier by occludin loop peptides activates beta-catenin/TCF/LEF-mediated transcription. *EMBO Reports*, *2*, 306–312.

Wan, H., Winton, H. L., Soeller, C., Tovey, E. R., Gruenert, D. C., Thompson, P. J., Stewart, G. A., Taylor, G. W., Garrod, D. R., Cannell, M. B., et al. (1999). Der 1 facilitates transepithelial allergen delivery by disruption of tight junctions. *The Journal of Clinical Investigation*, *104*, 123–133.

Wang, W., Uzzau, S., Goldblum, S. E., & Fasano, A. (2000). Human zonulin, a potential modulator of intestinal tight junctions. *Journal of Cell Science*, *24*(113), 4435–4440.

Wong, V., & Gumbiner, B. M. (1997). A synthetic peptide corresponding to the extracellular domain of occludin perturbs the tight junction permeability barrier. *The Journal of Cell Biology*, *136*, 399–409.

Yang, Y., Estrada, E. Y., Thompson, J. F., Liu, W., & Rosenberg, G. A. (2007). Matrix metalloproteinase-mediated disruption of tight junction proteins in cerebral vessels is reversed by synthetic matrix metalloproteinase inhibitor in focal ischemia in rat. *Journal of Cerebral Blood Flow and Metabolism: Official Journal of the International Society of Cerebral Blood Flow and Metabolism*, *27*, 697–709.

Yeh, T. H., Hsu, L. W., Tseng, M. T., Lee, P. L., Sonjae, K., Ho, Y. C., & Sung, H. W. (2011). Mechanism and consequence of chitosan-mediated reversible epithelial tight junction opening. *Biomaterials*, *32*, 6164–6173.

Zhang, Y., Sivasankar, S., Nelson, W. J., & Chu, S. (2009). Resolving cadherin interactions and binding cooperativity at the single-molecule level. *Proceedings of the National Academy of Sciences of the United States of America*, *106*, 109–114.

Zhou, L., Yang, B., Wang, Y., Zhang, H. L., Chen, R. W., & Wang, Y. B. (2014). Bradykinin regulates the expression of claudin-5 in brain microvascular endothelial cells via calcium-induced calcium release. *Journal of Neuroscience Research*, *92*, 597–606.

Zwanziger, D., Hackel, D., Staat, C., Bocker, A., Brack, A., Beyermann, M., Rittner, H., & Blasig, I. E. (2012). A peptidomimetic tight junction modulator to improve regional analgesia. *Molecular Pharmaceutics*, *9*, 1785–1794.

Chapter 10

Paracellular Channel Evolution

10.1 CELL JUNCTION IN INVERTEBRATE

10.1.1 Ultrastructure, Function, and Phylogeny

The septate junction (SJ) is a specialized intercellular junction in invertebrate epithelia, which appears as ladder-like septa of ~15 nm in length, spanning the paracellular space between adjacent cells and located basolaterally to the adherens junction (AJ) (Fig. 10.1). SJ functions as a permeation barrier to restrict the free diffusion of ions and solutes through the paracellular pathway (Baldwin, Loeb, & Riemann, 1987; Szollosi & Marcaillou, 1977). Phylogenetic tracing reveals that the lowest animal phylum with SJ is Porifera (Banerjee, Sousa, & Bhat, 2006). The phyla of Coelenterata, Platyhelminthes, Annelida, Arthropoda, Mollusca, and Echinodermata all make SJs. One notable exception is the phylum of Nematoda, which expresses a single combined apical junction (Fig. 10.2) (Banerjee et al., 2006).

10.1.2 Apicobasal Polarity

The establishment of apicobasal polarity is intimately connected to the formation of cell junction (Meng & Takeichi, 2009). A regulatory hierarchy dictated by the interaction and repulsion among three conserved protein complexes underlies the apicobasal polarization process (Bilder, Schober, & Perrimon, 2003). In *Drosophila*, these three protein complexes each contain one of the *p*ostsynaptic density 95/*d*iscs large/ZO-1 (PDZ) domain proteins—Bazooka (Baz), Stardust (Sdt), and Scribble (Scrib). The Baz complex is first recruited to the AJ by cell adhesion-dependent mechanisms. The Scrib complex represses apical domain expansion toward basolateral direction by antagonizing Baz-initiated apical polarity. The Sdt complex is then recruited by the Baz complex to counteract the activity of Scrib (Bilder et al., 2003). Vertebrate cells utilize a similar mechanism to establish the apicobasal polarity with three orthologous proteins—Par3 (Baz), PALS1 (Sdt), and Scrib (Shin, Fogg, & Margolis, 2006).

10.2 APICAL JUNCTION IN *CAENORHABDITIS ELEGANS*

C. elegans belongs to the phylum of Nematoda. Only one type of intercellular junction, the *C. elegans* apical junction (CeAJ) being a hybrid form of AJ and SJ, has been described on the ultrastructural level (McMahon, Legouis,

The Paracellular Channel. http://dx.doi.org/10.1016/B978-0-12-814635-4.00010-3

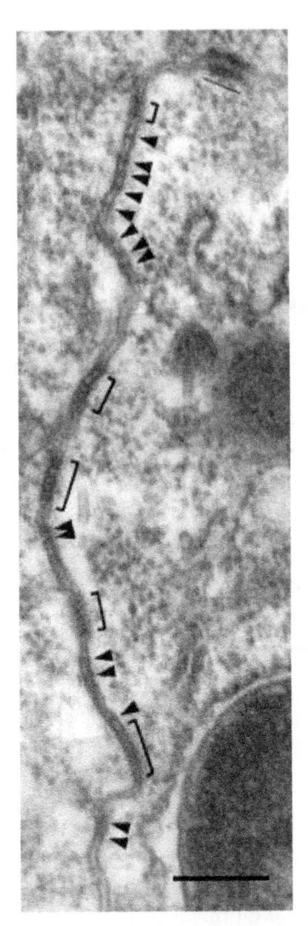

FIGURE 10.1 ***Drosophila*** **septate junction.** Transmission electron micrograph of late stage 17 *Drosophila* epidermal epithelium showing the AJ (*thin line*) and the SJ. *Brackets* indicate clustered groups of septa and *arrowheads* point to individual septa. Bar: 200 nm. *(Reproduced with permission from Wu, V.M., Schulte, J., Hirschi, A., Tepass, U., & Beitel, G.J. (2004). Sinuous is a Drosophila claudin required for septate junction organization and epithelial tube size control. The Journal of Cell Biology, 164, 313–323).*

Vonesch, & Labouesse, 2001). To date, five claudin-like proteins, CLC-1 to -4 and VAB-9, have been identified in *C. elegans* by homology to vertebrate claudins (Asano, Asano, Sasaki, Furuse, & Tsukita, 2003; Simske et al., 2003). Phylogenetic analyses reveal that these *C. elegans* claudins belong to a distant clade (Fig. 10.3). The vertebrate claudin proteins carry a common motif (-GLWCC; PROSITE ID: PS01346) in the first extracellular loop domain (Chapter 2, Section 2.3). This motif is conserved in *C. elegans* claudins (Asano et al., 2003; Simske et al., 2003). The carboxyl-terminal domains of vertebrate claudins contain a PDZ binding motif (YV) that is critical for interaction with the PDZ domain protein ZO-1 (Itoh et al., 1999). While CLC-3 and CLC-4

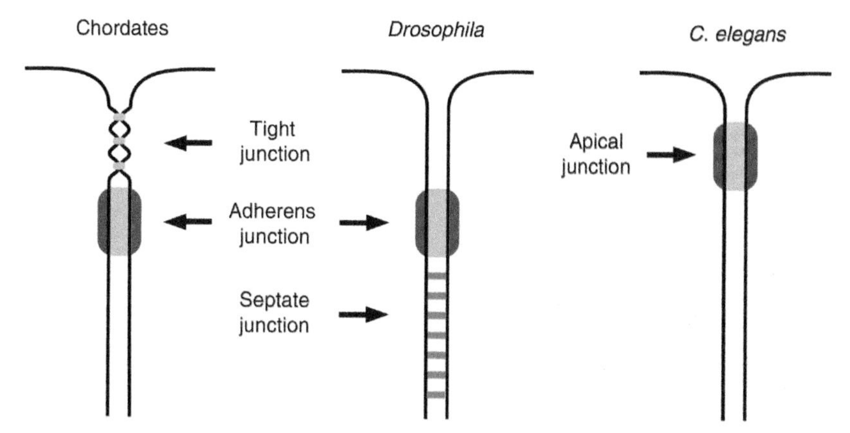

FIGURE 10.2 **Apical junctional complexes in chordates, *Drosophila* and *C. elegans*.** TJs and AJs are found in chordates (desmosomes that form spots rather than circumferential bands, are omitted). AJs and SJs are found in most nonchordate animals, including *Drosophila*. *C. elegans* shows only a single element in its apical junctional complex, which is often referred to as the *C. elegans* apical junction (CeAJ). *(Reproduced with permission from Tepass, U. (2003). Claudin complexities at the apical junctional complex. Nature Cell Biology, 5, 595–597).*

have putative PDZ binding motifs, CLC-1, -2 and VAB-9 clearly lack this sequence (Asano et al., 2003; Simske et al., 2003). The *C. elegans* ZO-1 orthologue, ZOO-1, anchors the CeAJ onto actin cytoskeleton during morphogenesis (Lockwood, Zaidel-Bar, & Hardin, 2008). Whether ZOO-1 directly interacts with CLCs or VAB-9 is unknown at present. CLC-1 is found at the CeAJ of the epithelial cells in the pharyngeal portion of *C. elegans* digestive organ (Asano et al., 2003). RNA interference experiments demonstrated that depletion of the CLC-1 gene transcripts in *C. elegans* increased the paracellular permeability to 10-kDa dextran across the pharyngeal epithelial cells (Asano et al., 2003). VAB-9 is expressed at the CeAJ by all epithelia in *C. elegans*. The *vab-9* mutant worm developed morphological defects due to disorganization of the actin cytoskeleton underneath the CeAJ (Simske et al., 2003). VAB-9 also recruits ZOO-1 to the CeAJ (Lockwood et al., 2008). It appears that VAB-9 regulates the interaction between CeAJ and actin cytoskeleton rather than the biophysical property of CeAJ itself.

10.3 SEPTATE JUNCTION IN *DROSOPHILA*

10.3.1 Pleated Septate Junction

Arthropods have two types of SJs: the pleated SJ (pSJ) and the smooth SJ (sSJ). The pSJ is found in ectoderm-derived epithelia, such as the epidermis, the foregut, the hindgut, the salivary glands, the trachea and the imaginal discs. The sSJ is found in endoderm-derived epithelia including the midgut and the

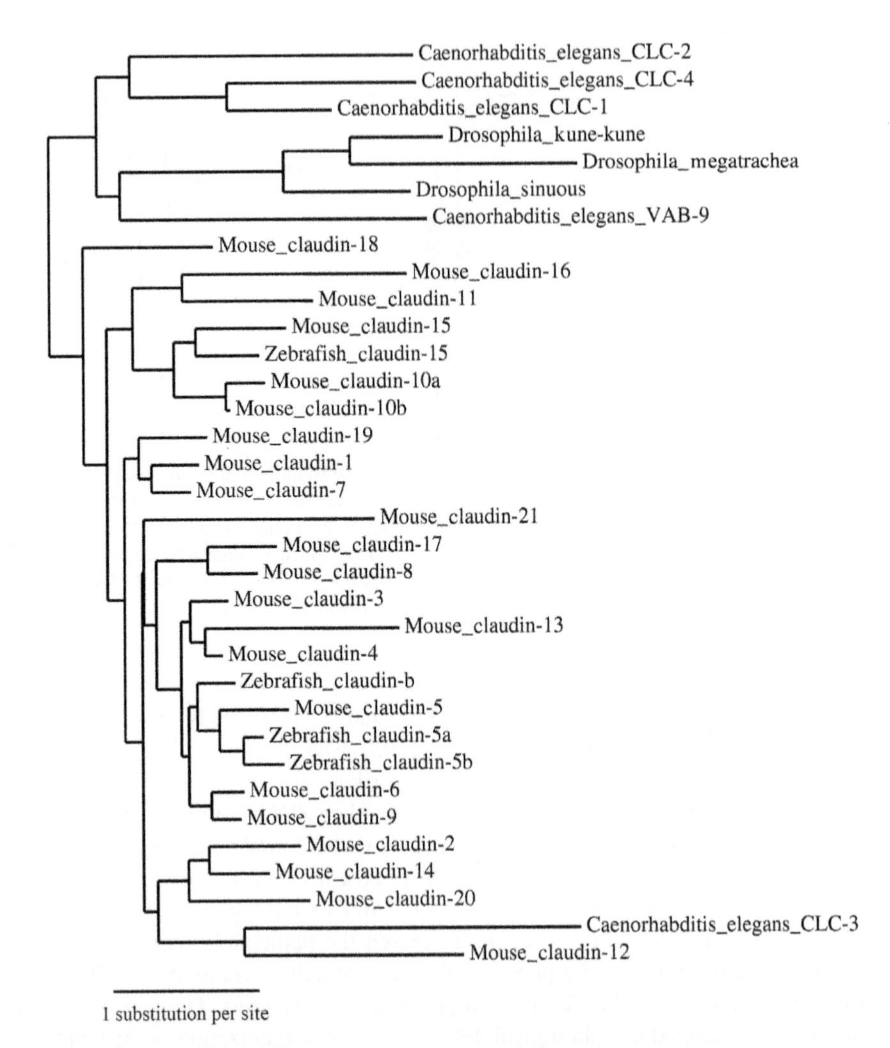

FIGURE 10.3 Phylogenetic tree of *C. elegans*, *Drosophila*, zebrafish, and mouse claudin proteins. The tree is constructed by the neighbor-joining method. The length scale indicates the rate of amino acid substitution per site.

gastric cecum. The criterion for distinguishing these two types of SJs is ultrastructural. The septa of pSJ form undulating rows while those in sSJ are arranged in parallel lines (Tepass & Hartenstein, 1994). The pSJ is common in many other invertebrate species (Jonusaite, Donini, & Kelly, 2016). *Drosophila* pSJs express three claudin-like proteins, Megatrachea (Mega), Sinuous (Sinu), and Kune-kune (Kune) (Behr, Riedel, & Schuh, 2003; Nelson, Furuse, & Beitel, 2010; Wu, Schulte, Hirschi, Tepass, & Beitel, 2004). All three *Drosophila* claudins display key features of vertebrate claudins, including the

GLWCC motif and the PDZ binding motif located in the first extracellular loop domain and the carboxyl-terminal domain respectively. From the phylogenetic tree, it is clear that *Drosophila* claudins cluster more closely to *C. elegans* claudins than to vertebrate claudins (Fig. 10.3). All three *Drosophila* claudins play vital roles in the regulation of epithelial tube size and paracellular barrier function in the tracheal system. Mutating any single *Drosophila* claudin gene caused tracheal tube elongation and dilation due to altered cell shape and increased paracellular permeability (Behr et al., 2003; Nelson et al., 2010; Wu et al., 2004).

10.3.2 Tricellular Septate Junction

At invertebrate tricellular junction, the intercellular space is sealed by a specialized SJ known as the diaphragm or tricellular septate junction (tSJ) (Graf, Noirot-Timothee, & Noirot, 1982; Noirot-Timothee, Graf, & Noirot, 1982). Three protein components of tSJ have been identified in *Drosophila*, which are Gliotactin (Gli), Anakonda (Aka), and Sidekick (Sdk) (Fig. 10.4). Gli is a cholinesterase-like transmembrane protein and belongs to a class of adhesion molecules termed electrotactins (Auld, Fetter, Broadie, & Goodman, 1995).

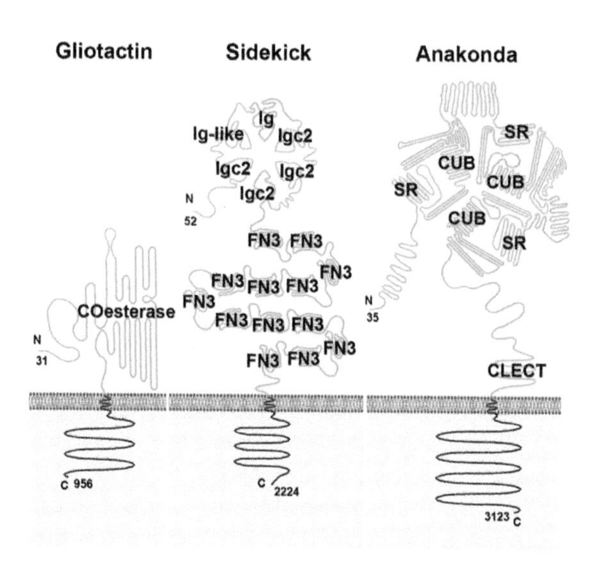

FIGURE 10.4 Diagram of tSJ protein topology. *Drosophila* Gliotactin, Sidekick, and Anakonda proteins are shown. *Top*, extracellular; *bottom*, intracellular. *CLECT*, C-type lectin domain; *COesterase*, carboxylesterase domain; *CUB*, complement C1r/C1s/Uegf/Bmp1 domain; *FN3*, fibronectin type III domain; *Ig*, Ig domain; *Igc2*, Ig domain C2-set type; *Ig-like*, Ig-like domain; *SR*, scavenger receptor cysteine-rich domain. *(Reproduced with permission from Higashi, T., & Miller, A.L. (2017). Tricellular junctions: How to build junctions at the TRICkiest points of epithelial cells. Molecular Biology of the Cell, 28, 2023–2034).*

Gli was initially discovered from the glial cells in the peripheral nervous system and considered essential for the formation of glial ensheathment of axons (Auld et al., 1995). In epithelia, Gli is localized specifically to the tSJ (Schulte, Tepass, & Auld, 2003). *Gli* mutants showed increased paracellular permeability to molecules with sizes up to 10 kDa (Schulte et al., 2003). There were several prominent ultrastructural changes in bicellular septate junction (bSJ) and tSJ in *Gli* mutants. The septa in bSJ were present but not compacted into clusters, whereas the septa in tSJ were completely lost in *Gli* mutants (Schulte et al., 2003). The tSJ also acts as a signaling hub to regulate intestinal stem cell behavior in *Drosophila*. Acute loss of Gli in enterocytes stimulated proliferation whilst suppressed differentiation of intestinal stem cells in young flies, which represents a hallmark of accelerated ageing process (Resnik-Docampo et al., 2017). Aka is a transmembrane protein with an unusual tripartite repeat structure in the extracellular domain and a PDZ-binding motif in the carboxyl-terminal domain (Byri et al., 2015). The tripartite repeat structure is speculated to maintain the tSJ architecture via *trans* Aka interaction. Mutation of the Aka gene in *Drosophila* caused paracellular barrier defects due to the loss of stacked diaphragm in the tSJ (Byri et al., 2015). Aka is also important for recruiting the Gli protein to the tSJ, suggesting that Aka may mediate the tSJ assembly via its *cis*-interaction with other tSJ proteins (Byri et al., 2015). Sdk is an immunoglobulin-like transmembrane protein required for the pattern formation of photoreceptor cells in *Drosophila* eyes (Nguyen, Liu, Litsky, & Reinke, 1997). In a systematic protein trap experiment, Sdk was found exclusively localized to the tSJ in *Drosophila* embryos (Lye, Naylor, & Sanson, 2014). Whether Sdk plays a role in tSJ structure and function has not been determined. Notably, Gliotactin, Anakonda, and Sidekick show no homology to tricellulin or angulins, the tricellular tight junction (tTJ) proteins, which suggests that tSJ and tTJ are evolutionarily divergent cellular organelles.

10.4 TIGHT JUNCTION IN ZEBRAFISH

Fifteen zebrafish (*Danio rerio*) claudin genes were initially identified by homology to mammalian claudin genes (Kollmar, Nakamura, Kappler, & Hudspeth, 2001). A recent genome-wide search has expanded the list to 54 claudins in zebrafish, which include the orthologues to all mammalian claudins (Baltzegar, Reading, Brune, & Borski, 2013). Zebrafish and mammalian claudins share a common phylogeny (Fig. 10.3). Claudin-b, the zebrafish orthologue to human claudin-4, is expressed at the TJ by the epithelial cells in the skin, the gill and the kidney (Kwong, Kumai, & Perry, 2013; Kwong & Perry, 2013). Transgenic knockdown of claudin-b in zebrafish caused edema in the pericardial cavity and the yolk sac owing to increased paracellular permeability to Na^+ (Kwong & Perry, 2013). Zebrafish expresses two claudin-5 genes, claudin-5a and -5b. Claudin-5a assumes a strong neuroepithelial expression whereas claudin-5b is endothelium-specific (Zhang et al., 2010). Loss of claudin-5a in zebrafish disrupted the neuroepithelial paracellular barrier, which resulted in decreased brain ventricular volume (Zhang et al., 2010). The paracellular pathway also underlies the mechanism responsible for single-lumen specification in zebrafish gut tubes. The claudin-15 knockdown zebrafish developed multiple lumens in the gut, due to reductions in paracellular ionic permeability and luminal fluid accumulation (Bagnat, Cheung, Mostov, & Stainier, 2007).

10.5 SPECIAL VERTEBRATE CELL JUNCTION

10.5.1 Paranodal Junction

The axon in vertebrate nervous system is myelinated by the surrounding glial cells. Neuron-glia interactions establish three membrane domains along the myelinated axon, including the node of Ranvier, the paranodal junction and the juxtaparanode. The paranodal junction is essential for separating the periaxonal from the extracellular space to fulfill myelin's insulating role (Rosenbluth, 2009). Under transmission electron microscopy, the paranodal junction appears as septum-like structure of approximately 2–4 nm in length, also known as "transverse band," which connects the myelin lamella to the axolemma (Fig. 10.5). Functionally, the paranodal junction acts as a barrier to impede the permeation of molecules to and from the periaxonal space (Mierzwa, Shroff, & Rosenbluth, 2010). The paranodal junction shares a core adhesive protein complex with the SJ in invertebrates (Hortsch & Margolis, 2003). Nonetheless, claudin or its related protein has not been found in the paranodal junction, which makes it difficult to conclude the paracellular channel exists in this type of cell junction.

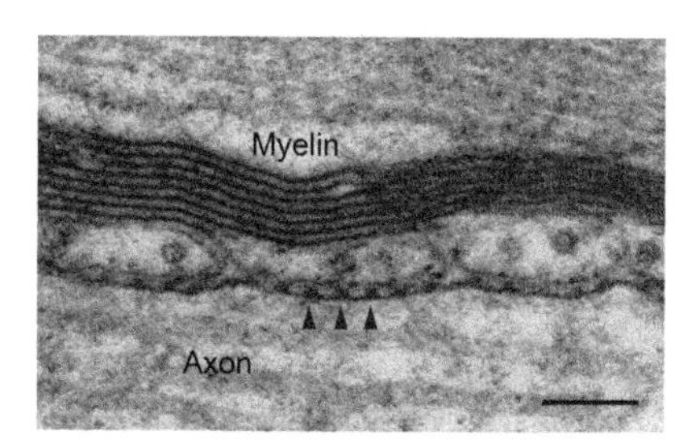

FIGURE 10.5 Mouse paranodal junction. Transmission electron micrograph of adult mouse optic nerve longitudinal section showing the sequential termination of myelin lamellae in the paranodal region. *Arrowheads* indicate the paranodal junction as electron-dense transverse bands between the paranodal loop and the axonal membrane. Bar: 100 nm. *(Reproduced with permission from Chang, K.J., Zollinger, D.R., Susuki, K., Sherman, D.L., Makara, M.A., Brophy, P.J., et al. (2014). Glial ankyrins facilitate paranodal axoglial junction assembly. Nature Neuroscience, 17, 1673–1681).*

10.5.2 Slit Diaphragm

The glomerulus in vertebrate kidney functions as a molecular sieve capable of filtering small plasma molecules but excluding large plasma proteins such as albumin. The glomerulus consists in a filtration unit made of the endothelial cell, the podocyte, and the glomerular basement membrane. The slit diaphragm (SD) is a specialized cell junction formed between podocyte foot processes, which appears as a continuous band of 30–45 nm in length (Fig. 10.6). Transmission electron microscopy reveals rectangular pores of approximately 4 × 14 nm in cross section in the SD (Rodewald & Karnovsky, 1974). The key protein component of SD is nephrin, a single-pass transmembrane protein with a large extracellular structure comprising a fibronectin type III domain and eight immunoglobulin domains (Kestila et al., 1998). Nephrin is structurally similar to cadherin, and *trans* nephrin interactions have been visualized in the SD (Wartiovaara et al., 2004). Several TJ proteins including ZO-1, occludin, and JAM-A are localized to the SD (Fukasawa, Bornheimer, Kudlicka, & Farquhar, 2009; Schnabel, Anderson, & Farquhar, 1990). Claudin and its related protein TM4SF10, the vertebrate orthologue to *C. elegans* VAB-9, also play important roles in SD structure and function. For example, claudin-1 interacts with nephrin in both *cis* and *trans* and destabilizes the SD architecture. Transgenic overexpression of claudin-1 in mouse podocytes caused microalbuminuria due to dissolution of the SD (Gong, Sunq, Roth, & Hou, 2017). TM4SF10 was found to loosen the nephrin anchorage to actin cytoskeleton in response to multiple nephrotic insults (Azhibekov, Wu, Padiyar, Bruggeman, & Simske, 2011).

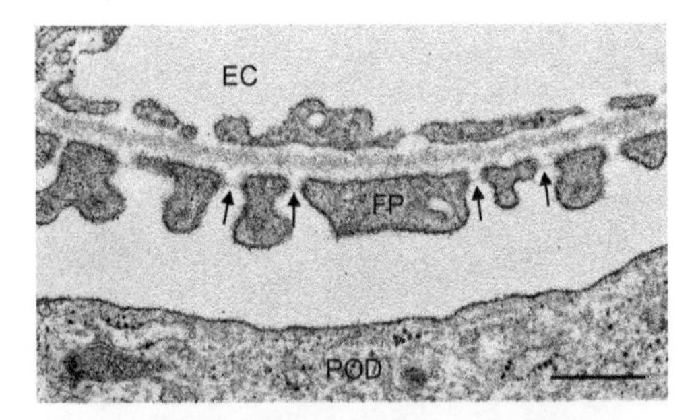

FIGURE 10.6 Mouse slit diaphragm. Transmission electron micrograph of adult mouse kidney section showing the glomerular filtration unit. *Arrows* indicate the slit diaphragm as electron-dense continuous bands connecting podocyte foot processes. *EC*, Endothelial cell; *FP*, foot process; *POD*, podocyte. Bar: 500 nm. *(Reproduced with permission from Gong, Y., Sunq, A., Roth, R.A., & Hou, J. (2017). Inducible expression of Claudin-1 in glomerular podocytes generates aberrant tight junctions and proteinuria through slit diaphragm destabilization. Journal of the American Society of Nephrology, 28, 106–117).*

REFERENCES

Asano, A., Asano, K., Sasaki, H., Furuse, M., & Tsukita, S. (2003). Claudins in *Caenorhabditis elegans*: Their distribution and barrier function in the epithelium. *Current Biology, 13*, 1042–1046.

Auld, V. J., Fetter, R. D., Broadie, K., & Goodman, C. S. (1995). Gliotactin, a novel transmembrane protein on peripheral glia, is required to form the blood-nerve barrier in Drosophila. *Cell, 81*, 757–767.

Azhibekov, T. A., Wu, Z., Padiyar, A., Bruggeman, L. A., & Simske, J. S. (2011). TM4SF10 and ADAP interaction in podocytes: Role in Fyn activity and nephrin phosphorylation. *American Journal of Physiology Cell Physiology, 301*, C1351–C1359.

Bagnat, M., Cheung, I. D., Mostov, K. E., & Stainier, D. Y. (2007). Genetic control of single lumen formation in the zebrafish gut. *Nature Cell Biology, 9*, 954–960.

Baldwin, K. M., Loeb, M. J., & Riemann, J. G. (1987). A novel occluding junction which lacks membrane fusion in insect testis. *Tissue and Cell, 19*, 413–421.

Baltzegar, D. A., Reading, B. J., Brune, E. S., & Borski, R. J. (2013). Phylogenetic revision of the claudin gene family. *Marine Genomics, 11*, 17–26.

Banerjee, S., Sousa, A. D., & Bhat, M. A. (2006). Organization and function of septate junctions: An evolutionary perspective. *Cell Biochemistry and Biophysics, 46*, 65–77.

Behr, M., Riedel, D., & Schuh, R. (2003). The claudin-like megatrachea is essential in septate junctions for the epithelial barrier function in drosophila. *Developmental Cell, 5*, 611–620.

Bilder, D., Schober, M., & Perrimon, N. (2003). Integrated activity of PDZ protein complexes regulates epithelial polarity. *Nature Cell Biology, 5*, 53–58.

Byri, S., Misra, T., Syed, Z. A., Batz, T., Shah, J., Boril, L., et al. (2015). The triple-repeat protein anakonda controls epithelial tricellular junction formation in drosophila. *Developmental Cell, 33*, 535–548.

Chang, K. J., Zollinger, D. R., Susuki, K., Sherman, D. L., Makara, M. A., Brophy, P. J., et al. (2014). Glial ankyrins facilitate paranodal axoglial junction assembly. *Nature Neuroscience, 17*, 1673–1681.

Fukasawa, H., Bornheimer, S., Kudlicka, K., & Farquhar, M. G. (2009). Slit diaphragms contain tight junction proteins. *Journal of the American Society of Nephrology, 20*, 1491–1503.

Gong, Y., Sunq, A., Roth, R. A., & Hou, J. (2017). Inducible expression of claudin-1 in glomerular podocytes generates aberrant tight junctions and proteinuria through slit diaphragm destabilization. *Journal of the American Society of Nephrology, 28*, 106–117.

Graf, F., Noirot-Timothee, C., & Noirot, C. (1982). The specialization of septate junctions in regions of tricellular junctions. I. Smooth septate junctions (=continuous junctions). *Journal of Ultrastructure Research, 78*, 136–151.

Higashi, T., & Miller, A. L. (2017). Tricellular junctions: How to build junctions at the TRICkiest points of epithelial cells. *Molecular Biology of the Cell, 28*, 2023–2034.

Hortsch, M., & Margolis, B. (2003). Septate and paranodal junctions: Kissing cousins. *Trends in Cell Biology, 13*, 557–561.

Itoh, M., Furuse, M., Morita, K., Kubota, K., Saitou, M., & Tsukita, S. (1999). Direct binding of three tight junction-associated MAGUKs, ZO-1, ZO-2, and ZO-3, with the COOH termini of claudins. *The Journal of Cell Biology, 147*, 1351–1363.

Jonusaite, S., Donini, A., & Kelly, S. P. (2016). Occluding junctions of invertebrate epithelia. *Journal of Comparative Physiology B, Biochemical, Systemic, and Environmental Physiology, 186*, 17–43.

Kestila, M., Lenkkeri, U., Mannikko, M., Lamerdin, J., McCready, P., Putaala, H., et al. (1998). Positionally cloned gene for a novel glomerular protein--nephrin--is mutated in congenital nephrotic syndrome. *Molecular Cell, 1*, 575–582.

Kollmar, R., Nakamura, S. K., Kappler, J. A., & Hudspeth, A. J. (2001). Expression and phylogeny of claudins in vertebrate primordia. *Proceedings of the National Academy of Sciences of the United States of America, 98*, 10196–10201.

Kwong, R. W., & Perry, S. F. (2013). The tight junction protein claudin-b regulates epithelial permeability and sodium handling in larval zebrafish, Danio rerio. *American Journal of Physiology Regulatory, Integrative and Comparative Physiology, 304*, R504–R513.

Kwong, R. W., Kumai, Y., & Perry, S. F. (2013). Evidence for a role of tight junctions in regulating sodium permeability in zebrafish (Danio rerio) acclimated to ion-poor water. *Journal of Comparative Physiology B, Biochemical, Systemic, and Environmental Physiology, 183*, 203–213.

Lockwood, C., Zaidel-Bar, R., & Hardin, J. (2008). The C. elegans zonula occludens ortholog cooperates with the cadherin complex to recruit actin during morphogenesis. *Current Biology, 18*, 1333–1337.

Lye, C. M., Naylor, H. W., & Sanson, B. (2014). Subcellular localisations of the CPTI collection of YFP-tagged proteins in Drosophila embryos. *Development (Cambridge, England), 141*, 4006–4017.

McMahon, L., Legouis, R., Vonesch, J. L., & Labouesse, M. (2001). Assembly of C. elegans apical junctions involves positioning and compaction by LET-413 and protein aggregation by the MAGUK protein DLG-1. *Journal of Cell Science, 114*, 2265–2277.

Meng, W., & Takeichi, M. (2009). Adherens junction: Molecular architecture and regulation. *Cold Spring Harbor Perspectives in Biology, 1*, a002899.

Mierzwa, A., Shroff, S., & Rosenbluth, J. (2010). Permeability of the paranodal junction of myelinated nerve fibers. *The Journal of Neuroscience, 30*, 15962–15968.

Nelson, K. S., Furuse, M., & Beitel, G. J. (2010). The Drosophila Claudin Kune-kune is required for septate junction organization and tracheal tube size control. *Genetics, 185*, 831–839.

Nguyen, D. N., Liu, Y., Litsky, M. L., & Reinke, R. (1997). The sidekick gene, a member of the immunoglobulin superfamily, is required for pattern formation in the Drosophila eye. *Development (Cambridge, England), 124*, 3303–3312.

Noirot-Timothee, C., Graf, F., & Noirot, C. (1982). The specialization of septate junctions in regions of tricellular junctions. II. Pleated septate junctions. *Journal of Ultrastructure Research, 78*, 152–165.

Resnik-Docampo, M., Koehler, C. L., Clark, R. I., Schinaman, J. M., Sauer, V., Wong, D. M., et al. (2017). Tricellular junctions regulate intestinal stem cell behaviour to maintain homeostasis. *Nature Cell Biology, 19*, 52–59.

Rodewald, R., & Karnovsky, M. J. (1974). Porous substructure of the glomerular slit diaphragm in the rat and mouse. *The Journal of Cell Biology, 60*, 423–433.

Rosenbluth, J. (2009). Multiple functions of the paranodal junction of myelinated nerve fibers. *Journal of Neuroscience Research, 87*, 3250–3258.

Schnabel, E., Anderson, J. M., & Farquhar, M. G. (1990). The tight junction protein ZO-1 is concentrated along slit diaphragms of the glomerular epithelium. *The Journal of Cell Biology, 111*, 1255–1263.

Schulte, J., Tepass, U., & Auld, V. J. (2003). Gliotactin, a novel marker of tricellular junctions, is necessary for septate junction development in Drosophila. *The Journal of Cell Biology, 161*, 991–1000.

Shin, K., Fogg, V. C., & Margolis, B. (2006). Tight junctions and cell polarity. *Annual Review of Cell and Developmental Biology, 22*, 207–235.

Simske, J. S., Koppen, M., Sims, P., Hodgkin, J., Yonkof, A., & Hardin, J. (2003). The cell junction protein VAB-9 regulates adhesion and epidermal morphology in C. elegans. *Nature Cell Biology*, *5*, 619–625.

Szollosi, A., & Marcaillou, C. (1977). Electron microscope study of the blood-testis barrier in an insect: Locusta migratoria. *Journal of Ultrastructure Research*, *59*, 158–172.

Tepass, U. (2003). Claudin complexities at the apical junctional complex. *Nature Cell Biology*, *5*, 595–597.

Tepass, U., & Hartenstein, V. (1994). The development of cellular junctions in the Drosophila embryo. *Developmental Biology*, *161*, 563–596.

Wartiovaara, J., Ofverstedt, L. G., Khoshnoodi, J., Zhang, J., Makela, E., Sandin, S., et al. (2004). Nephrin strands contribute to a porous slit diaphragm scaffold as revealed by electron tomography. *The Journal of Clinical Investigation*, *114*, 1475–1483.

Wu, V. M., Schulte, J., Hirschi, A., Tepass, U., & Beitel, G. J. (2004). Sinuous is a Drosophila claudin required for septate junction organization and epithelial tube size control. *The Journal of Cell Biology*, *164*, 313–323.

Zhang, J., Piontek, J., Wolburg, H., Piehl, C., Liss, M., Otten, C., et al. (2010). Establishment of a neuroepithelial barrier by Claudin5a is essential for zebrafish brain ventricular lumen expansion. *Proceedings of the National Academy of Sciences of the United States of America*, *107*, 1425–1430.

Chapter 11

Perspective

11.1 STRUCTURAL ORGANIZATION OF PARACELLULAR CHANNEL

11.1.1 *De Novo* Assembly

A simplified model of paracellular channel structure is viewed as a high-order protein complex made of multiple claudin subunits. Even though the structure of claudin monomer has been resolved, it remains to be seen how claudins are assembled in the extracellular space. Claudin polymerization is part of a sophisticated cellular process known as TJ assembly, which involves hundreds of proteins via several key steps such as cell adhesion, polarity establishment, and cytoskeletal anchorage. Cell-free systems such as artificial membranes may allow bypass the regulatory hierarchy required for paracellular channel biogenesis. Peripheral myelin protein 22 (PMP22), a protein structurally related to claudin, spontaneously organized the membrane into myelin-like lamellae when reconstituted into lipid bilayers (Fig. 11.1A) (Mittendorf et al., 2017). VE-cadherin self-assembled into an artificial adherens junction when reconstituted into liposomes (Fig. 11.1B) (Taveau et al., 2008). The extracellular domains in PMP22 or VE-cadherin are vital to the respective assembly process. Because the extracellular domains in claudin also mediate cell adhesion (Kubota et al., 1999), liposomes reconstituted with claudins may form TJ-like structures. These artificial TJs, while not identical to endogenous TJs in cells, will still provide critical insights to *trans* claudin interaction, spatial organization of paracellular channel, and protein to lipid relationship.

11.1.2 *In Situ* Visualization

Endogenous TJs contain hundreds of proteins (Tang, 2006). The paracellular channels are in reality bundled with many other TJ proteins such as those in the junctional plaque or performing signaling functions. Molecular imaging tools are essential for measuring paracellular channel density in the TJ and for revealing the spatial relationship between paracellular channels and other TJ components. Single-molecule super-resolution microscopic methods, more commonly known as stochastic optical reconstruction microscopy (STORM) and photoactivated localization microscopy (PALM), are based on localizing individual photoactivatable fluorophores by fitting their point-spread functions (Rust, Bates, & Zhuang, 2006). TJs are well suited to STORM/PALM

The Paracellular Channel. http://dx.doi.org/10.1016/B978-0-12-814635-4.00011-5

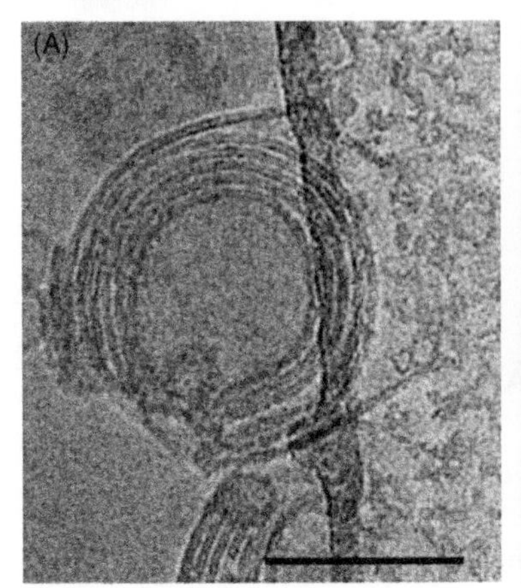

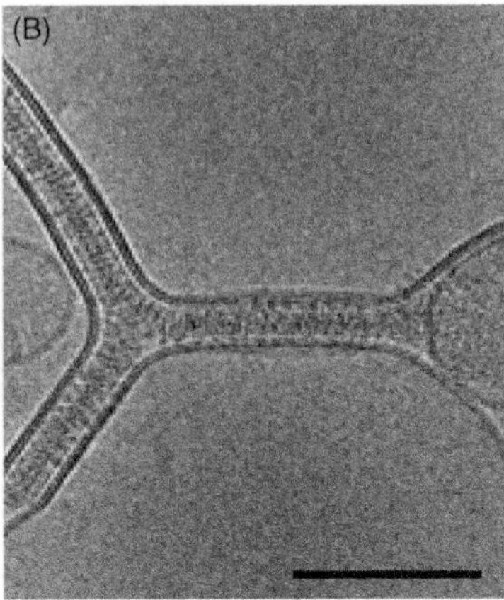

FIGURE 11.1 Cryo-electron micrographs of artificial myelin and adherens junction. (A) Vitrified liposomes containing PMP22 proteins spontaneously formed myelin-like structures. (B) Vitrified liposomes containing VE-cadherin proteins established cell junctions resembling the adherens junction. Bar: 100 nm. *(Reproduced with permission from Mittendorf, K.F., Marinko, J.T., Hampton, C.M., Ke, Z., Hadziselimovic, A., Schlebach, J.P., et al. (2017). Peripheral myelin protein 22 alters membrane architecture. Science Advances, 3, e1700220; Taveau, J.C., Dubois, M., Le Bihan, O., Trepout, S., Almagro, S., Hewat, E., et al. (2008). Structure of artificial and natural VE-cadherin-based adherens junctions. Biochemical Society Transactions, 36, 189–193).*

because they contain clustered claudin proteins sufficient to provide submicron spatial resolution according to the Nyquist sampling theory (molecular density $\sim 10^4/\mu m^2$ for achieving ~20 nm resolution). Kaufmann and coworkers estimated the claudin-5 molecular density in the TJ to be $\sim 4000/\mu m^2$ using spectral position determination microscopy, which in theory would allow STORM/PALM to achieve ~50 nm resolution (Fig. 11.2) (Kaufmann et al., 2012). These methods are expected to reveal the local architecture of claudin molecular assembly. As a proof of principle, Wu and coworkers have imaged the adherens junction with STORM/PALM and determined how E-cadherin molecules are assembled, that is, a median number of six E-cadherin molecules make a cluster; the mean diameter of each cluster is 60 nm; and the median spacing between two clusters is 157 nm (Wu, Kanchanawong, & Zaidel-Bar, 2015).

Cryo-electron microscopy of vitreous section (CEMOVIS) is a new method for visualizing the ultrastructure of subcellular organelle with nanometer resolution and in close to native state. With CEMOVIS, Al-Amoudi and coworkers have revealed the architecture of desmosome in human epidermis (Al-Amoudi, Dubochet, & Norlen, 2005). The molecular interfaces of desmosomal cadherin assembly are seen as electron-dense, transverse lines with ~5 nm periodicity (Fig. 11.3A). These transverse lines are molecular clusters of cadherins undergoing *trans* W-like and *cis* V-like interactions (Fig. 11.3B) (Al-Amoudi, Diez, Betts, & Frangakis, 2007). If CEMOVIS is applied to the TJ, then potential

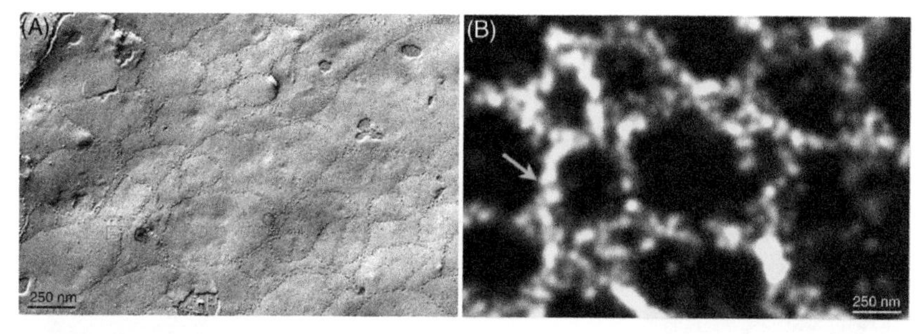

FIGURE 11.2 Correlative study of TJ by freeze fracture electron microscopy and spectral position determination microscopy. (A) Freeze fracture electron micrograph of Cldn5-YFP/Cldn3 cotransfected HEK293 cells. (B) Spectral position determination micrograph of Cldn5-YFP transfected HEK293 cells. Under freeze fracture electron microscopy, the TJ strands appear as chains of intramembranous particles of ~10 nm in diameter (*green arrow* in A) on the exoplasmic (E) and protoplasmic (P) faces of the plasma membrane. Due to lower spatial resolution of light microscopy, the TJ strands appear thicker. It is not known whether the structure with reinforced signals (*green arrow* in B) corresponds to the intramembranous particle in A. Bar: 250 nm. *(Reproduced with permission from Kaufmann, R., Piontek, J., Grull, F., Kirchgessner, M., Rossa, J., Wolburg, H., et al. (2012). Visualization and quantitative analysis of reconstituted tight junctions using localization microscopy. PLoS One, 7, e31128).*

structural information of claudin assembly will be obtained. It will even be possible to visualize how paracellular channels are opened and closed due to claudin interactions or in response to various physiological and pathological stimuli.

11.2 SEPARATION OF PARACELLULAR CONDUCTANCE FROM TRANSCELLULAR CONDUCTANCE

Current electrophysiological approaches are unable to discern paracellular conductance from transcellular conductance on the basis of channel behavior, circuit configuration, or signal analysis. Scanning ion conductance microscopy (SICM) has proven to be an effective tool to isolate the paracellular conductance from surrounding transcellular pathways by filtering the transcellular conductance signals according to their spatial proximity to the TJ (Chapter 3, Section 3.2.2.2) (Chen, Zhou, Morris, Hou, & Baker, 2013). The spatial resolution alone is, however, not sufficient to allow delineating discrete paracellular channel conductance. In the configuration of SICM, application of a transepithelial potential (V_{TE}) drives ions to move across the epithelial layer through both paracellular and transcellular pathways, which generates two currents, indicated as i_{para} and i_{trans} respectively. An unknown fraction of i_{trans} will inevitably be picked up by the recording pipet in SICM. Hou and coworkers have proposed an improved design of SICM to electrically offset the contribution of i_{trans} by incorporating whole-cell patch clamps into SICM (Fig. 11.4) (Zhou, Zeng, Baker, & Hou, 2015). The purpose of this method is to "zero out" the driving force for

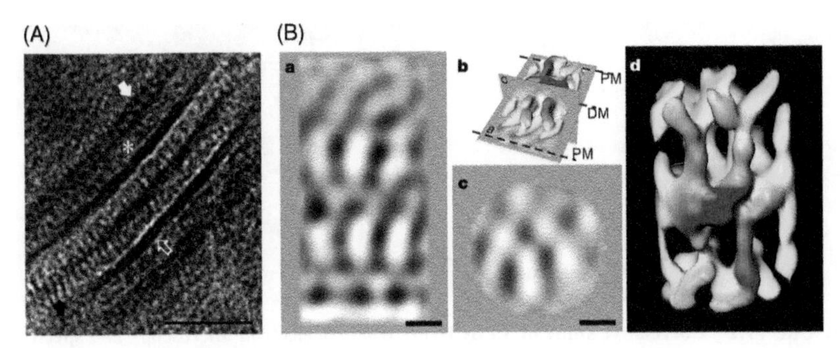

FIGURE 11.3 Cryo-electron micrograph of desmosome in human epidermis. (A) The desmosome is ~33 nm wide and contains transverse electron-dense lines with ~5 nm periodicity (*black arrow*). On the cytoplasmic side, an opaque zone of medium electron density (*white asterisk*) separates the electron-dense plasma membrane (*hollow white arrow*) from the electron-dense layer of desmosomal plaque (*solid white arrow*). Bar: 50 nm. (B) Visualization of the average of the subtomograms of desmosome. a. Coronal slice through the average of the subtomograms, which are extracted from the extracellular space only. The plasma membranes are not shown but their position is indicated in b. The elongated curved densities arrange in a periodic manner and their dimensions compare remarkably well with the W-like conformation observed in the C-cadherin X-ray structure. b. Cartoon image showing the relative orientations of the slices visualized in a and c and the isosurface in d. The position of the coronal slice visualized in a, the axial slice in c, and their location with respect to the isosurface shown in d are indicated. The dashed lines indicate the location of the dense midline (DM) as well as the position of the plasma membrane (PM). The latter is removed from the subtomograms and thus not visible in the average structure or shown in a. c. Axial slice through the averaged subtomograms. Distinct cadherin molecules can be recognized as dense blobs of approximately 3 nm in diameter. d. Isosurface visualization of the organization of the cadherin molecules. The threshold is chosen so that the thickness of the central cadherin molecule is approximately 3 nm. The color coding varies as a function of the depth from *red* (close) to *blue* (far). Two cadherin dimers arranged in a *trans* W-like manner are shown in the foreground in *orange*, with their concave side facing left. One layer deeper, shown in *green*, four cadherin *trans* W-like dimers have their concave side oriented to the right. The *orange* and the *green* molecules emanating from the same cell surface interact to form V-like *cis* dimers. Two additional W-like cadherin dimers have their concave side oriented to the left (*blue*) and interact with the *green* molecules to form V-like *cis* dimers. Bar: 7 nm. *(Reproduced with permission from Al-Amoudi, A., Diez, D.C., Betts, M.J., & Frangakis, A.S. (2007). The molecular architecture of cadherins in native epidermal desmosomes. Nature, 450, 832–837; Al-Amoudi, A., Dubochet, J., & Norlen, L. (2005). Nanostructure of the epidermal extracellular space as observed by cryo-electron microscopy of vitreous sections of human skin. The Journal of Investigative Dermatology, 124, 764–777).*

i_{trans} (due to the application of V_{TE}) by controlling the intracellular potentials of the cells on either side of the TJ with whole-cell patch clamps. Baker and coworkers undertook a different direction to resolve multiple transport pathways with SICM. They automated the SICM measurements to map the conductance levels over the entire epithelial surface (Fig. 3.7) (Zhou, Gong, Hou, & Baker, 2017). While the spatial resolution over each recorded location remains the same, that is, approximately one micron in diameter, the conductance heat map allows direct visualization of the transport heterogeneity within an epithelial layer. Not only are differences between paracellular and transcellular pathways

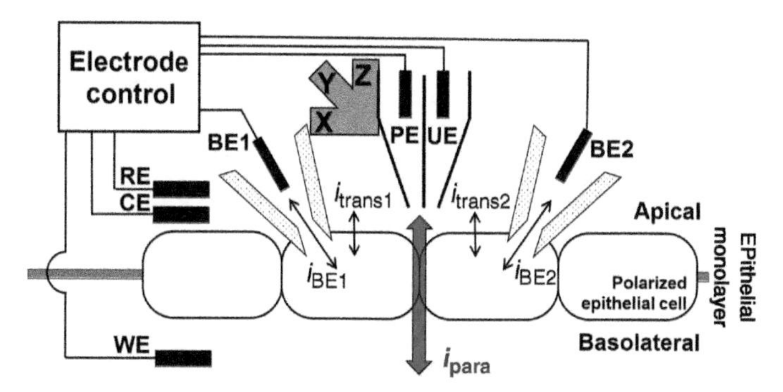

FIGURE 11.4 **Patch clamp integrated SICM.** A double-barrel pipette for topographical imaging (PE) and local potential measurement (UE) is attached to a piezo positioner. The electric circuit from working electrode (WE) to reference electrode (RE) applies the transepithelial potential (V_{TE}) across the epithelial monolayer. Two balance electrodes (BE1 and BE2) control the intracellular potentials of two clamped cells and record the current flowing to and from BE1 or BE2 (i_{BE1} or i_{BE2}) respectively. Upon the application of V_{TE}, ion current flows across the epithelial cell layer through both paracellular pathway (i_{para}) and transcellular pathway (i_{trans1} and i_{trans2}). When the intracellular potentials of the two clamped cells are held at their resting membrane potentials to cancel out the apical membrane currents, the majority of V_{TE} induced basolateral membrane currents are drawn through BE1 and BE2. *(Reproduced with permission from Zhou, L., Zeng, Y., Baker, L.A., & Hou, J. (2015). A proposed route to independent measurements of tight junction conductance at discrete cell junctions. Tissue Barriers, 3, e1105907).*

compared side by side, but also conductance hot spots are revealed for distinctive locations in the epithelium. These hot spots may represent transient open states of paracellular or transcellular channels.

11.3 SPATIAL AND CELLULAR HETEROGENEITY IN PARACELLULAR CHANNEL

Different binding affinity of claudin proteins partitions the paracellular pathway into several compartments along the lateral membrane. In the inner ear, the TJ between the sensory hair cell and the supporting cell contains two subdomains that are distinguishable by ultrastructural morphology and claudin composition (Nunes et al., 2006). Claudin-14 preferably assembles into the apical subdomain of TJ whereas claudin-6 and claudin-9 co-assemble into the basal subdomain of TJ (Fig. 11.5). The overall paracellular permeability is determined as a composite function of the paracellular channels made of claudin-14 and of claudin-6 and claudin-9. New advancement in single-molecule super-resolution microscopy may reveal new cases of molecular heterogeneity in the paracellular pathway. Intriguingly, claudin binding affinity difference also partitions the paracellular pathway based upon cellular heterogeneity. In the TJs of kidney thick ascending limb cells, the localization of claudin-10b and claudin-16 was mutually exclusive whereas the localization of claudin-16 and claudin-19 was universally

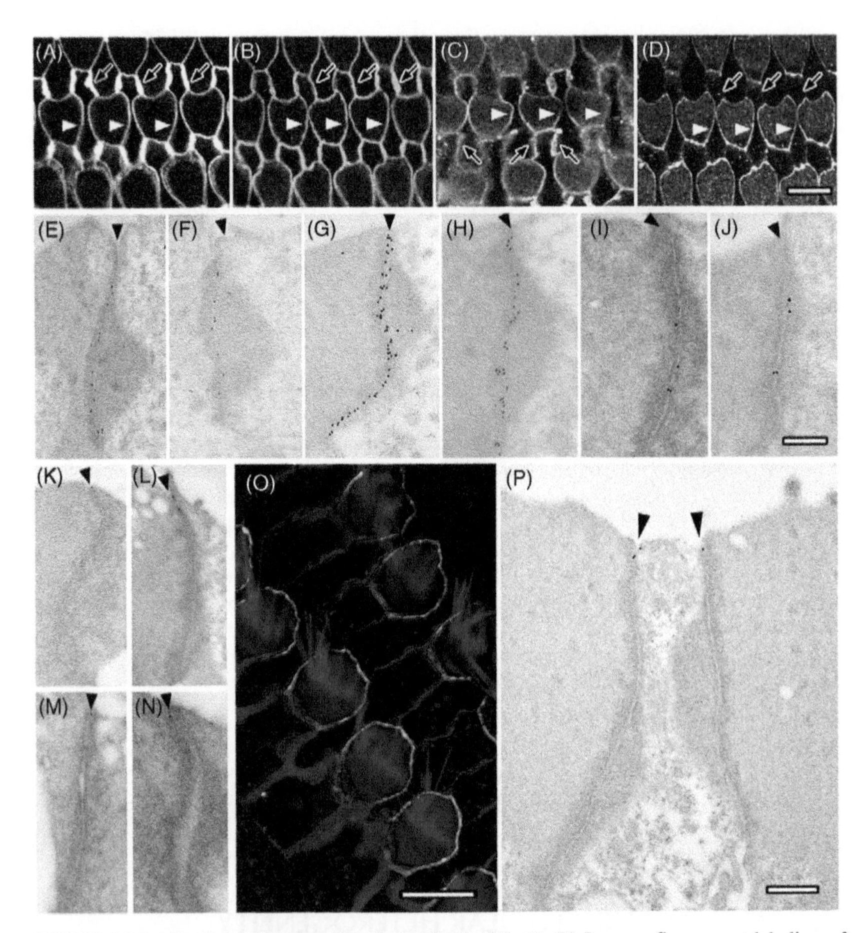

FIGURE 11.5 Spatial heterogeneity of claudin in TJ. (A–D) Immunofluorescent labeling of ZO-1 (A), claudin-9 (B), claudin-6 (C), and claudin-14 (D) in the cell junction between supporting cell and outer hair cell (*white arrowheads*) or between supporting cells (*black arrows*) in the organ of Corti. (E–N) Immunogold electron microscopy showing ZO-1 (E), occludin (F), claudin-9 (G and H), claudin-6 (I and J) and claudin-14 (K–N) in the cell junction between supporting cell and outer hair cell. Outer hair cells are to the right. *Black arrowheads*: apices of cell junctions. (O) Single confocal optical section near the apical surface of the reticular lamina showing immunofluorescence for claudin-14 in the cell junction between supporting cell and outer hair cell (*green*), ZO-1 in the cell junction between supporting cells or between supporting cell and outer hair cell (*blue*), and actin (labeled with rhodamine phalloidin, *red*) in the stereocilia. (P) Immunogold electron microscopy showing additional claudin-14 labeling (*arrowheads*) in the cell junction between supporting cell and outer hair cell. Tissue samples are from: guinea pig (A–F, O and P); rat (G, I, K, and L) and mouse (H, J, M and N). Bar: 10 μm (A–D); 0.4 μm (E–N); 10 μm (O); 0.3 μm (P). *(Reproduced with permission from Nunes, F.D., Lopez, L.N., Lin, H.W., Davies, C., Azevedo, R.B., Gow, A., et al. (2006). Distinct subdomain organization and molecular composition of a tight junction with adherens junction features. Journal of Cell Science, 119, 4819–4827).*

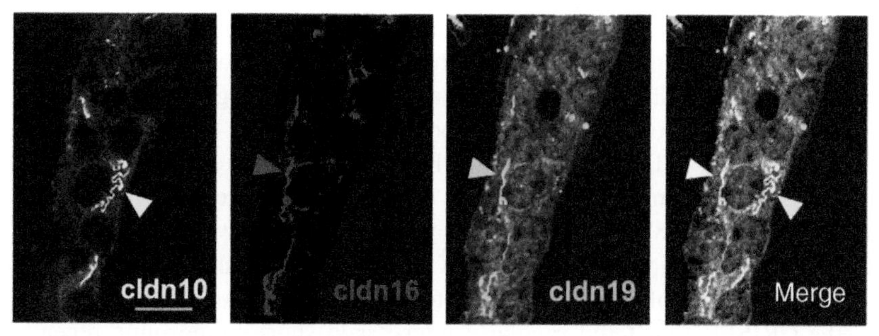

FIGURE 11.6 Mosaic claudin localization in TJ. Single isolated mouse thick ascending limb tubules are immunostained for cldn10b, cldn16, and cldn19. In contrast to immunostaining in sections, intracellular claudin expression is not distinguishable from background. Cldn16 (*magenta*) and cldn19 (*green*) are co-localized within the same TJ (shown as *white* in the merged image, *white arrowhead*), but cldn10b (*yellow arrowhead*) never co-localizes with cldn16 or cldn19 in the TJ. Bar: 10 μm. *(Reproduced with permission from Milatz, S., Himmerkus, N., Wulfmeyer, V.C., Drewell, H., Mutig, K., Hou, J., et al. (2017). Mosaic expression of claudins in thick ascending limbs of Henle results in spatial separation of paracellular* Na^+ *and* Mg^{2+} *transport. Proceedings of the National Academy of Sciences of the United States of America, 114, E219–E227).*

overlapping (Fig. 11.6) (Milatz et al., 2017). The separation of claudin-10b from claudin-16 or claudin-19 may create two distinct cell populations with different paracellular permeabilities. Whether claudin segregation arises from random interaction or cellular regulation will be the most important question to address. Modern techniques such as single-cell RNA sequencing and single-cell proteomics may reveal how claudin mosaics are established on cell and organ levels, and what role the mosaics play in paracellular channel physiology.

11.4 NEW ASPECT OF PARACELLULAR CHANNELOPATHY

Conventional wisdom has it that loss of a gene function, due to either genetic mutation in human patients or transgenic deletion in experimental mice, results in a phenotype that is attributed to the gene, the so called monogenic effect. The claudin genes, however, do not always follow this principle. Mechanistically, there are two structural models to explain the paracellular channel function in general. *A*, claudins form parallel channels each with its own physiological signature and the overall paracellular permeability reflects the combination of the properties of the claudins expressed; *B*, claudins form hybrid channels with novel properties that require their *cis*- and *trans*- interactions. Under either circumstance, the phenotype incurred by removal of a claudin will be highly dependent upon the remaining claudins in the TJ. Sometimes, cells may compensate for the loss of a claudin by synthesizing a new claudin that is not normally present in the TJ. For example, deletion of claudin-10b from the mouse kidney stimulated an over 20-fold increase in claudin-14 gene expression (Breiderhoff et al., 2012). Therefore, interpretation of the phenotype of claudin deficiency must take into account the background claudins or the compensating claudins. The knockout approach itself could be problematic in the study of human disease. After analyzing 29 different claudin-16 mutations from the familial hypomagnesemia with hypercalciuria and nephrocalcinosis (FHHNC) syndrome, Konrad and coworkers noticed that not all of these mutations resulted in a loss of function (Chapter 8, Section 8.2.5) (Konrad et al., 2008). Some retained a residual function, which may explain the phenotypic variation in FHHNC. Clearly, these variables are not easily addressed by a simple genetic knockout animal. The knock-in approach aided by the CRISPR (clustered regularly interspaced short palindromic repeats) technology will likely provide a bona fide view of how genetic mutations cause paracellular channelopathy.

11.5 COUPLING OF PARACELLULAR PATHWAY WITH TRANSCELLULAR PATHWAY

The concept of transport coupling between paracellular and transcellular channels is not new. Numerous studies have revealed that the paracellular permeability can be regulated by the membrane channels or transporters such as Na^+/K^+–ATPase, Na^+-glucose cotransporter 1 (SGLT1), Na^+/H^+ exchanger 3 (NHE3), chloride channels, transient receptor potential channels, and aquaporins (Rajasekaran, Beyenbach, & Rajasekaran, 2008). Intracellular Na^+, Ca^{++}, Cl^-, and volume levels are important second messengers. What remains intriguing is the fact that some of these membrane channels or transporters are located in the TJ. For example, the hypotonicity-activated chloride channel, ClC-2 is found predominantly in the TJ of intestinal epithelial cells (Mohammad-Panah et al., 2001). The *Drosophila* Na^+/K^+–ATPase is part of the protein complex making the septate junction, the insect orthologue to mammalian TJ (Genova & Fehon, 2003). Some of the membrane channels or transporters could interact with claudins and modulate their function. Reciprocally, claudins may regulate the membrane channels directly or indirectly. Phylogenetic analyses have suggested that claudin and voltage-gated calcium channel γ subunit (CACNG) share a common ancestral root (Anderson & Van Itallie, 2009). Both classes of proteins carry the same characteristic motif (-GLWCC; PROSITE ID: PS01346) in the first extracellular loop domain (Chu, Robertson, & Best, 2001). CACNG2, also known as stargazin, has been shown to mediate cell adhesion when transfected into mouse L fibroblasts (Price, Davis, Deng, & Burgess, 2005). On the basis of claudin homology to CACNG, one might speculate that claudin participates in membrane calcium channel conductance. Interestingly, claudin-16 has been found to increase the transcellular Cl^- permeability in MDCK cells via interaction with the calcium-activated chloride channel (Gunzel et al., 2009). From an evolutionary point of view, the paracellular channel may have derived from an event of gene duplication, which created a new type of ion channel bearing traditional ion channel structure.

11.6 STRUCTURE AND FUNCTION OF TRICELLULAR TIGHT JUNCTION

Tricellular tight junction (tTJ) differs significantly from bicellular tight junction (bTJ) in molecular content, ultrastructural architecture, and paracellular permeability. Yet, tTJ and bTJ are interdependent. For example, deletion of occludin, a bTJ protein from mouse inner ear caused dislocalization of tricellulin, a tTJ protein to cytoplasm (Kitajiri et al., 2014). Reciprocally, knockdown of tricellulin reduced the abundance level of occludin in the bTJ (Ikenouchi et al., 2005). These results indicate that tTJ and bTJ proteins directly interact. The interaction might be vital to the trafficking of tTJ proteins. In fact, how proteins are targeted to the tTJ remains largely unknown. They are, perhaps, initially recruited to the bTJ, where additional mechanisms cluster them to the tTJ. The tTJ permeability is the most important topic for investigation. Under electron microscopy, tTJ appears as a hollow tube of ~1 μm in length and ~10 nm in diameter. This ultrastructure has fueled the speculation that tTJ makes a different type of paracellular channel, which allows macromolecular permeation owing to its large dimension. The paracellular channel in the tTJ will less likely hold structural rigidity or electrostatic field to establish size and charge selectivity. On the other hand, when a tTJ protein, angulin-2 was removed, the tTJ became highly permeable to water (Gong et al., 2017). It appears that tTJ consists in a paracellular water channel. The molecular nature of the paracellular water channel is not clear. Apart from tricellulin and angulins, there may exist yet unidentified proteins in the tTJ. A proximity-based proteomic approach may allow identifying new proteins located in the tTJ or interacting with the tTJ components (Van Itallie et al., 2013).

REFERENCES

Al-Amoudi, A., Dubochet, J., & Norlen, L. (2005). Nanostructure of the epidermal extracellular space as observed by cryo-electron microscopy of vitreous sections of human skin. *The Journal of investigative dermatology*, *124*, 764–777.

Al-Amoudi, A., Diez, D. C., Betts, M. J., & Frangakis, A. S. (2007). The molecular architecture of cadherins in native epidermal desmosomes. *Nature*, *450*, 832–837.

Anderson, J. M., & Van Itallie, C. M. (2009). Physiology and function of the tight junction. *Cold Spring Harbor Perspectives in Biology*, *1*, a002584.

Betzig, E., Patterson, G. H., Sougrat, R., Lindwasser, O. W., Olenych, S., Bonifacino, J. S., et al. (2006). Imaging intracellular fluorescent proteins at nanometer resolution. *Science (New York, NY)*, *313*, 1642–1645.

Breiderhoff, T., Himmerkus, N., Stuiver, M., Mutig, K., Will, C., Meij, I. C., et al. (2012). Deletion of claudin-10 (Cldn10) in the thick ascending limb impairs paracellular sodium permeability and leads to hypermagnesemia and nephrocalcinosis. *Proceedings of the National Academy of Sciences of the United States of America*, *109*, 14241–14246.

Chen, C. C., Zhou, Y., Morris, C. A., Hou, J., & Baker, L. A. (2013). Scanning ion conductance microscopy measurement of paracellular channel conductance in tight junctions. *Analytical Chemistry*, *85*, 3621–3628.

Chu, P. J., Robertson, H. M., & Best, P. M. (2001). Calcium channel gamma subunits provide insights into the evolution of this gene family. *Gene, 280*, 37–48.

Genova, J. L., & Fehon, R. G. (2003). Neuroglian, gliotactin, and the Na^+/K^+ ATPase are essential for septate junction function in Drosophila. *The Journal of Cell Biology, 161*, 979–989.

Gong, Y., Himmerkus, N., Sunq, A., Milatz, S., Merkel, C., Bleich, M., et al. (2017). ILDR1 is important for paracellular water transport and urine concentration mechanism. *Proceedings of the National Academy of Sciences of the United States of America, 114*, 5271–5276.

Gunzel, D., Amasheh, S., Pfaffenbach, S., Richter, J. F., Kausalya, P. J., Hunziker, W., et al. (2009). Claudin-16 affects transcellular Cl- secretion in MDCK cells. *The Journal of Physiology, 587*, 3777–3793.

Ikenouchi, J., Furuse, M., Furuse, K., Sasaki, H., Tsukita, S., & Tsukita, S. (2005). Tricellulin constitutes a novel barrier at tricellular contacts of epithelial cells. *The Journal of Cell Biology, 171*, 939–945.

Kaufmann, R., Piontek, J., Grull, F., Kirchgessner, M., Rossa, J., Wolburg, H., et al. (2012). Visualization and quantitative analysis of reconstituted tight junctions using localization microscopy. *PloS One, 7*, e31128.

Kitajiri, S., Katsuno, T., Sasaki, H., Ito, J., Furuse, M., & Tsukita, S. (2014). Deafness in occludin-deficient mice with dislocation of tricellulin and progressive apoptosis of the hair cells. *Biology Open, 3*, 759–766.

Konrad, M., Hou, J., Weber, S., Dotsch, J., Kari, J. A., Seeman, T., et al. (2008). CLDN16 genotype predicts renal decline in familial hypomagnesemia with hypercalciuria and nephrocalcinosis. *Journal of the American Society of Nephrology, 19*, 171–181.

Kubota, K., Furuse, M., Sasaki, H., Sonoda, N., Fujita, K., Nagafuchi, A., et al. (1999). $Ca(2^+)$-independent cell-adhesion activity of claudins, a family of integral membrane proteins localized at tight junctions. *Current Biology, 9*, 1035–1038.

Milatz, S., Himmerkus, N., Wulfmeyer, V. C., Drewell, H., Mutig, K., Hou, J., et al. (2017). Mosaic expression of claudins in thick ascending limbs of Henle results in spatial separation of paracellular Na^+ and Mg^{2+} transport. *Proceedings of the National Academy of Sciences of the United States of America, 114*, E219–E227.

Mittendorf, K. F., Marinko, J. T., Hampton, C. M., Ke, Z., Hadziselimovic, A., Schlebach, J. P., et al. (2017). Peripheral myelin protein 22 alters membrane architecture. *Science Advances, 3*, e1700220.

Mohammad-Panah, R., Gyomorey, K., Rommens, J., Choudhury, M., Li, C., Wang, Y., et al. (2001). ClC-2 contributes to native chloride secretion by a human intestinal cell line, Caco-2. *The Journal of Biological Chemistry, 276*, 8306–8313.

Nunes, F. D., Lopez, L. N., Lin, H. W., Davies, C., Azevedo, R. B., Gow, A., et al. (2006). Distinct subdomain organization and molecular composition of a tight junction with adherens junction features. *Journal of Cell Science, 119*, 4819–4827.

Price, M. G., Davis, C. F., Deng, F., & Burgess, D. L. (2005). The alpha-amino-3-hydroxyl-5-methyl-4-isoxazolepropionate receptor trafficking regulator "stargazin" is related to the claudin family of proteins by its ability to mediate cell-cell adhesion. *The Journal of Biological Chemistry, 280*, 19711–19720.

Rajasekaran, S. A., Beyenbach, K. W., & Rajasekaran, A. K. (2008). Interactions of tight junctions with membrane channels and transporters. *Biochimica et Biophysica Acta, 1778*, 757–769.

Rust, M. J., Bates, M., & Zhuang, X. (2006). Sub-diffraction-limit imaging by stochastic optical reconstruction microscopy (STORM). *Nature Methods, 3*, 793–795.

Tang, V. W. (2006). Proteomic and bioinformatic analysis of epithelial tight junction reveals an unexpected cluster of synaptic molecules. *Biology Direct, 1*, 37.

Taveau, J. C., Dubois, M., Le Bihan, O., Trepout, S., Almagro, S., Hewat, E., et al. (2008). Structure of artificial and natural VE-cadherin-based adherens junctions. *Biochemical Society Transactions*, *36*, 189–193.

Van Itallie, C. M., Aponte, A., Tietgens, A. J., Gucek, M., Fredriksson, K., & Anderson, J. M. (2013). The N and C termini of ZO-1 are surrounded by distinct proteins and functional protein networks. *The Journal of Biological Chemistry*, *288*, 13775–13788.

Violette, M. I., Madan, P., & Watson, A. J. (2006). Na^{+}/K^{+}-ATPase regulates tight junction formation and function during mouse preimplantation development. *Developmental Biology*, *289*, 406–419.

Wu, Y., Kanchanawong, P., & Zaidel-Bar, R. (2015). Actin-delimited adhesion-independent clustering of E-cadherin forms the nanoscale building blocks of adherens junctions. *Developmental Cell*, *32*, 139–154.

Zhou, L., Zeng, Y., Baker, L. A., & Hou, J. (2015). A proposed route to independent measurements of tight junction conductance at discrete cell junctions. *Tissue Barriers*, *3*, e1105907.

Zhou, L., Gong, Y., Hou, J., & Baker, L. A. (2017). Quantitative visualization of nanoscale ion transport. *Analytical Chemistry*, *89*, 13603–13609.

Davies, C., [illegible] M., Le Bihan, Y., Tarmann, B., Abreu, S., Howe, [illegible], 2018. [illegible] of anti-[illegible] and natural [illegible]-based alternatives [illegible] Society [illegible]

[illegible] The [illegible] [illegible] 179, [illegible]

Vicente, [illegible] [illegible] and [illegible] in [illegible] [illegible]

Wu, Y., [illegible], P., R. [illegible], [illegible] R. [illegible] [illegible] [illegible]

[illegible] [illegible] [illegible] Figure [illegible]

Zhu, L., [illegible] [illegible] [illegible] 2017. [illegible] [illegible]

Index

G

H

I

K

L

V

Z

Printed and bound by CPI Group (UK) Ltd, Croydon, CR0 4YY
08/06/2025
01896869-0002